The Challenges of Dam Removal and River Restoration

edited by

Jerome V. De Graff
U.S. Department of Agriculture Forest Service
Clovis, California 93611
USA

James E. Evans
Department of Geology
190 Overman Hall
Bowling Green State University
Bowling Green, Ohio 43403
USA

THE GEOLOGICAL SOCIETY OF AMERICA®

Reviews in Engineering Geology XXI

3300 Penrose Place, P.O. Box 9140 ▪ Boulder, Colorado 80301-9140, USA

2013

Copyright © 2013, The Geological Society of America (GSA), Inc. All rights reserved. Copyright is not claimed on content prepared wholly by U.S. government employees within the scope of their employment. Individual scientists are hereby granted permission, without fees or further requests to GSA, to use a single figure, a single table, and/or a brief paragraph of text in other subsequent works and to make unlimited photocopies of items in this volume for noncommercial use in classrooms to further education and science. Permission is also granted to authors to post the abstracts only of their articles on their own or their organization's Web site providing the posting cites the GSA publication in which the material appears and the citation includes the address line: "Geological Society of America, P.O. Box 9140, Boulder, CO 80301-9140, USA (http://www.geosociety.org)," and also providing the abstract as posted is identical to that which appears in the GSA publication. In addition, an author has the right to use his or her article or a portion of the article in a thesis or dissertation without requesting permission from GSA, provided the bibliographic citation and the GSA copyright credit line are given on the appropriate pages. For any other form of capture, reproduction, and/or distribution of any item in this volume by any means, contact Permissions, GSA, 3300 Penrose Place, P.O. Box 9140, Boulder, Colorado 80301-9140, USA; fax +1-303-357-1073; editing@geosociety.org. GSA provides this and other forums for the presentation of diverse opinions and positions by scientists worldwide, regardless of their race, citizenship, gender, religion, sexual orientation, or political viewpoint. Opinions presented in this publication do not reflect official positions of the Society.

Published by The Geological Society of America, Inc.
3300 Penrose Place, P.O. Box 9140, Boulder, Colorado 80301-9140, USA
www.geosociety.org

Printed in U.S.A.

GSA Reviews in Engineering Geology Series Editor: Syed E. Hasan

Library of Congress Cataloging-in-Publication Data

The challenges of dam removal and river restoration / edited by Jerome V. De Graff, James E. Evans.
    p. cm. — (Reviews in engineering geology ; 21)
  Includes bibliographical references.
  Summary: "River restoration is a societal goal in the United States. This collection of research articles focuses on our current understanding of the impacts of removing dams and the role of dam removal in the larger context of river restoration. The papers are grouped by topic: (1) assessment of existing dams, strategies to determine impounded legacy sediments, and evaluating whether or not to remove the dam; (2) case studies of the hydrologic, sediment, and ecosystem impacts of recent dam removals; (3) assessment of river restoration by modifying flows or removing dams; and (4) the concept of river restoration in the context of historical changes in river systems"—Provided by publisher.
  ISBN 978-0-8137-4121-5 (pbk.)
  1. Stream restoration—United States. 2. Dam retirement—United States. I. De Graff, Jerome V. II. Evans, James E. (James Erwin), 1954– III. Series: Reviews in engineering geology ; v. 21.
  TA705.C45 2013
  627'.12—dc23

                                                                                           2012038995

Cover: Looking upstream through the breach of the IVEX dam (Chagrin River, Ohio), which failed in 1994 because of seepage piping failure. In the background is the reservoir fill that accumulated between 1842 and 1994. The breach was at the contact of the earthen dam (to the right) and the masonry spillway, still standing to the left. An earlier seepage piping failure repair can be observed on the left. The dam failure was a combination of 86% loss of storage capacity due to sedimentation, lack of a hydraulically designed spillway, lack of an emergency spillway, and poor maintenance (trees were allowed to grow on the dam itself). This is a common scenario for thousands of older dams in the United States. Photograph from 15 August 1994 appears courtesy of James E. Evans.

10 9 8 7 6 5 4 3 2 1

Contents

Foreword v

Preface vii

1. *Dam removal: A history of decision points* 1
   Laura Wildman

2. *Engineering considerations for large dam removals* 11
   Thomas E. Hepler

3. *Assessing sedimentation issues within aging flood-control reservoirs* 25
   Sean J. Bennett, John A. Dunbar, Fred E. Rhoton, Peter M. Allen, Jerry M. Bigham,
   Gregg R. Davidson, and Daniel G. Wren

4. *Using ground-penetrating radar to determine the quantity of sediment stored behind
   the Merrimack Village Dam, Souhegan River, New Hampshire* 45
   David J. Santaniello, Noah P. Snyder, and Allen M. Gontz

5. *Prediction of sediment erosion after dam removal using a one-dimensional model* 59
   Blair Greimann

6. *Sediment management at small dam removal sites* 67
   James G. MacBroom and Roy Schiff

7. *Multiyear assessment of the sedimentological impacts of the removal of the Munroe
   Falls Dam on the middle Cuyahoga River, Ohio* 81
   John A. Peck and Nicholas R. Kasper

8. *Sediment impacts from the Savage Rapids Dam removal, Rogue River, Oregon* 93
   Jennifer A. Bountry, Yong G. Lai, and Timothy J. Randle

9. *Changes in biotic and habitat indices in response to dam removals in Ohio* 105
   Kenneth A. Krieger and Bill Zawiski

10. *Using airborne remote-sensing imagery to assess flow releases from a dam in order to
    maximize renaturalization of a regulated gravel-bed river* 117
    M.S. Lorang, F.R. Hauer, D.C. Whited, and P.L. Matson

11. *Assessing stream restoration potential of recreational enhancements on an urban stream,
    Springfield, Ohio* 133
    John Ritter, Kelly Shaw, Aaron Evelsizor, Katherine Minter, Chad Rigsby, and Kristen Shearer

12. ***Effects of multiple small stock-pond dams in a coastal watershed in central California: Implications for removing small dams for restoration*** 149
    J.L. Florsheim, A. Chin, and A. Nichols

13. ***The shortcomings of "passive" urban river restoration after low-head dam removal, Ottawa River (northwestern Ohio, USA): What the sedimentary record can teach us*** 161
    J.E. Evans, N. Harris, and L.D. Webb

14. ***The rise and fall of Mid-Atlantic streams: Millpond sedimentation, milldam breaching, channel incision, and stream bank erosion*** 183
    Dorothy Merritts, Robert Walter, Michael Rahnis, Scott Cox, Jeffrey Hartranft, Chris Scheid, Noel Potter, Matthew Jenschke, Austin Reed, Derek Matuszewski, Laura Kratz, Lauren Manion, Andrea Shilling, and Katherine Datin

Foreword

This is the first volume in the *Reviews in Engineering Geology* series that has been prepared under the aegis of the Environmental and Engineering Geology Division, which in 2011, updated its 64-year-old name—the Engineering Geology Division—to better reflect the profession's growing involvement in protection and preservation of the Earth's environment.

While engineering geologists have been providing indispensable contributions to site selection, design, construction, and maintenance of civil engineering structures for more than a century, the ever-increasing engagement of engineering geologists in environmental projects for the past several decades led to the addition of the word "environmental" to the division's name.

It is quite a coincidence that this Reviews in Engineering Geology volume titled *The Challenges of Dam Removal and River Restoration* highlights the expanded scope of our profession in the evolving field of dam removal for ecological restoration. Several contributions in this volume are state-of-the-art papers that outline the key geologic and hydrologic considerations that should form the basis for site investigations; they set the standard of operation for projects related to dam removal and river restoration.

The system approach to the study of Earth emphasizes the role of the biosphere in maintenance of environmental quality. This volume can also be considered "a first" in the sense that it marks the close collaboration of geologists with biologists in determining the most suitable method of dam removal aimed at optimal restoration of the fluvial ecosystem. Papers by Krieger and Zawiski (Chapter 9) and Florsheim et al. (Chapter 12) highlight the collective input of earth and life science professionals in arriving at workable solutions to the complex issue of restoration of the fluvial ecosystem. These papers underscore the fact that the "living component" of the environment is a major factor in successful dam removal and restoration efforts.

I would like to take this opportunity to extend our appreciation to my predecessor, Dr. Charles W. Welby, who served as the Division's Publication Committee chair for many years and was responsible for preparation of several volumes in the Reviews in Engineering Geology series.

Syed E. Hasan
Chair, Publications Committee
Environmental and Engineering Geology Division
Department of Geosciences, University of Missouri–Kansas City
Kansas City, Missouri 64110-2499, USA

June 2012

Preface

As the manuscripts for this volume were about to be submitted for publication, U.S. Bureau of Reclamation workers turned off the generators at the Elwha Dam in Washington State. This is the first concrete step in removing both the Elwha and Glines Canyon dams and beginning restoration of the Elwha River on Washington's Olympic Peninsula (http://www.nps.gov/olym/naturescience/elwha-ecosystem-restoration .htm, accessed 10 December 2012). It seems fitting that publication of this volume and the actual start of restoration on the Elwha River should coincide. The idea for this volume was sparked by a session at the 2005 Geological Society of America Annual Meeting in which several papers pertaining to dam removal and restoration of the Elwha River were prominently featured.

As a Geological Society of America publication series sponsored by the Engineering Geology Division, Reviews in Engineering Geology might seem an unlikely venue for a collection of papers on dam removal and river restoration. This perception has more to do with the past association of engineering geology with dam building and other civil engineering efforts. In recent decades, engineering geology practice is increasingly involved with a broad range of applied geology topics. The Engineering Geology Division (EGD) of the Geological Society of America, the oldest Division in the Society, changed its name in 2011 to the Environmental and Engineering Geology Division. This is to better reflect the state of current practice in our field. This volume displays some of the diversity within engineering geology practice as well as interaction with other geologic specialties that share a common interest in the fluvial environment.

It would be easy for people who are only familiar with dam removal and river restoration from the popular media perspective to see these issues as a new controversy arising from environmental concerns. The first paper in the volume, "Dam removal: A history of decision points" by Laura Wildman, convincingly demonstrates that controversy over dam removal and construction has a very long history in the United States. While contemporary dam removal controversies tend to involve large dams, many of the arguments for and against dam removal were part of the early controversies involving small dams. "Engineering considerations for large dam removals" by Thomas Hepler explores the unique aspects of removing large dams.

Having broadly set the scene for dam removal and related fluvial issues, four papers examine various aspects of dealing with a primary issue, sedimentation. "Assessing sedimentation issues within aging flood-control reservoirs" by Sean Bennett and others speaks to a primary concern in the decision to remove a dam. The amount of sediment available in the reservoir and its ultimate fate within the downstream environment once released is both a complex and important consideration. Determining the quantity of sediment present is essential to predicting sediment impacts. "Using ground-penetrating radar to determine the quantity of sediment stored behind the Merrimack Village Dam, Souhegan River, New Hampshire" by David Santaniello and others illustrates how a fairly new geophysical technique can be employed to improve estimates of available sediment. Blair Greimann in his paper, "Predictions of sediment erosion after dam removal using a one dimensional model," describes how modeling can yield insights into expected sediment movement. Such analytical forecasting of post-removal sediment transport enables better identification of where monitoring and mitigation measures should be placed well before the actual dam removal. While Greimann's paper offers insights for large dam removals contemplated in the future, James MacBroom and Roy Schiff's paper "Sediment management at small dam removal sites" uses the experience gained from removal of many small dams to suggest a procedure to evaluate and select sediment management strategies. It describes a practical approach to this key design issue.

Four case studies from Ohio, Oregon, and Idaho explore actual dam removals and document the sedimentological impacts associated with them. Sediment impacts are at the heart of John Peck and Nicholas Kasper's paper, "Multiyear assessment of the sedimentological impacts of the removal of the Munroe Falls Dam on the Middle Cuyahoga River, Ohio" and Jennifer Bountry and others' paper, "Sediment impacts from the Savage Rapids Dam removal, Rogue River, Oregon." Both papers offer unique perspectives given differences in geographic locations, dam size, and number of years of observation. Kenneth Krieger and Bill Zawiski examine indirect measures suitable for determining the biological consequences of sediment after dam removal in their paper entitled "Changes in biotic and habitat indices in response to dam removals in Ohio." Their study links the physical transport of sediment following dam removal to its impact on aquatic species. M.S. Lorang and others examine the broad physical changes in channels using remote sensing and geographical information systems data in their paper to assess flow releases from a dam in order to maximize renaturalization of a regulated gravel-bed river. Their insights into flows and natural channel conditions are beneficial to successful dam removal and general river restoration.

The last four papers in this volume take a broader view of both dam removal and river restoration. John Ritter and others in their paper entitled "Assessing stream restoration potential of recreational enhancements on an urban stream, Springfield, Ohio" take a look at how river restoration can improve or degrade post-removal recreational uses. They highlight the importance of stream grade in the design of instream recreational features. "Effects of multiple small stock-pond dams in a coastal watershed in central California: Implications for removing small dams for restoration" by J.L. Florsheim and others emphasizes the importance of basin-scale management approaches in watersheds with a high density of small dams. They highlight the importance of re-establishing connectivity within the watershed under any restoration strategy being considered. J.E. Evans and others in their paper, "The shortcomings of 'passive' urban river restoration after low-head dam removal, Ottawa River (northwestern Ohio, U.S.A.): What the sedimentary record can teach us," demonstrate that the truism, "let nature takes its course" is not a good guide for river restoration in an urban river setting. Understanding the river's flow regime prior to alteration as well as current conditions allows a more desirable restoration outcome that requires some level of "active" restoration for a river within an urban setting. The last contribution returns the volume to where it started on the eastern seaboard of the United States with its long history of small dams. Dorothy Merritts and others examine the erosion in streams resulting from breached millpond dams in their paper entitled "The rise and fall of Mid-Atlantic streams: Millpond sedimentation, milldam breaching, channel incision, and stream bank erosion." While the study looks specifically at incision of streams and resulting sedimentation in southeastern Pennsylvania impacting Chesapeake Bay, the study results have relevance to watersheds in New York, Maryland, and Virginia.

For engineering geologists, this volume explores aspects of traditional practices related to dam construction, geophysical exploration methods, and modeling as well as current practices focused on sediment management, "passive" versus "active" restoration measures, and regulation by flow releases. There should also be a significant amount of material of equal interest to fluvial geomorphologists and other geologic practitioners. Finally, these papers collectively illustrate that construction, modification, and removal of dams represent a continuum of overlapping human interactions with the environment. The longevity of most dams, and the magnitude of the hydrological and sedimentological response to these human interactions with the environment, make these case studies outstanding examples of the concept of humans as geological agents.

Jerome V. De Graff
James E. Evans

Dam removal: A history of decision points

Laura Wildman

Princeton Hydro LLC, New England Regional Office, 931 Main Street, Suite 2, South Glastonbury, Connecticut 06073, USA

ABSTRACT

In the northeastern United States, we have been removing dams for almost as long as we have been building them, yet many communities involved in current decisions to repair, replace, or remove a dam are not aware of this. This paper highlights some of the stories that have been recorded regarding the history of decision points for dams, including the colorful history of the Billerica Dam in Massachusetts, which has been removed and rebuilt numerous times and is now under consideration for removal for at least the sixth time in its 300 yr history.

By understanding that dam removal is just one of the potential dam safety decisions that needs to be analyzed over the life cycle of a dam, and that dams are manmade structures with finite life spans, we can deconstruct the notion of dam removal as a radical concept. Dam removal is just one of many dam safety options that may be discussed over the course of a dam's history. It is most commonly implemented when a dam no longer serves any economic purpose that justifies the expense of maintaining the dam structure.

In the past, dams have been removed for many of the same reasons that we remove dams today; however, the procedures currently required to remove a dam are far more complex and highly regulated. This has led to increased documentation of dam removal efforts and now allows us to compare and categorize dam removal projects, such that the lessons learned from these projects can be incorporated into a more informed decision-making process in the future.

INTRODUCTION

The issue of dam removal is not as new as many today might assume. In fact, we have been removing dams for almost as long as we have been building them. Historic decisions as to whether a community should build or remove a dam were often controversial, as they can be today, and included battles waged to stop the construction of a dam or to remove a dam recently constructed, often due to unacceptable impacts such as flooding or blockage of fish runs. Dam removal is just one decision made in a long history of decision points in time regarding the use and management of river resources. These decision points most commonly center on issues pertaining to safety, liability, economics, the dam's uses and condition, as well as riverine resources, such as fisheries and flow rights. A critical decision point for a dam may arise every few years or every few decades, but it will most commonly arise when the economic use of the dam changes, the dam is modified, or the dam is in need of significant repair.

The communities involved in current debates regarding dam removal often are unaware of this long history of decision points.

This chapter documents many examples that demonstrate the series of decisions typically made regarding a dam. At first, a community must decide if the benefits of constructing a dam outweigh the multiple impacts to the riverine resources. Typically, multiple dams have been constructed at or near the same site, as the design life of the original dam ends or a different configuration is needed to serve a new economic purpose. As the uses of the dam change over time, the community must reassess whether the dam's value outweighs its continued impacts. This decision is often economically based and depends heavily on the current benefits of the dam and its condition. Because rivers impacted by dams are shared resources, the decision often involves the entire community and not just the dam owner.

Historically, dams have been removed for a wide variety of reasons. They have been removed to make way for the construction of a new dam, to eliminate impacts such as upstream flooding of private lands or blocked fish runs, to restore flowage rights, to eliminate dam safety concerns, to do away with maintenance costs, or because the dam no longer serves a purpose that justifies its upkeep and coverage of potential liability. Sometimes, dams are removed following a significant structural failure, when replacing the dam is not economically feasible. However, since the removal of a dam, even a failing dam, is not a mandatory requirement at the end of its useful life cycle, our landscape is dotted with derelict dams that have not been maintained for years and no longer serve an economically justifiable purpose. Often, the general public mistakenly equates dams (manmade structures requiring maintenance) with waterfalls (relatively structurally stable natural features) and, therefore, fails to understand the reasons for maintaining or removing a dam. As manmade structures with finite life spans, dams pose a risk of failure and a threat to public safety that waterfalls do not. Dam removal is just one tool available to help dam safety officials and local communities come to a decision regarding dam safety and the continued use of riverine resources.

DAM REMOVAL DECISIONS: THEN AND NOW

Significance of the Dam Site

The history of dams and dam removal is also a history of rivers and their management. In New England, dams have existed in our landscape for a little over 300 yr. In geologic terms, dams have existed for only a brief period of time, while the rivers they modify have scoured out their paths on the landscape for thousands of years or more. The diverse assemblages of species that utilize rivers have also evolved over this extended history in conjunction with the rivers. While many people in communities with existing dams remember only the dammed river, the locations where dams were constructed often have a long history that points to the site's valued river resources. Many of the early colonial villages, and later industrial towns along rivers in the northeastern United States, were sited on the exact locations that were once Native American encampments. These locations were selected by the tribes to gain easy access to the river's flow and fisheries resources. Frequently, the sites were chosen because of bedrock pinch points or falls that created excellent fishing sites for harvesting the diadromous fish species of the river. Later, dams were constructed on these same sites, in order to take advantage of the underlying bedrock for the dam's foundation and to maximize the available riverhead. Often, the Native American name for a river described the river's geologic features and thereby foretold which rivers would later become the most heavily dammed. Rivers such as the Penobscot ("the rocky part" or a place "where the white rocks are"), Presumpscot ("river of many falls"), Merrimac ("swift stream"), Pawtucket ("at the falls in the river"), Neversink ("continuously flowing" or "place of big rock"), and Pootatuck ("river with falls," and now named the Housatonic) were destined to someday be dammed to utilize their swift, continuous flow, and steep, rocky riverbeds. Historically, these same rivers were also home to some of the most productive diadromous fish runs in the northeastern United States, thus initiating many of the early conflicts and decision points regarding dams, dam removals, and riverine resources.

Historic Dam Removal Decisions

Many communities who are deciding today whether or not to keep a dam are the same communities that originally fought to stop the dam from being constructed, or fought to have the dam removed just after construction, often on the basis of upstream flooding, flowage rights, and damage to fisheries resources (Kulik, 1985).

Many colorful stories exist chronicling the long history of decision points regarding dams and dam removal in the northeastern United States. One of the most fascinating and well documented of these stories revolves around the history of the Billerica Dam on the Concord River.

Billerica Dam

Early colonists settled in Billerica, Massachusetts, in 1653, where the Pennacook people of the Abenaki Confederation had lived and fished for generations along the shores of the Concord River, which flowed swiftly over a series of small waterfalls in the area. The falls were last visible in 1710, when the Billerica Dam was constructed, among much local controversy. Proponents of the dam lost their first legal contest in 1711, when the dam owner was ordered to pay restitution to a flooded upstream land owner. In 1722, the dam was removed by court order, but the dam owner rebuilt the dam later that same year. The case went back to court, but before the judge could make a decision, an angry band of farmers, whose fields had been flooded, removed the dam by force. Due to mill damage that resulted from that incident, the dam owner was allowed to rebuild the dam a third time (Kulik, 1985; Donahue, 1989; Ingraham, 2009; Wildman, 2003).

The Billerica Dam was back in court in 1809 and 1815, due to additional upstream flooding caused by the raising of the dam to divert water to a newly constructed canal system. This time the court sided with the dam owner and the dam remained. In 1839, Henry David Thoreau commented on the dam's potential removal while paddling the Concord River with his brother: "…mere Shad, armed only with innocence and a just cause…I for one am with thee, and who knows what may avail a crow-bar against that Billerica dam?" (Thoreau, 1849). When the canal charter was revoked in 1859, the flooded farmers hired Thoreau to survey the river, gathering evidence for the defendants in their newest highly publicized and controversial court case. The dam was again ordered to be removed, with full compensation to the dam owner through the construction of a new steam-powered generator for the mill. However, in 1861, the dam owner filed a petition for repeal of this act. The lawyer for the defense pleaded, "For generations, a painful and expensive controversy has existed in relation to [the Billerica Dam], and if [not removed now], the children and children's children of these parties will be cursed with strife and contention." The dam was ordered removed again, but, by April of that year, the country headed into civil war, and all efforts to remove the dam were put aside (Kulik, 1985; Donahue, 1989; Wildman, 2003). One hundred and fifty years later, there are still efforts under way to remove the dam, and controversy still surrounds the decision between maintaining or removing the dam. Figure 1 demonstrates the complex time line of decision points for the Billerica Dam.

Coonamesset Dams

Another early-documented dam removal effort was the "herring war" of 1798, in Falmouth, Massachusetts, which was initiated after three mill dams were constructed, resulting in a decline of the anadromous herring populations and the passing of a new bylaw giving an elected Herring Committee the right to remove any obstructions to restore fish passage to the Coonamesset River. Features at the core of this decision point were conflicting rights of property owners versus public rights of access to the common fisheries resource. At one point, members of the anti-herring party filled a cannon on the town green with alewife (a species of anadromous herring) in an act of protest and lit the fuse; the explosion of the cannon resulted in the death of the gunner. Records of the town's bylaws demonstrate that the discussion over the removal of the dam was still active in 1865 (Anonymous, 2010). The mill impoundments were later turned into cranberry bogs, and the river and its floodplain were significantly modified. The history of decision points for the Coonamesset dams continues to this day, with a request for proposals being ordered by the town in January 2011 to study the potential to remove four dams from the river and restore the site's wetland and fisheries functions and values.

Presumpscot Falls Dam (aka Smelt Hill Dam)

In 1736, local native tribes in Maine protested the 1734 construction of a mill dam at Presumpscot Falls on the Presumpscot River almost immediately after the Smelt Hill Dam was constructed (Friends of Sebago Lake et al., 2002). The mill dam was later replaced by a hydroelectric dam. In 2002, after the hydroelectric dam had been significantly damaged by a large flood in 1996, the dam was removed as part of a larger effort to assess the removal or installation of fishways at nine dams along the Presumpscot River. The larger restoration effort is continuing, utilizing a staged approach, with the decision to initiate fish passage at each upstream dam once passage has been established at the base of each dam.

Pawtucket Falls Dams (aka Slater Mill and Main Street Dams)

In 1748, local residents went to court to try to get a furnace dam removed that blocked fish passage at the Pawtucket Falls on the Blackstone River, in Rhode Island. The judge agreed with the plaintiffs and ordered that "said dam should be broken and a way made through the same," but the dam owners petitioned the General Assembly and were allowed to keep the dam. The controversy at Pawtucket Falls was reinitiated in 1792, when the owners of the water privileges at the falls "did…utterly subvert, pull down and destroy" another partially finished cotton mill dam built 200 yards above the falls. At their trial, they freely admitted to destroying the dam "as lawfully they might," since they considered the dam a nuisance that diverted the natural stream (Kulik, 1985). At that time, common law supported the removal of any dam at will, if the dam was defined as a public nuisance. Dam owners could appeal to local courts following the dam's removal, but not preceding the removal (Kulik, 1985). This law was later changed with legislation that protected dam construction and was more sympathetic to industry and the dam owners. Figure 2 shows one of the historic dams at the Pawtucket Falls site. Two dams still remain on the Pawtucket Falls, the Slater Mill Dam and Main Street Dam. The Slater Mill Dam no longer serves an economic purpose, but its associated mill building has been converted into a museum highlighting the industrial history of the Blackstone River. The museum owners are currently discussing the option of adding a small hydroelectric turbine to the dam. The Main Street Dam has already been retrofitted with a hydroelectric turbine; however, it is unclear if it is still actively used. Both dams are currently under consideration for the addition of fishways, while multiple dams further upstream are being considered for removal.

Selected Historic Dam Removals

Many additional reports of historic dam removal efforts can be found by reviewing historic records, court documents, and newspaper archives. A selection of some of these accounts has been compiled here:
- New Milford Dam: In 1799, Elijah Boardman, who later became a U.S. Senator from Connecticut, led angry citizens to remove a dam in New Milford on the Housatonic River by force, convinced that the dam was causing

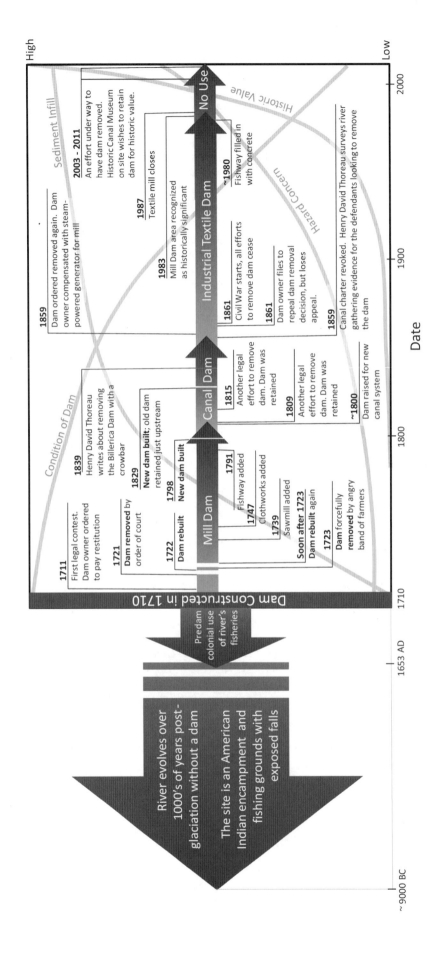

Figure 1. The history of decision points for a New England dam, based on the story of the Billerica Dam in Billerica, Massachusetts.

Figure 2. Photograph of Pawtucket Falls with dam in place. Photograph source: Accession: 2009–11, Rhode Island State Archives.

the repeated fever outbreaks in their town (Gordon and Raber, 2000).
- East Dam: In 1856, shad fishermen and some local property owners attempted to blow up the East Dam on the Connecticut River in Windsor Locks, Connecticut (Kulik, 1985).
- Stow's Pond Dam: A 1934 *Hartford Courant* newspaper article refers to the removal of a dam at Stow's Pond in Plantsville, Connecticut, but provides no reason for the dam's removal, nor does the brief article refer to any controversy surrounding the dam's removal (Harford Courant, 1934). Many dams were removed due to safety concerns over the course of history, and it is likely that large numbers of these decision points were not heavily contested.
- Connecticut Dams: The 1983 Connecticut Department of Environmental Protection's *Map of Connecticut Dams* lists 131 dams as "removed" but gives no reasons for the removals, so it is unclear how many of these "removals" were actual dam removals as opposed to uncontrolled dam failures.
- Valley of the Tigris Dams: One of the earliest-documented series of deliberate dam removals occurred in 331 B.C., when "Alexander the Great led his forces into the Valley of the Tigris. The records of his campaign indicate that dams on the river had to be partially removed to permit passage of his fleet." The dams "have been described as massive rubble-masonry weirs which served as diversion works for canal intakes." The dams were presumably later repaired, since subsequent reports chronicle the irrigation systems of the lands he conquered (Jansen, 1980). It is interesting that these dams were removed to restore passage for boats, since one of the earliest and most common uses of rivers was as transportation corridors.

LEGACY DAMS

When dams with strong economic justification and purpose fall into disrepair or become antiquated in their design, the decision is typically made to construct a new dam. Historically, if the original dam was not removed, it was commonly decided to submerge the older dam within the new impoundment. This decision was often more cost effective for the dam owner. Submerging the original dam saved on demolition costs and provided water control during the construction of the new dam. However, the historic decision points that led to the submergence of an antiquated upstream dam have often complicated recent decisions to remove current dams. We refer to the submerged dam as a legacy dam. An 1829 detail of the Billerica Dam (Fig. 3), previously described in this paper, demonstrates a well-documented example of a legacy dam left in place just upstream of the current Billerica Dam. The two dams were then structurally connected by filling the interstitial void between them. It is unusual to find documentation of this type for a legacy dam.

In some cases, such as the Great Works and Veazie Dams on the Penobscot River in Maine, the legacy dams submerged within the impoundment are significantly larger in mass than the current dams downstream. The Great Works and Veazie Dams are scheduled to be removed as part of a large-scale restoration of the Penobscot River. In order to obtain the restoration goals of fish passage and free-flowing riverine conditions, a portion of the legacy dams will need to be removed along with the primary dams, greatly adding to the complexity and cost of this restoration effort.

The need for the removal of an upstream legacy dam as part of a downstream dam removal project has become common in the field of dam removal. Some legacy dams can be easily identified utilizing aerial photographs prior to the dam removal, or are

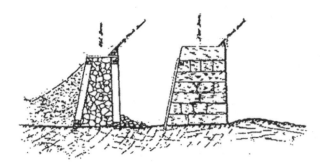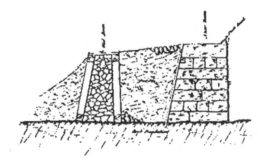

Figure 3. Example of a legacy dam shown on a detail of the design of the fifth Billerica Dam just downstream from the fourth Billerica Dam, drawn in ca. 1829. (Source: Ingraham, 2009.)

visible when water surface elevations within the impoundment are low. However, other legacy dams have become completely submerged by impounded sediment and can be difficult to detect prior to dam removal, even when numerous sediment probes or borings have been conducted throughout the impoundment.

Another example of a legacy dam exists within the impoundment of the current Rainbow Dam on the Farmington River in Connecticut. This largely intact timber crib legacy dam became visible in 2010 when the Rainbow Dam was drawn down for repair. The photographs shown in Figure 4 demonstrate the significant scale of this particular legacy dam.

The photographs in Figure 5 show yet another significant legacy dam clearly visible behind the Bear Lake Dam in California, when the impoundment was drawn down for repairs.

Legacy dams are commonly uncovered during dam removal efforts and impoundment drawdowns; however, they rarely are identified as separate dams on state and federal dam inventories. Based on the frequency with which legacy dams are encountered during current dam removal efforts nationwide, it is likely that many more of these legacy dams exist but have yet to be identified. Future decisions to remove dams will therefore need to include both a determination as to whether a legacy dam exists and whether the removal of the legacy dam will need to be incorporated into the downstream dam removal effort.

CURRENT DAM REMOVAL DECISIONS

Current decisions to remove dams are a continuation of the long history of decisions points regarding dams and river resources. While many dams still serve valuable societal functions and are actively maintained, the vast majority of dam removals involve dams that have no economic purpose and are not being maintained. Dams continue to be removed for many reasons, similar to historic dam removal efforts. Dams today are removed to increase public safety, decrease economic burdens, reduce liability, dewater properties submerged by a dam, reduce upstream flooding, restore a navigation pathway, make way for a new dam, reestablish more natural flow patterns, restore passage for aquatic organisms, manage contaminated sediment in a controlled manner, reestablish a more natural stream-temperature regime, improve water quality, uncover submerged habitats, and restore riverine connectivity. Pohl (2002) found

Figure 4. 2010 photograph and aerial photograph of the timber crib legacy dam behind the Rainbow Dam, exposed during downstream dam repairs. (Photo source: Joe Spirko.)

Figure 5. (Left) Bear Lake Dam looking upstream with legacy dam fully submerged. (Right) Bear Lake Dam looking downstream with arched legacy dam exposed. (Photo source: Rick Keppler Collection.)

that in the 1990s, almost half of the dams recently dismantled nationwide were removed for environmental reasons (Mullens, 2003). However, recent dam removal experience has also demonstrated that, while a dam removal project may commonly be initiated for its environmental benefits, such as system defragmentation and fish passage, it is often the dam safety and economic arguments that are the final determining factors when deciding if a particular dam will be removed. Communities, while sympathetic to environmental arguments, are more commonly influenced by the potential dam safety and economic benefits of dam removal, especially if a dam safety order requiring costly repairs to a dam has been issued or if there has been any loss of life as a direct result of the dam.

Today's dams are not removed with the intention of restoring "presettlement" conditions to a river. The numerous impacts of development on a river system, including changes in flow regimes and base-level elevations, stabilization of river banks and beds, encroachment from buildings, railways, roadway, bridges, and culverts, the removal of riverine vegetated buffers, the introduction of invasive species, and the disruption in sediment transport regimes, cannot be reversed or eliminated through the removal of a dam. The numerous environmental organizations and state and federal agencies leading the current effort to defragment rivers and remove obsolete dams are looking only to restore as many riverine functions as possible with the intention of placing the river back on a trajectory toward some level of rehabilitation, reflecting the given current societal and environmental constraints. Therefore, when dams are removed, it is understood that the historic channel, series of channels, floodplain, or wetland complex that existed prior to the dam's construction can often not be fully reestablished in the exact form and with the same functions and values it once provided or offered. Even when it is possible to more fully rehabilitate the site, it often becomes cost prohibitive to attempt a large-scale restoration effort. For these reasons, dam removals more often focus on removing the barrier, and potentially managing the mobile portion of the impounded sediment, while allowing the natural riverine and vegetative processes to take over and continue along the trajectory of restoration for the site. By keeping barrier removal projects cost effective, more barriers can be removed, and longer reaches of rivers can be defragmented.

Some decisions to remove dams are made to eliminate the "time bomb" of dam failures. While dam removal projects include detailed study and management of any potential impacts that the dam removal might have, such as management of contaminated sediment, sensitive species relocation, infrastructure protections, and water-quality protections, dam failures include none of these precautions. Dam failures can release large quantities of contaminated sediment and are often responsible for severe environmental and private property damage, as well as potential loss of life. The removal of an unmaintained dam prior to its failure eliminates these "time bombs" and protects the environment, private property, and lives. Many dams under consideration for removal have already failed or partially failed prior to the dam being removed. In some cases, the dam failed while dam removal was under consideration, such as the Baker Chocolate Dam on the Neponset, where a partial failure allowed for the uncontrolled release of polychlorinated biphenyl (PCB)–contaminated sediment, and the Anaconda Dam on the Naugatuck River, where a partial failure undermined an active sewer line downstream, instigating an accelerated removal of the dam under an emergency order from the state dam safety office.

The number of dams in the U.S. ranges from 99,000 dams, documented in the U.S. Fish and Wildlife Service Barrier Inventory, to estimates of millions of dams, although no fully inclusive inventory of dams exists. The American Society of Civil Engineers (ASCE) Infrastructure Report Card has given dams nationally a grade of "D" (poor) ever since it initiated the score for dams in 1998 (ASCE, 2009, 2011). Most of the dams in the northeastern United States were built during the Industrial Revolution. However, dams were designed with finite life spans, based on the longevity of the materials from which they were built and the sediment loads of the particular river systems that flow into them. Many dams have far outlived their design lives and no longer serve an economic purpose that can support their ongoing maintenance. According to data collected from state dam safety offices by the American Association of State Dam Safety Officials (ASDSO), the number of high-hazard dams in need of repair has increased from 488 in 2001 to 1826 in 2007. The total number of deficient dams on record has increased from 1348 to 4095 over the same period of record (ASCE, 2009). In addition, according to records compiled by ASDSO, the rate of reported dam failures has increased more than 175% from 2000 to 2006, as compared to the reported dam failure rate from 1960 to 2000. This does not include the many unreported or undetected dam failures. The gap between dams needing repair and those actually repaired is growing significantly (ASCE, 2009). The cost of repairing and maintaining obsolete dams can be exceedingly high and thus overly burdensome for dam owners (Whitelaw and MacMullan, 2002; Mullens, 2003). In 2009, the ASDSO estimated that the total cost to repair the nation's dams totaled $50 billion and the needed investment to repair high-hazard-potential dams totaled $16 billion (ASCE, 2009). As more dams approach the end of their functional lives, state dam offices and communities will be facing an increasing number of decisions regarding dam structures (Mullens, 2003) and their potential removal.

In the northeastern states, the most typical dams being removed are low-head mill/industrial dams, less than 25 ft (7.62 m) in height, that have commonly not been actively used for over 30–50 yr. Most are relics of an industrial past where mechanical power was used or water was diverted to canals that fed multiple industrial sites within a town. Some of these sites were later converted to hydroelectric power but were then found to be economically infeasible, and the sites were abandoned. Smaller numbers of the dams removed were recreational dams, hydroelectric dams, cranberry bog dams, or farm ponds. Many of the dams no longer serve their original purpose but are now incidentally used for recreation or water diversion. There is no record of an active flood-control dam being removed, and with respect to the few dams that have been removed that served as water-supply dams, either they had already been decommissioned or suitable water supply alternatives were available. The Edwards Dam in Maine, removed in 1999, is the only documented case where the Federal Energy Regulatory Commission (FERC) actually ordered the removal of an active hydroelectric facility against the wishes of the dam owner, due to the fact that the economic value of the dam did not compensate for the environmental impacts of the dam on the Kennebec River's diadromous fish populations, and the installation of effective fishways would be more costly than dam removal (American Rivers et al., 1999). In all other cases, the consent of the dam owner was required in order to remove a dam, unless the dam was abandoned and a dam owner could not be identified. Several high-profile large dams have been removed, such as the 160-ft-long (49 m) and 120-ft-high (37 m) Occidental Chem Pond Dams in Tennessee in 1995, and the 700-ft-long (213 m) and 65-ft-high (20 m) Birch Run Dam in 2005 in Pennsylvania. Others are under consideration for removal include the 185 ft (56 m) Glines Canyon Dam in Washington and the 110 ft (34 m) Elwha Dam in Washington (American Rivers et al., 1999). However, the number of large dams removed is far less than the number of small dams removed. Large dam removals involve a much more significant effort when determining if the environmental and economic benefits for the dam's removal outweigh the current uses and economic benefits of the dams. The decision-making process can typically take 10 to 20 yr, as opposed to the more typical 2 to 3 yr process to remove smaller dams.

CURRENT DAM REMOVAL PROCEDURE

Unlike some of the colorful stories of dam removals in the 1700s, in which angry mobs removed the dam by force, the removal of a dam today is a carefully regulated process. Typically, anywhere from three to ten permits are required, along with multiple consultations with regulators, depending upon the state in which the dam is located. On almost all dam removal projects, impounded sediments are tested, the site is surveyed, hydrologic and hydraulic analyses are completed, and engineering design plans are prepared, showing the deconstruction sequence and the proposed site protections, prior to initiation of the dam's removal. In addition, on many removal projects, analyses and studies are completed regarding economic issues, specific ecological concerns, historic/archaeological considerations, hydrogeologic interconnections, channel morphology and restoration potential, sediment transport assessments, water-quality and flood-capacity issues, ice-jam potential, and infrastructure stability/protection. This significant amount of analysis relating to dam removal is often difficult to obtain through a traditional literature search since most of the analysis was typically completed by consultants for a project, remains unpublished, and is considered "gray" literature. In 2006, in response to a recommendation made in the Aspen Institute report titled *Dam Removal: A New Option for a New Century* (Aspen Institute, 2002), the Clearinghouse on Dam Removal Information (CDRI) was created at the Water Resource Center Archives at the University of California, Berkley, to address this deficiency. The CDRI now acts as the only national repository for the "gray" literature relating to dam removal, collecting grant proposals, settlement agreements, engineering design plans, cost estimates, modeling data and analyses, monitoring reports, presentations, photos, videos, and newspaper articles, as well as information on upcoming events, conferences, and training.

LEARNING FROM RECENT DAM REMOVAL EXPERIENCES

While almost all modern dam removals have documented and assessed existing conditions, and have analyzed the proposed conditions, the monitoring of dam removal sites subsequent to removal is far less common. In addition, a great deal of monitoring data goes unpublished, as it is held by state agencies that tend to collect data without publishing. The primary reason for lack of monitoring is limited available funding. While currently there are multiple funding sources to remove a dam, there are fewer funding sources that cover dam removal design, and even fewer still that cover the monitoring of a dam removal project. Recently, there has been an increase in monitoring efforts, based primarily on new federal funding requirements that require monitoring to be included in the proposal for dam removal funding in order to receive a grant for design or construction. The monitoring shortfalls of river restoration projects have been well documented in a series of papers produced by the National River Restoration Science Synthesis Project (NRRSS). As more funding becomes available for monitoring, more post–dam removal sites will be assessed, and this will help to improve the outcome of future dam removal projects.

In 2007, the Stream Barrier Removal Monitoring Guide was created by the Gulf of Maine Council on the Marine Environment and the National Oceanic and Atmospheric Administration's (NOAA) Restoration Center, in cooperation with multiple state agencies and the input from a diverse River Restoration Monitoring Steering Committee, a workshop with 70 restoration professionals, and 25 document reviewers. The guide attempts to standardize monitoring protocols for dam removals, and it developed a framework of monitoring parameters that can be used for stream barrier removal projects throughout the Gulf of Maine watershed, although it is clear that the guide has the potential for national implementation. The steering committee selected eight parameters that, when analyzed collectively, are expected to provide valuable data that will adequately characterize the physical, chemical, and biological response of a given stream to a barrier removal project. These eight critical monitoring parameters include: monumented cross sections; longitudinal stream profile; streambed sediment grain-size distribution; photo stations; water quality; riparian plant community structure; macroinvertebrates; and fish passage assessment. The monumented cross sections act as the skeletal framework for the other monitoring parameters and are permanently established and georeferenced (Collins et al., 2007).

Growing numbers of dam removal projects have been completed in recent years, and some obvious patterns and similarities in responses, among specific "categories" of dam removals, are starting to emerge. In 2010, Wildman and MacBroom proposed a Broad Classification for Dam Removal Projects, based on their experience and observations, in conjunction with the output of traditional empirical and analytical techniques, on over 100 dam removal projects, as well as multiple historical undocumented dam removals, dam breaches/failures, reservoir drawdowns, and postglacial lakes and dams. The classification system allows its user to compare similar dam removal sites, create better conceptual models at the initiation of a project and better predict the complexity of a project and the trends, such as whether postdam sediments will erode, if postdam channels will likely be straight or sinuous, the likely mechanism of erosion, and the project's potential recovery rates (Wildman and MacBroom, 2010).

The classification system was also developed to assist dam owners, stakeholders, and restoration professionals to better understand the risks and rewards of specific dam removal projects at the onset of a project. By categorizing dam removal projects on a sliding scale from relatively "easy" to complex, the classification system can help frame regulatory concerns, such that regulators better understand when to be comfortable with a dam removal and when additional questions, relevant to that specific dam category, still need to be addressed while avoiding overgeneralization relating to the potential impacts of dam removal. The classification can also be used to strategically monitor different dam removal categories that are likely to have similar responses and response periods, or as a simple framework to help determine the type of numerical modeling analysis that might be needed (Wildman and MacBroom, 2010). Future decision points regarding whether a dam should or should not be removed will be made easier with data from these documented dam removal examples.

CONCLUSIONS

Dams are manmade structures that require regular maintenance to remain intact and safe, similar to buildings and bridges. This paper helps to document the series of decision points that have been made regarding dams, dam removal, and riverine resources over the past 300 yr in the northeastern United States, as well as the current reasons for dam removals and the procedures typically required to complete a removal. In the end, any proposal to remove a dam today or in the future is just a continuation of a discussion initiated when the dam was first proposed. What decision we make now, through carefully balancing today's issues, will directly affect the decisions to be made in the future.

REFERENCES CITED

American Rivers, Friends of the Earth, and Trout Unlimited, 1999, Dam Removal Success Stories: http://www.americanrivers.org/assets/pdfs/reports-and-publications/SuccessStoriesReport6f14.pdf (accessed 31 July 2011).

American Society of Civil Engineers (ASCE), 2009, 2009 Report Card for America's Infrastructure: Washington, D.C., American Society of Civil Engineers, 162 p.

American Society of Civil Engineers (ASCE), 2011, Report Cards for America's Infrastructure: American Society of Civil Engineers, http://www.infrastructurereportcard.org/report-cards (accessed 31 July 2011).

Anonymous, 2010, Rivers of Barnstable County, Massachusetts: Coonamesset River, Pamet River, Bass River, Quashnet River, Pocasset River, and Childs River: General Books LLC, 34 p.

Aspen Institute, 2002, Dam Removal—A New Option for a New Century: Washington, D.C., Aspen Institute Publications Office, 68 p.

Collins, M., Lucey, K., Lambert, B., Kachmar, J., Turek, J., Hutchins, E., Purinton, T., and Neils, D., 2007, Stream Barrier Removal Monitoring Guide: Gulf of Maine Council on the Marine Environment, http://www.gulfofmaine.org/streambarrierremoval/Stream-Barrier-Removal-Monitoring-Guide-12-19-07.pdf (accessed 31 July 2011).

Donahue, B., 1989, Dammed at both ends and cursed in the middle: The flowage of the Concord River Meadows, 1798–1862: Environmental Review, v. 13, no. 3/4, p. 46–67, doi:10.2307/3984390.

Friends of Sebago Lake, Friends of the Presumpscot River and American Rivers, 2002, Historic Records Related to the Anadromous Fisheries of the Presumpscot River and Sebago Lake: http://friendsofsebago.org/historicreport.html (accessed 31 July 2011).

Gordon, R., and Raber, M., 2000, Industrial Heritage in Northwest Connecticut, a Guide to History and Archaeology: New Haven, Connecticut Academy of Arts and Sciences, p. 188.

Hartford Courant, 1934, September 30th 1934: ProQuest Historical Newspapers Hartford Courant, 1764–Current, p. C6.

Ingraham, A., 2009, A Short History of the Milldam at North Billerica, 1653–1995: Billerica, Massachusetts, Billerica Historical Society, 25 p.

Jansen, R., 1980, Dams and Public Safety: Water & Power Resources Service (SW/74), U.S. Department of the Interior, 332 p.

Kulik, G., 1985, Dams, fish, and farmers: Defense of public rights in eighteenth-century Rhode Island, in Hahn, S., and Prude, J., eds., The Countryside in the Age of Capitalist Transformation: Chapel Hill, University of North Carolina Press, p. 25–50.

Mullens, J., 2003, An examination of dam removal in New England: Proceedings of the New England–St. Lawrence Valley Geographical Society, v. XXXII, p. 51–62.

Pohl, M., 2002, Bringing down our dams: Trends in American dam removal rationales: Journal of the American Water Resources Association, v. 38, no. 6, p. 1511–1519, doi:10.1111/j.1752-1688.2002.tb04361.x.

Thoreau, H., 1849, A Week on the Concord and Merrimack Rivers: Penguin Classics, 384 p.; available at www.tower.com/a-week-on-concord-merrimack-rivers-h-daniel-peck-paperback/wapi/100539304 (accessed 9 October 2012).

Whitelaw, E., and MacMullan, E., 2002, A framework for estimating the costs and benefits of dam removal: Bioscience, v. 52, p. 724–730, doi:10.1641/0006-3568(2002)052[0724:AFFETC]2.0.CO;2.

Wildman, L., 2003, Cursed on Both Ends and Dammed in the Middle: Exploring the Human Dimensions in the Efforts to Remove Dams and Restore Rivers—The Billerica Dam Case Study: New Haven, Connecticut, Yale University, School of Forestry & Environmental Studies, Course Paper FES 565, 24 p.

Wildman, L., and MacBroom, J., 2010, A Broad Level Classification System for Dam Removals, in Joint 9th Federal Interagency Sedimentation Conference and 4th Federal Interagency Hydrologic Modeling Conference Proceedings, 27 June–1 July 2010, Las Vegas, Nevada: http://acwi.gov/sos/pubs/2ndJFIC/Contents/10B_Wildman_100416_paper.pdf (accessed 31 July 2011).

Manuscript Accepted by the Society 29 June 2012

Engineering considerations for large dam removals

Thomas E. Hepler

U.S. Bureau of Reclamation, P.O. Box 25007, Denver Federal Center, Denver, Colorado 80225-0007, USA

ABSTRACT

The removal of a large dam requires special engineering considerations not normally required for the removal of smaller dams. Large dams generally provide greater project benefits, but they represent a higher downstream hazard in the event of failure and greater challenges to fish passage. The reservoirs associated with large dams can impound more sediment and affect downstream water quality to a greater degree. The environmental compliance and decision-making process for a large dam removal project can take many years and will normally require the evaluation of a full range of project alternatives with estimated costs, the performance of a comprehensive environmental impact analysis, and the identification and implementation of extensive mitigation measures. Streamflow diversion and demolition plans for the removal of a large dam must ensure acceptable construction risks from start to finish, and produce reservoir drawdown at a controlled rate for sediment management purposes and to prevent instability of natural or embankment slopes. Large dams require more time for removal, at a higher cost, and contain greater volumes of materials for which disposal sites must be found.

INTRODUCTION

Although few large dams, defined as having a height of greater than 15 m,[1] have been removed to date, there are several river restoration projects throughout the western United States that, if implemented, would each include the removal of one or more large dams. This paper describes the special engineering considerations for large dam removals that are normally not required for the removal of smaller dams. Examples are provided from current projects involving the removal of two large concrete dams on the Elwha River in Washington, the removal of a large concrete dam on the Carmel River in California, and the potential removal of three large dams (plus one smaller diversion dam) on the Klamath River in Oregon and California.

Large dams generally provide a variety of project benefits, including the provision of storage for municipal and domestic water supply, flood control, recreation, fish and wildlife, hydroelectric power generation, and irrigation. Most large dams in the United States are owned and operated by federal or state agencies, or by public utilities (for additional details, refer to the National Inventory of Dams [NID] through https://nid.usace.army.mil), and must follow federal or state regulations for dam safety and environmental compliance. Large privately owned dams that generate power are licensed by the Federal Energy Regulatory Commission (FERC), and must be operated and maintained in accordance with FERC requirements as well as

---

[1]International Commission on Large Dams (ICOLD) definition of a large dam. Other qualifications include a height greater than 10 m with a crest length greater than 500 m, or a storage capacity greater than 1,000,000 m³, or a spillway discharge capacity greater than 2000 m³/s. Height is measured from the streambed at the downstream toe to the dam crest.

those of the state agency having regulatory jurisdiction. Since large dams would normally pose a danger to human life and/or produce serious economic damage in the event of failure, most carry a high hazard classification and must meet a higher level of engineering review and inspection than for smaller, low-hazard dams. Regulatory agencies can order the emergency evacuation and necessary repairs or removal of any dams found to pose an unacceptable risk to the public.

The primary reason for the removal of large dams is environmental, resulting from adverse impacts due to blockage of anadromous fish passage, changes in water quality, changes in hydrologic regime, and downstream degradation due to changes in the sediment regime. These impacts are generally much lower or nonexistent for smaller dams, which are more likely to be removed for public safety and economic reasons (Pohl, 2002). The reservoirs associated with large dams, with storage volumes greater than 1,000,000 m^3, can impound more sediment and affect downstream water quality to a greater degree. Ensuring the passage of anadromous fish around large dams can be particularly challenging and expensive. Regulatory agencies such as FERC can require the removal of dams that do not meet environmental requirements for which other alternatives are not selected. In the case of the Klamath River dams, FERC established mandatory conditions for fish passage at each of four hydroelectric dams owned by PacifiCorp (FERC, 2007), which led the owner to consider dam removal as a potentially more economical alternative than extensive structural modifications and a change in reservoir operations. The required fish passage measures would have also presented greater uncertainties than dam removal regarding their effectiveness for upstream passage of adults and downstream passage of juveniles, considering the size of the dams and reservoirs.

DECISION-MAKING PROCESS

The decision-making process for a large dam removal project often begins at the local level with the dam owner; stakeholder groups, including Native American tribes and nongovernmental organizations having an environmental or commercial fishing focus; or at the state level with regulatory agencies. Although studies performed in 1990 for the owners of the Elwha River dams concluded dam removal and sediment management would be extremely costly, studies completed the following year for the Lower Elwha Klallam tribe identified alternative methods of dam removal and sediment management that were less costly and had fewer environmental impacts. Early studies for the potential removal of four hydroelectric dams on the Klamath River were funded by the California State Coastal Conservancy (SCC), a state agency that works with other public agencies and the private sector to help preserve California's coastal resources. The SCC was also instrumental in the decision to remove San Clemente Dam (Hepler et al., 2011).

The National Environmental Policy Act (NEPA) of 1969 established a U.S. national policy to promote the environment, including an interdisciplinary approach to decision making, and it established the President's Council on Environmental Quality (CEQ). NEPA requirements, as defined by CEQ regulations (Title 40, Code of Federal Regulations, Chapter V, Sections 1500–1508), are generally applied to any project, whether federal, state, or local, that involves federal funding, regulatory control, or work performed by or for the federal government. The primary purpose of NEPA is to ensure that environmental factors are weighted equally when compared to other factors in the decision-making process undertaken by federal agencies. The removal of a large dam will normally qualify as a major federal action, for which an Environmental Impact Statement (EIS) will be required. A federal lead agency will be identified for purposes of development and approval of the EIS. An Environmental Assessment (EA) and a Finding of No Significant Impact (FONSI) may be sufficient for a smaller project.

The primary purpose of an EIS is to help public officials make informed decisions that are a reflection of an understanding of the environmental consequences of a proposed action (such as dam removal) and of the reasonable alternatives available. An EIS is required to describe:

(1) the environmental impacts of the proposed action;
(2) any adverse environmental effects that cannot be avoided should the proposal be implemented;
(3) the reasonable alternatives to the proposed action;
(4) the relationship between short-term uses of man's environment and the maintenance and enhancement of long-term productivity; and
(5) any irreversible and irretrievable commitments of resources that would be involved in the proposed action should it be implemented.

A detailed study is required to evaluate the environmental impacts of the proposed action, as well as those of a full range of project alternatives consistent with the purpose and need statement and with project objectives. The affected environment and environmental consequences of the proposed action and of reasonable alternatives should be presented in comparative form to provide a clear basis for choice among options by the decision maker and by the public. For a large dam removal project, the alternatives will include the "No Action" alternative, which would consist of "reasonably foreseeable" actions without the project, and alternatives that would remove or modify the structure to meet dam safety and/or environmental requirements, and would be considered both technically and economically feasible. Alternatives to the proposed removal of San Clemente Dam included provisions for fish passage and potential structural modifications to either: (1) strengthen the dam for seismic loads and increase the spillway discharge capacity, or (2) reduce the dam height to either accommodate the maximum loads or remove the downstream hazard in the event of dam failure (Hepler et al., 2011). Alternatives to the proposed removal of four dams on the Klamath River included: (1) the removal of the two larger dams and provisions for fish passage at the two smaller dams, and (2) provisions for fish passage at all four dams (U.S. Bureau of Reclamation, 2012).

NEPA requires that the EIS present the environmental consequences and their significance for each action alternative, considering the "context and intensity" of each action when compared to the baseline conditions for a wide range of resources that could be affected. The "No Action" alternative is commonly used as the baseline for determining the significance of the environmental impacts within an EIS. The affected environment and environmental consequences for each of the identified resources are discussed in detail. The affected environment includes the environmental setting (the physical environmental conditions in the vicinity of the project as they existed prior to preparation of the EIS) and the regulatory setting (the applicable laws, regulations, permits, and policies associated with each resource), as appropriate. The environmental consequences include an assessment of impacts, as well as a discussion of the respective mitigation, compensation, or restoration for each resource. Typical categories of resources include biological resources, such as fish, aquatic communities, and terrestrial species (wildlife and plants); physical resources, such as surface water hydrology, groundwater, water quality, water supply, air quality, and geology; and the social environment, including land use, population, socioeconomics, aesthetics, transportation, noise, public health and safety, public services and utilities, solid and hazardous waste, recreation, cultural and historical resources, power generation, environmental justice, and Indian Trust assets.

The significance of an impact under NEPA generally relies upon professional judgment and knowledge of the context within which the impact would occur. Significant, less-than-significant, and beneficial impacts, along with cumulative impacts, are identified for each resource being evaluated:

1. A significant impact would cause a substantial adverse change in the environment. Mitigation measures are normally required for all significant impacts.
2. A less-than-significant impact would cause an adverse, but not a substantially adverse, change in the environment. Compensation or restoration measures are recommended for all less-than-significant impacts.
3. A beneficial impact would cause a change in the environment for the better.
4. Cumulative impacts are impacts on the environment that result from the incremental impacts of a proposed action when added to other past, present, and reasonably foreseeable future actions. Cumulative impacts can result from individually minor, but collectively significant actions taking place over a period of time.

Mitigation measures are methods and techniques that can be implemented under a project alternative to reduce the amount of adverse environmental impacts during and after construction. The following potential mitigation measures are listed in the order in which they would be applied:

(1) avoid the impact by not taking a certain action or parts of an action;
(2) minimize the impact by limiting the degree or magnitude of the action and its implementation;
(3) rectify the impact by repairing, rehabilitating, or restoring the affected environment;
(4) reduce or eliminate the impact over time by preservation and maintenance operations during the life of the action; and
(5) compensate for the impact by replacing or providing substitute resources or environments.

The loss of reservoir storage may produce numerous impacts that need to be identified and evaluated as part of a dam removal project. Obviously, the original benefits for which the dam was authorized and constructed may be permanently lost, which could include water supply, flood control, power generation, and recreation. Legal rights to water diversions may need to be addressed. In addition to the loss of water storage, lower groundwater levels may result, which in turn could impact local wells and springs. Downstream water quality may be impacted by the passage of natural sediments (either as suspended solids or bed load) that had previously been contained within the reservoir. The coarser sediments may be deposited along the downstream channel, producing higher river stages and greater flooding potential. Downstream water intakes may also be adversely affected by sediment deposition.

The removal of a dam and loss of reservoir storage may also produce significant impacts to infrastructure within the reservoir area. The loss of channel depths may affect river navigation, and the removal of the dam may eliminate an important river crossing. Bridge piers, roadway and railroad embankments, levees, drainage culverts, and buried utilities (such as water and natural gas pipelines) within the reservoir area may become subjected to higher flow velocities, potential scour, and surface erosion. Previously inundated cultural and archaeological sites may become exposed and subject to erosion or human disturbance. Property values along the former lake shoreline may be affected. The lakebed exposures may produce a dust hazard that requires abatement, erosion control, and vegetation. Mitigation for any upstream and downstream impacts may be required as part of the permitting process.

The EIS for a large dam removal project should be prepared concurrently with and integrated with environmental impact analyses and related surveys required by the Endangered Species Act (Section 7), the Fish and Wildlife Coordination Act, the National Historic Preservation Act (Section 106), and other applicable federal and state environmental review laws and executive orders. In addition, all federal permits, licenses, and other entitlements that must be obtained in implementing the proposed action should be listed in the EIS. These would normally include several permits under the Clean Water Act, including a Section 404 permit from the U.S. Army Corps of Engineers (USACE) for the discharge of dredged or fill materials into waters of the United States, including wetlands; a state-issued water-quality certification, or Section 401 permit; and a state-issued National Pollutant Discharge Elimination System (NPDES), or Section 402, permit. Various other state and local permits may also be required for dam removal.

If a cost-benefit analysis relevant to the choice among environmentally different alternatives is considered for the proposed action, it must discuss the relationship between that analysis and any analyses of unquantified environmental impacts and other important qualitative considerations. Many potential benefits associated with large dam removal projects may be difficult to assess monetarily, especially those associated with Native American tribal fishing rights and cultural resources. The Elwha River Restoration Project and the Klamath River Dams Removal Study each used a nonuse value survey to collect information pertaining to the value that people derive from public goods or natural resources (such as for a reservoir vs. a free-flowing river) that are independent of any use, present or future, that people might make of those goods.

The decision-making process for a major federal action under NEPA would include the following steps:
(1) Issuance of a Notice of Intent (NOI) for preparation of the EIS.
(2) Public scoping for the EIS and receipt of public and agency comments.
(3) Preparation of a draft EIS.
(4) Issuance of a Notice of Availability of the draft EIS, and circulation of the draft EIS for a 60 day public and agency review and comment period.
(5) Preparation of a final EIS (including responses to all comments received) and identification of the preferred alternative.
(6) Filing of the final EIS with the U.S. Environmental Protection Agency (EPA) and publication of the Notice of Availability of the final EIS in the Federal Register.
(7) Final EIS 30 day no action period.
(8) Filing of a federal Record of Decision (ROD) regarding the project alternative to be implemented.

The final approval, design, and implementation of a large dam removal project can take many years. This may be due to various legal, environmental, political, social, and economic reasons. FERC began preparing an EIS for the Elwha and Glines Canyon Hydroelectric Projects in 1989 to evaluate the potential impacts of the hydroelectric projects for re-licensing and of potential alternative actions. With strong stakeholder support, the Elwha River Ecosystem and Fisheries Restoration Act (Elwha Act) was enacted in 1992 to suspend FERC's authority to issue licenses for the projects, and to require the Secretary of the Interior to propose a plan to fully restore the Elwha River ecosystem and native anadromous fisheries. The Secretary of the Interior determined in 1994 that removal of both dams was both feasible and necessary to meet restoration goals. The National Park Service (NPS) was assigned lead responsibility for preparation of an interagency, programmatic (or policy) EIS for the evaluation of several alternatives, including no action, dam retention with fish passage, removal of Elwha Dam, removal of Glines Canyon Dam, and removal of both dams, including the environmental impacts of each alternative. Following public comment and the issuance of the Final Programmatic EIS in June 1995, a Record of Decision was prepared by NPS recommending removal of both dams. The Final Implementation EIS was issued in November 1996 to address the specific methods and mitigating measures necessary to drain the reservoirs, remove the dams, manage the accumulated sediments, restore the reservoir areas, improve the fisheries, and protect downstream water quality. The Final Supplemental EIS was issued in July 2005 to account for changes since 1996, including fish species newly listed under the Endangered Species Act (ESA) and the final water-quality mitigation plans. The design and construction of significant mitigation measures had to be completed before the dams could be removed. Additional project funding was approved in 2009, and a contract for removal of both dams was awarded by NPS in 2010, or 18 years after the Elwha Act was enacted (Hepler, 2010).

ENGINEERING DESIGN AND CONSTRUCTION

Design Level and Scope

Engineering design for large dam removal projects includes planning, design, and implementation, and should incorporate the overall site restoration goals and project objectives. Natural physical and biological processes of the river should be incorporated to the extent possible to facilitate final site restoration. The application of current technologies and engineering practices by competent and experienced design engineers increases the likelihood of site-appropriate designs, realistic cost estimates, reduced risk and uncertainty, and optimum results (Aspen Institute, 2002). Individuals with a full range of relevant expertise and experience should be considered for the design team, including civil, geotechnical, mechanical, and electrical engineers; cost estimators; hydrologists; geologists; geomorphologists; land-use planners; safety professionals; construction and hazardous material specialists; fishery biologists; chemists; economists; ecologists; archaeologists; and historians.

For projects directly undertaken by or for federal agencies, engineering designs are prepared at the feasibility level for the EIS, and are often based on previous preliminary designs. Feasibility-level designs and cost estimates utilize information and data obtained during pre-authorization investigations, from which feasibility-level quantities for each kind, type, or class of material, equipment, or labor may be obtained. Feasibility-level cost estimates are used to assist in the selection of the preferred alternative, to determine the economic feasibility of a project, and to support seeking project authorization. Engineering designs and cost estimates will normally be prepared for one or more project alternatives in addition to dam removal. Mitigation measures may also require engineering designs and cost estimates. The mitigation measures for the Elwha River Restoration Project included the design and construction of flood-control levees, two new water treatment plants, and a new fish hatchery, which ultimately cost many times more than the removal of the two dams alone (Hepler, 2010).

Final designs and specifications will be prepared for the preferred alternative based on the feasibility design, following an expression of approval or support from the regulatory agencies and key stakeholders, and the identification of a funding source. The final design specifications for a large dam removal project will include details of the existing site conditions and the project requirements to achieve the desired results. Design assumptions will be made for development of the owner's cost estimates. The specific methods to meet the project requirements, such as for streamflow diversion and structure demolition, are normally left to the contractor to determine, for approval through a specified submittal and review process. Any special site restrictions, such as reservoir drawdown rate limitations and the prohibition of blasting, or any environmental constraints, such as in-water work periods and noise restrictions, should be included in the specifications. Strict compliance with the specifications and with applicable project permit requirements and conditions will be required of the contractor by the owner and regulator.

A negotiated procurement process, in which contractors are requested to submit formal proposals for evaluation using predefined technical and price criteria, and are given the opportunity to respond to questions and revise their proposals, is an effective tool for the selection of a contractor and award of a large dam removal contract. By allowing the contractor to plan and implement the dam removal based on performance specifications, efficiencies can be realized during construction. Potential contractors for large dam removal projects should be bonded and meet minimum experience requirements for the proposed work. The use of a prequalified contracting process can facilitate the review of technical proposals, improve the accuracy of cost estimates, and ultimately result in a contract award to the best qualified contractor.

Dam Type and Structure Removal Limits

The type of dam and discharge capacities of appurtenant structures will help determine the engineering design required for its removal. Large dams include one or more appurtenant structures for control of reservoir releases (i.e., spillways and outlet works) and are constructed of concrete or earth materials, or both. Embankment dams consist of earth and/or rock, and normally include an impermeable barrier (either an impervious central core or upstream membrane) and outer slope protection (commonly rock riprap on upstream face, and rockfill or grass on downstream face). Concrete dams may be of the gravity, arch, or buttress type, based on how the reservoir loads are carried. Composite dams consist of both embankment and concrete sections, and will generally include the spillway and outlet works within the concrete section. Spillways may consist of a free overflow crest, or include gates or flashboards to regulate flood releases to the downstream channel. The outlet works may consist of a low-level pipe, conduit, or tunnel constructed through or around the dam, and will include gates or valves for controlled releases for power generation, water supply, emergency evacuation, or diversion during construction. Diversion outlets may have been sealed following construction of the dam and may require modifications to restore.

The limits of structure removal should be identified as part of the planning and design process. Full removal generally means removal of the dam and appurtenant structures such that no visible evidence of the original structures will remain. The stream channel is normally restored to its original alignment and grade following full removal, with natural-looking ground contours provided on the abutments. Portions of the structures may remain buried on site, especially below the original ground surface. Partial removal will allow some structures, or portions of structures, to remain in place, either within or outside of the original stream channel, but no significant reservoir impoundment will remain. Remaining structures must be capable of performing satisfactorily in the future under anticipated potential loading conditions and may have to be modified for public safety (either to prevent entry or accidents) or for the safe passage of large floods. The stream channel may be restricted or diverted to a new alignment following a partial breach of the dam, as for the case of a large composite dam for which the embankment portion only is removed to reduce project costs.

Structure removal limits will generally be based on a wide range of factors, including regulatory requirements, public safety and liability issues, operation and maintenance requirements, landowner preferences, sediment management issues, aesthetic factors, historical and cultural values, and costs. The type of dam will play an important role in the establishment of removal limits. If partial removal is considered, the construction materials used in the dam (concrete, earth, or rockfill) can determine the extent of structure removal required to meet project requirements and ensure the future performance of any remaining portions. Concrete gravity dams can generally be excavated to vertical surfaces, and can usually withstand overtopping during floods, unless foundation scour presents a stability concern. Modifications to a concrete arch dam may be limited to ensure structural stability of any remaining cantilever sections under potential hydraulic loads. Embankment dams will require excavation to much flatter slopes for stability (generally 2H:1V or flatter, depending on the materials) and may require some degree of erosion and scour protection to withstand any potential overtopping. A notch through an embankment dam for a partial breach is generally designed so that the perimeter of the notch is resistant to erosive flows for the design flood.

A partial breach to retain the lower portion of a dam to some height above streambed may be desirable to serve as a silt retention barrier and for stabilization of upstream sediments, or to reduce the potential for dam failure. Such was the case at Agency Dam, a 10-m-high concrete arch dam on the Rocky Boys Indian Reservation in north-central Montana, which was notched to retain the existing impounded sediments and reduce downstream risk in the event of dam failure. A partial breach of San Clemente Dam in California was considered as an alternative to dam removal (Hepler et al., 2011). Public safety and liability issues would require consideration of the potential hazards to the public represented by the remaining portions of the dam, such as an attractive nuisance on the stream or an abandoned gate tower or surge tank that could topple during a large earthquake. Remaining structures may require some long-term operation and maintenance activities that should be considered in any decision for partial removal. Aesthetic factors may affect a decision for partial removal, especially in areas subject to public view. Historical and cultural values may also play a role in the removal of older structures. The dam and appurtenant structures may be protected under the National Historic Preservation Act, and retention of portions of the dam may be desirable to preserve a particular element of the original design for posterity. Finally, land-use and landowner/stakeholder preferences must be taken into consideration when establishing removal limits. Full removal was required at Elwha Dam in Washington to fulfill federal trust obligations and to meet the cultural resource needs of the Lower Elwha Klallam tribe. Partial removal was acceptable to the NPS for Glines Canyon Dam, also located on the Elwha River, to provide for public overlook areas and to help document the original hydroelectric project dating back to 1927 (Hepler, 2010).

Design Data Collection

A complete and thorough understanding of the existing structures is important to the success of any dam removal project (Aspen Institute, 2002). All existing data appropriate to the scale of a project should be collected prior to the final design stage. This will include all available construction drawings and any supporting design calculations and project specifications. Older structures may have little or no existing information, in which case new drawings reflecting current conditions may have to be prepared based on field surveys and measurements. Operating entities and regulatory agencies should be consulted for any pertinent information available in their records, including inspection reports. Modifications performed over the years should be identified and incorporated into the design records. Local historical societies, archives, and public libraries may have additional project information, including project histories, construction photographs, and newspaper reports. An Internet search may also provide some useful information. Of particular importance would be construction materials testing data (such as concrete compressive strengths or embankment gradations), and details of embedded metalwork, post-tensioned anchors, and steel reinforcement in concrete structures. Preconstruction topographic maps would be useful to determine the original ground surfaces within the reservoir for estimating sediment volumes (when used with current bathymetric data) and at the dam site for developing removal limits.

After collecting all available engineering data, additional fieldwork consistent with the project goals and objectives may be required to confirm the site conditions. This may include field measurements of the structural dimensions of all major project features for the specifications drawings and quantity estimates; documentation of the operating history and present condition of gated spillways and outlet works to be used for streamflow diversion; and a complete inventory of mechanical and electrical equipment to be removed, including weights (if available). Drawings prepared for the removal specifications may be useful for Historic American Engineering Record (HAER) documentation purposes. Offsite locations and haul distances for disposal of waste materials will normally need to be identified. Additional site surveys may be required to develop dam site ground surface contours (generally at 0.6 m intervals); river cross sections, including plunge pools and tailrace areas; and reservoir bathymetry, to develop a sediment profile at the dam and reservoir cross sections. Field test data could include concrete compressive strengths and unit weights; index and engineering properties of embankment materials and reservoir sediments (including gradation, plasticity, and density); and the identification of onsite hazardous materials, including asbestos, coating contaminants (such as lead-based paint), batteries, treated wood, chemicals, petroleum products, polychlorinated biphenyls (PCBs), and mercury. Hazardous materials are often associated with power plants and mechanical or electrical equipment, and may affect a decision to fully or partially remove a structure. Materials determined to be hazardous will require special handling and disposal at approved facilities, as sampling, testing, labeling, manifesting, transporting, and disposal of hazardous materials are regulated by law. Testing of the surrounding soils and impounded reservoir sediments should be performed for possible contamination. Historic industrial or mining operations upstream of the site and unrelated to the dam project may have contaminated the reservoir sediments. Septic systems and underground storage tanks should also be located and tested (ASCE, 1997).

The evaluation of geologic conditions for larger dam removal projects could include an assessment of slope stability of the dam abutments and upstream slopes during reservoir drawdown and following dam removal; determination of the erosion resistance of the dam abutments and foundation for flood flows; subsurface explorations for potential diversion channels or tunnels; and estimation of foundation permeability and groundwater levels for unwatering the site excavations. Geologic data may also be required for sediment transport and river erosion models to predict important changes to the streambed profile (deposition and erosion) and alignment (river meander or avulsion). A sediment management plan will be required when the impoundment contains large quantities of sediment (both clean and contaminated), to provide for the natural erosion, or proper handling and disposal, of the sediment. Potential earthquake loadings and seismic stability should be considered for the retention of any large, tower-like structures.

The evaluation of the hydrology of the site would be required for a partial dam removal to ascertain the appropriate size for a safe breach, particularly if there would be downstream consequences in the event of dam failure. Many sources of hydrologic, hydraulic, and sediment data are available for gauged streams within the United States (Aspen Institute, 2002). Streamflow records should be obtained for the site and be evaluated to determine frequency flood peaks for diversion purposes during dam removal. For larger projects, water surface profiles should be prepared along the stream for various discharges to develop stage-discharge relationships both with and without the dam. If the intention of the breach is to render the dam a low-hazard structure, flood routings and dam break studies may be required to verify that the consequences of failure of the lower structure are acceptable. If the portions remaining after the breach are expected to withstand flood flows, hydraulic and armoring analysis and design would be needed. Remaining site features exposed to flood loadings should remain stable up to some prescribed minimum return period (such as 100 years), depending upon the potential consequences of failure or as required by the regulatory agencies.

Reservoir Drawdown and Streamflow Diversion

The reservoir size and streamflow characteristics of the site, and the available capability for drawing the reservoir down safely while passing normal flows are critical in planning for the removal of a large dam. The size and location of existing release facilities should be determined, and discharge curves should be developed or obtained from available records to plan for a reservoir drawdown. These curves will indicate the expected discharge for a given reservoir level, based on a specific gate opening. Operating restrictions may be in place to limit gate openings to avoid cavitation damage or excessive flow velocities or air demand; however, since the releases are to facilitate dam removal, some damage to the release facilities may be tolerable. The downstream channel capacity should also be evaluated to determine whether any damage would result from maximum reservoir releases during drawdown. Emergency Action Plans (EAPs) or downstream inundation maps may contain information pertaining to safe channel capacities, based on a stage-discharge relationship.

Reservoir drawdown requires the release of reservoir volume as well as the downstream passage of inflow to the reservoir. A storage-elevation curve or reservoir capacity tables should be obtained from project records when available, or developed using site topography or bathymetry. Streamflow records for the upstream basin should be evaluated when available, or developed based on the records of a nearby gauged basin. Mean, minimum, and maximum daily flows for the construction period would be most useful, with some additional information as to the potential for general rainfall or thunderstorm events. Critical operations during dam removal may require some degree of flood protection, such as a 5- or 10-year event, or greater, depending upon the estimated duration for the work and the potential consequences in the event of failure. Timing the key demolition activities for low-flow periods should be considered when possible; however, environmental constraints may restrict the dam removal schedule. A large concrete dam on the Klamath River (Copco No. 1 Dam) would have to be removed during the winter (between 1 January and 15 March) to minimize downstream impacts on fish due to the release of fine sediments (U.S. Bureau of Reclamation, 2012).

Ideally, the outlet works for the dam would be located near the original streambed and be large enough to drain the reservoir at an optimum rate, without exceeding drawdown rate limitations established for the site (ASCE, 1997). The rate of reservoir drawdown must be evaluated pertaining to the potential instability of upstream embankment slopes or landslides along the reservoir rim. Average drawdown rates may have to be limited to between 0.3 and 0.6 m per day to prevent a rapid drawdown condition that could cause slope instability. Even this might not be slow enough if the slopes are composed of slow-draining materials and are marginally stable. Drawdown rate limitations may also be established for sediment management, to ensure sufficient time for erosion of impounded sediments downstream during dam removal, rather than leaving large volumes of sediment within the reservoir area for future erosion. Reservoir drawdown for the removal of Elwha and Glines Canyon Dams in Washington, for example, was limited to a maximum of 0.9 m every 48 hours, with hold periods of 14 days for every 4.6 m of drawdown, for sediment management considerations. Environmental impacts to fish related to water quality may require the establishment of "fish windows," during which time no reservoir drawdown producing elevated turbidity levels from the release of sediments would be allowed. The Elwha River Restoration Project identified three such "fish windows" covering a total period of five and a half months per year (Hepler, 2010).

The ability to safely pass streamflow during dam removal is often critical to the dam removal process, especially for an embankment dam, which is much more vulnerable than a concrete dam during floods that can cause overtopping. When the existing release facilities are incapable of passing anticipated flows, or are not low enough to draw the reservoir down sufficiently, alternative diversion methods must be employed. For large embankment dams, a diversion tunnel may be excavated through one of the abutments, requiring a costly lake tap excavation beneath the reservoir surface. Limited overtopping protection may be provided for an embankment dam by armoring the flow surfaces with large or grouted riprap, or with precast concrete blocks or mats, but the potential risk of failure must be taken into consideration. Concrete dams can typically accommodate streamflow diversion more easily by passing flows over the top of the dam. A series of notches can be excavated through a concrete dam on alternating sides to permit reservoir drawdown for removal of the remaining portions of the dam under dry conditions. Maximum notch depths for concrete arch dams may need to be determined by analysis to prevent potential sliding or overturning of concrete blocks during large flood events. Excavated notches up to 4.6 m deep were permitted for the removal of Glines Canyon Dam in Washington (Hepler, 2010). Alternatively, a diversion outlet can be provided through the dam, either by modification of an existing outlet conduit (such as a power penstock, by modification or removal of the turbine), by restoration of the original diversion outlet (such as by removing a concrete plug from a diversion tunnel), or by constructing a new outlet through a concrete dam (such as by excavating to an upstream underwater bulkhead). Surface diversion channels that pass streamflow around the dam, through an excavated rock channel, a lined earth channel, or a temporary flume or pipeline may also be employed.

The streamflow diversion method and release capacity will determine the actual reservoir levels and drawdown rates experienced during dam removal. The passage of flood flows that exceed the release capacity of the streamflow diversion facilities may cause a temporary suspension of work and impact the contractor's schedule. Some form of flow control such as a gate or valve may be provided for the diversion facilities, especially for those requiring streamflow to pass through the dam or abutment well below the reservoir surface. This provides a means by which to control the rate of reservoir drawdown for sediment management purposes, or to prevent downstream releases that could exceed the safe downstream channel capacity during construction. The construction of an upstream cofferdam can provide temporary storage to reduce the natural flow through the demolition site, or provide greater head on a temporary diversion pipeline to increase its diversion capacity. A downstream cofferdam can prevent backwater effects on the demolition site. Cofferdams can be constructed of earth or rockfill embankments using onsite materials (often with a geomembrane water barrier), large sandbags, precast concrete gravity blocks, sheet-piling, or various commercial products. Cofferdams that utilize in-stream materials, or clean gravels and cobbles, may require less effort for subsequent removal. The design flood for streamflow diversion facilities will be much more frequent for removal of a small, low-hazard dam than for removal of a large, high-hazard dam.

Demolition and Site Restoration

Large dams will require more time for removal than smaller dams, and contain greater volumes of materials for which disposal sites must be found. Although the volume of material in an earth or rockfill dam is generally much greater than that required for a concrete dam, the excavation and disposal of earth and rockfill materials is usually much easier. Concrete waste disposal may pose problems, depending upon the composition of the concrete and location and availability of suitable waste disposal sites. Concrete crushing and recycling may be an economical option. These factors will influence the choice of demolition methods and dam removal costs (ASCE, 1997).

Drilling and blasting is generally the most economical and effective method for concrete demolition, where permissible. Controlled blasting techniques may be required depending upon the site conditions, with limited load factors and the possible use of blasting mats, in order to control flyrock and prevent damage beyond the structure removal limits. Mechanical demolition methods are also common for dam removal projects, and include impact equipment such as boom-mounted hydraulic impact hammers (or hoe-rams), crane-operated wrecking balls, and jackhammers. Hydraulic splitters are effective for controlled excavation in plain and reinforced concrete and in masonry, and hydraulic shears are effective for removal of reinforced concrete walls and roof slabs. Earth and rockfill dams may be removed using common excavation methods and earth-moving equipment, and they can provide a source of clay, sand, gravel, cobbles, and rock for site restoration or for local commercial use. The removal sequence for a large embankment dam should generally be from the top down to prevent steep slopes. A controlled breach may be considered for removal of the lower portion of an embankment dam, provided the resulting erosion of earth materials would not harm the downstream channel or produce an excessive peak flow compared to the safe channel capacity.

Construction access to the project features and waste disposal sites is generally an important factor affecting dam removal costs. A broad canyon site should provide good access to one or both abutments and will facilitate dam removal operations and provide flexibility in demolition methods and equipment. A narrow, deep canyon site may limit access to the top of the dam, however, and may require the use of helicopters or the construction of a new access road. Other potential options include the erection of a large crane on one or both abutments, or the installation of an overhead cableway across the canyon, for movement of equipment and materials. Access roads used for the original construction of the dam may not be suitable for modern construction equipment or current safety requirements, considering roadway widths, grades, horizontal alignment (i.e., curves), and load capacities (including bridges, culverts, and retaining walls). Significant improvements to existing access roads and bridges may be required to perform the work, and should be considered in the project cost and schedule. Site restoration requirements should specify whether site access roads are to be preserved or obliterated following dam removal. This will be dependent upon the proposed future land use of the site.

Materials removed from a dam and its appurtenant structures may include concrete, reinforcing steel, mechanical and electrical equipment, miscellaneous metalwork, earth materials (including clay, sand, gravel, cobbles, and rock), timber (treated or untreated), and various other building materials. Suitable waste disposal sites should be identified within a reasonable haul distance for all of these materials. Some waste disposal sites may require special separation of materials prior to disposal, such as the removal of reinforcing steel from concrete, and the separation of combustible and noncombustible materials. Waste concrete can be crushed for reuse as a road base or other construction materials in areas where supplies are limited. Mechanical and electrical items should be evaluated for potential salvage value, either intact for its commercial or historic value, or recycled for its metallic components. Borrow areas used for the original construction of the dam should be considered for the disposal of earth materials and concrete rubble if available. Onsite disposal areas may require special permits and approvals from state and local entities. Waste earth and concrete materials from demolition activities may be suitable for use as backfill, as long as adequate compaction is provided to prevent the development of future surface subsidence or sinkholes. Topsoil and seeding should generally be provided over backfilled areas to promote revegetation. Materials determined to be hazardous will require special handling and disposal at approved off-site facilities.

Following removal of all or parts of the dam and appurtenant structures, the remaining features may have to be modified to ensure public safety or to minimize long-term operation and maintenance requirements. Retired buildings should be secured to prevent unauthorized entry by locking doors, covering windows, sealing access hatches and vents, removing access ladders, and posting signs. Roofing materials and drainage features may need to be maintained to prevent water damage. Site fencing may need to be added or extended at any remaining facilities. Tunnel portals may need to be screened, plugged, or backfilled, with possible special provisions for future inspection and drainage. Buried pipelines may need to be removed to prevent future deterioration and collapse. Historic properties may require other protections in accordance with state regulations. Sites that are to be opened to the public may require special accommodations for disabled visitors in accordance with the Americans with Disabilities Act (ADA), including ramps and accessible restroom facilities (ASCE, 1997). The sale of land or the release of easements may first require some level of site restoration to remove potential public safety or environmental hazards.

CASE STUDIES

The following case studies represent three large river restoration projects, either proposed or under way in the western United States, involving the removal of one or more large dams. Other large dam removal projects include the removal of Condit Dam from the White Salmon River in Washington, and the potential removal of Matilija Dam from Matilija Creek in California. Each of these large dams is being removed to facilitate anadromous fish passage.

Elwha River Restoration Project (from Hepler, 2010)

Elwha and Glines Canyon Dams were located on the Elwha River in northwestern Washington State, on the Olympic Peninsula. Elwha Dam (Fig. 1) was completed in 1913, ~7.9 km upstream from the mouth of the river, and included a 33-m-high concrete gravity section, gated spillways on both abutments, and a powerhouse with four generating units rated at a combined capacity of 12 megawatts (MW). Glines Canyon Dam (Fig. 2) was completed in 1927, ~14 km upstream from Elwha Dam, and included a 64-m-high concrete thin arch section, a gated spillway on the left abutment, a thrust block on the right abutment, and a powerhouse with a single generating unit rated at 16 MW. Both dams were owned by the U.S. Federal Government and were operated and maintained for the NPS by the Bureau of Reclamation. Glines Canyon Dam was located within Olympic National Park.

The removal of Elwha and Glines Canyon Dams was authorized and funded by Congress to provide anadromous fish passage to more than 112 km of the Elwha River and its tributaries, and for full restoration of the Elwha River ecosystem. A construction contract for the removal of both dams was awarded in August 2010, with site work beginning in June 2011 and to be completed by September 2014. The demolition work was required to accommodate river flows during dam removal, and facilitate sediment management through restricted construction schedules and controlled releases for reservoir drawdown, while achieving all project objectives including environmental compliance and impact mitigation. Work completed by others for project mitigation prior to dam removal included the construction or modification of flood-control levees along the downstream river channel for continued flood protection after sediment deposition, the construction of two water treatment plants for the protection of municipal and industrial water supplies from suspended sediments, and the construction of a new tribal fish hatchery. Natural erosion during reservoir drawdown was expected to remove up to one-half of the impounded reservoir sediment (estimated at over 18 million cubic meters), with the balance remaining stable along the reservoir margins outside the new river channel. Following completion of the dam removal contract, the former lake beds and remaining sediment terraces will be revegetated through natural processes and by a planting program of native species. Visitor improvements at the Glines Canyon Dam site will be completed to provide paved parking areas and interpretive signs on both abutments, overlooking the deep canyon. Total project costs for dam removal and restoration efforts are approximately $350 million, including costs for project acquisition ($29.5 million), dam removal ($26.9 million), and all mitigation measures.

San Clemente Dam (from Hepler et al., 2011)

San Clemente Dam (Fig. 3) was constructed in 1921 at the confluence of the Carmel River and San Clemente Creek, ~24 km southeast of Carmel, California. The 32-m-high thin arch concrete dam is owned and operated by California American Water (CAW), a regulated utility that serves the Monterey Peninsula. The dam was constructed as a point of diversion for surface water from the Carmel River, which would flow to the Carmel Valley Filter Plant and be distributed to CAW's customers. Since construction, more than 1.9 million cubic meters of sediment have accumulated behind the dam, nearly filling the reservoir and essentially eliminating its usable storage capacity. Engineering studies performed in 1992 concluded that the dam could sustain structural damage resulting in the potential loss of the reservoir during a maximum credible earthquake (MCE), and would be overtopped and possibly fail during a probable maximum flood (PMF). The dam currently serves no useful purpose.

The San Clemente Dam (SCD) Seismic Safety Project was originally established to meet current standards for maximum earthquake and flood loads for dam safety, and to provide for fish passage at the dam, while maintaining a point of diversion

Figure 1. Elwha Dam, Elwha River—Concrete gravity section and gated spillway.

Figure 2. Glines Canyon Dam, Elwha River—Concrete arch section and gated spillway.

for CAW. A joint Environmental Impact Report/Environmental Impact Statement (EIR/EIS) was prepared in 2006 to address the environmental effects of the project alternatives. The owner's proposed project was to strengthen the dam and construct a new fish ladder; however, several alternatives were evaluated. The Carmel River Reroute and Dam Removal (CRRDR) alternative would remove the concrete dam and fish ladder, and use the rubble to help stabilize sediment at the site. The lower portion of the Carmel River above the dam would be permanently bypassed by excavating a channel through the bedrock ridge that separates the Carmel River from San Clemente Creek. The bypassed portion of the Carmel River would be used as a sediment disposal site for accumulated sediments in both the Carmel River and San Clemente Creek. The lower San Clemente Creek channel would be reconstructed to handle the natural flows from both drainages, while accommodating upstream passage of native steelhead to over 40 km of spawning and rearing habitat.

Although the owner's proposed project was found to be the least-cost alternative, several organizations recognized that the CRRDR alternative would provide additional benefits to the public by removal of the dam, and began working with CAW to adopt this alternative as the preferred project. The SCC was appointed as the lead state agency and began spearheading supplemental technical studies to (1) confirm the feasibility of the project design, including fish passage within the combined flow reach and reroute channel; (2) further assess potential impacts associated with downstream sediment transport; (3) provide updated cost estimates based on more detailed designs; (4) provide sufficient information to enable consensus among the parties on a feasible strategy for removing the dam; and (5) prepare the CRRDR project for the permitting and final design phases.

Following completion of SCC's feasibility studies, CAW and SCC signed a Memorandum of Understanding (MOU) outlining a cooperative strategy by which CAW would undertake the dam removal project and contribute an amount equivalent to the cost of the dam strengthening project ($49 million) and SCC would lead the project design and permitting effort, and secure additional funding up to $35 million to cover the increased costs of the CRRDR. CAW also agreed to donate 375 ha of the project site to the Bureau of Land Management (BLM) for watershed conservation and public access. Final designs are to be completed under a design-build contract, with the construction activities scheduled from April 2013 to November 2015.

Klamath River Dams (from U.S. Bureau of Reclamation, 2012)

The Klamath Hydroelectric Project is owned by PacifiCorp, and it includes four hydroelectric dams along the main stem of the Upper Klamath River, located between 304 and 365 km upstream from the Pacific Ocean. The existing project has an installed generating capacity of 169 MW and generates an average of 716,800 MWh of electricity annually. PacifiCorp began relicensing proceedings with FERC in 2000. A Final EIS was issued by FERC in 2007, and it contained FERC staff evaluations of the proposal submitted by PacifiCorp for continued operation of the four hydroelectric dams with new environmental measures, in addition to alternatives developed by FERC for relicensing the project. Project alternatives included the Staff Alternative, which incorporated most of PacifiCorp's proposed environmental measures with some modifications; the Staff Alternative with Mandatory Conditions, which required the installation of fishways at each dam; and two Staff dam removal alternatives, which included (1) the removal of Copco No. 1 and Iron Gate Dams, and the installation of fishways at Copco No. 2 Dam and J.C. Boyle Dam, and (2) the removal of J.C. Boyle, Copco No. 1, Copco No. 2, and Iron Gate Dams.

Copco No. 1 Dam (Fig. 4) is a large concrete gravity arch dam with a gated spillway and a height of 41 m, completed in 1922 on the Klamath River in California, just south of the Oregon border. Copco No. 2 Dam is a gated concrete diversion dam with a height of 10 m, completed in 1925 ~0.5 km downstream from Copco No. 1. J.C. Boyle Dam (Fig. 5) is a large earthfill embankment with a gated spillway section and a height of 21 m, completed in 1958 ~42 km upstream from Copco No. 1, in Oregon. Iron Gate Dam (Fig. 6) is a large earthfill embankment with an ungated side channel spillway and a height of 58 m, completed in 1962 ~14 km downstream from Copco No. 1. Copco No. 1 and Iron Gate Dams include diversion tunnels that, with structural modifications, can be used to facilitate reservoir drawdown. J.C. Boyle Dam includes a double-box culvert used for diversion during construction that can be modified for reservoir drawdown.

The SCC funded additional technical studies in 2006 to characterize the sediment impounded by each of the four dams, evaluate the potential downstream effects of reservoir sediment erosion, and develop a feasible method of removing the dams, including the preparation of cost estimates and construction schedules. These additional studies were intended to provide an overview analysis of dam removal and its effects on downstream water quality, and served to advance dam removal as a potentially viable alternative.

The Klamath Hydroelectric Settlement Agreement (KHSA) was completed in February 2010 for the express purpose of resolving the pending FERC relicensing proceedings by establishing a process for potential removal of the four hydroelectric dams and for operation of the hydroelectric project until that time. The KHSA addresses the proposed determination by the Secretary of the Interior whether to proceed with facilities removal, defined as the "physical removal of all or part of each of the facilities to achieve at a minimum a free-flowing condition and volitional fish passage, site remediation and restoration, including previously inundated lands, measures to avoid or minimize adverse downstream impacts, and all associated permitting for such actions." The KHSA describes the process for studies, environmental review, and participation by the signatory parties (including PacifiCorp) and the public to inform the Secretary of the Interior's determination. As a part of the basis for the secretarial determination,

Figure 3. San Clemente Dam, Carmel River—Concrete arch dam and overflow spillway.

Figure 4. Copco No. 1 Dam, Klamath River—Concrete gravity dam.

Figure 5. J.C. Boyle Dam, Klamath River—Embankment and concrete sections.

Figure 6. Iron Gate Dam, Klamath River—Embankment dam and powerhouse.

a Detailed Plan to implement facilities removal was required, to include the following components:

(1) the physical methods to be undertaken to achieve facilities removal, including but not limited to a timetable for decommissioning and facilities removal;
(2) as necessary and appropriate, plans for management, removal, and/or disposal of sediment, debris, and other materials;
(3) a plan for site remediation and restoration;
(4) a plan for measures to avoid or minimize adverse downstream impacts;
(5) a plan for compliance with all applicable laws, including anticipated permits and permit conditions;
(6) a detailed statement of the estimated costs of facilities removal;
(7) a statement of measures to reduce risks of cost overruns, delays, or other impediments to facilities removal; and
(8) the identification, qualifications, management, and oversight of a non-federal Dam Removal Entity (DRE), if any, that the Secretary of the Interior may designate.

Feasibility-level designs and cost estimates for removal of the four dams and for the natural erosion of impounded sediment were completed in 2012 to support the secretarial determination. These studies confirmed that dam removal was technically feasible and could be accomplished within the $450 million funding limit established under the agreement. Environmental compliance activities included the preparation of a joint Environmental Impact Statement/Environmental Impact Report (EIS/EIR) intended to meet both federal and State of California requirements, and the identification of appropriate mitigation measures. If approved, the four dams would be removed in 2020.

ACKNOWLEDGMENTS

Review comments provided by Dennis Gathard and Wayne Edwards are gratefully acknowledged.

REFERENCES CITED

ASCE, 1997, Guidelines for Retirement of Dams and Hydroelectric Facilities: New York, American Society of Civil Engineers, 222 p.

Aspen Institute, 2002, Dam Removal—A New Option for a New Century: Washington, D.C., Aspen Institute Publications Office, 68 p.

Federal Energy Regulatory Commission (FERC), 2007, Final Environmental Impact Statement for Relicensing of the Klamath Hydroelectric Project No. 2082-027: Washington, D.C., Federal Energy Regulatory Commission.

Hepler, T.E., 2010, Large dam removals for restoration of the Elwha River, *in* Proceedings, Dam Safety, Seattle, Washington, 19–23 September 2010: Lexington, Kentucky, Association of State Dam Safety Officials, p. 1–15 (on CD-ROM).

Hepler, T., Greimann, B., Chapman, T., and Szytel, J., 2011, Remove or reinforce: Design alternatives to meet dam safety and fish passage requirements at San Clemente Dam, *in* Proceedings of the 21st Century Dam Design—Advances and Adaptations, 31st Annual U.S. Society on Dams Conference, San Diego, California, 11–15 April 2011: Denver, Colorado, U.S. Society on Dams, p. 779–794.

Pohl, M.M., 2002, Bringing down our dams: Trends in American dam removal rationales: Journal of the American Water Resources Association, v. 38, no. 6, p. 1511–1519, doi:10.1111/j.1752-1688.2002.tb04361.x.

U.S. Bureau of Reclamation, 2012, Detailed Plan for Dam Removal—Klamath River Dams: Denver, Colorado, Bureau of Reclamation, Technical Service Center, http://klamathrestoration.gov/keep-me-informed/secretarial-determination/role-of-science/secretarial-determination-studies (accessed September 2012).

Manuscript Accepted by the Society 29 June 2012

# Assessing sedimentation issues within aging flood-control reservoirs

Sean J. Bennett
Department of Geography, University at Buffalo, Buffalo, New York 14261-0055, USA

John A. Dunbar
Department of Geology, Baylor University, Waco, Texas 76798-7354, USA

Fred E. Rhoton
U.S. Department of Agriculture–Agricultural Research Service National Sedimentation Laboratory, P.O. Box 1157, Oxford, Mississippi 38655, USA

Peter M. Allen
Department of Geology, Baylor University, Waco, Texas 76798-7354, USA

Jerry M. Bigham
School of Environment and Natural Resources, Ohio State University, Columbus, Ohio 43210, USA

Gregg R. Davidson
Department of Geology and Geological Engineering, University of Mississippi, University, Mississippi 38677, USA

Daniel G. Wren
U.S. Department of Agriculture–Agricultural Research Service National Sedimentation Laboratory, P.O. Box 1157, Oxford, Mississippi 38655, USA

ABSTRACT

Flood-control reservoirs designed and built by federal agencies have been extremely effective in reducing the ravages of floods nationwide. Yet some structures are being removed for a variety of reasons, while others are aging rapidly and require either rehabilitation or decommissioning. The focus of the paper is to summarize collaborative research activities to assess sedimentation issues within aging flood-control reservoirs and to provide guidance on such tools and technologies. Ten flood-control reservoirs located in Oklahoma, Mississippi, and Wisconsin have been examined using vibracoring, stratigraphic, geochronologic, geophysical, chemical, and geochemical techniques and analyses. These techniques and analyses facilitated: (1) the demarcation of the pre-reservoir sediment horizon within the deposited reservoir sediment, (2) definition of the textural and stratigraphic characteristics of the sediment over time and space, (3) the accurate determination of the remaining reservoir storage capacity, (4) the quantification of sediment quality with respect to agrichemicals and environmentally important trace elements over both time and space, and (5) the determination of geochemical conditions within the deposited sediment and the potential mobility of associated elements. The techniques employed and discussed here have proven to be successful in the assessment of sediment deposited within aging flood-control reservoirs, and it is envisioned that these same approaches could be adopted by federal agencies as part of their national reservoir management programs.

INTRODUCTION

Flood-control reservoirs have been very effective in mitigating the adverse effects of flooding in nearly every region of the United States. This is especially true for the more than 11,000 U.S. Department of Agriculture–Natural Resources Conservation Service (USDA-NRCS) flood-control reservoirs built as a result of the Flood Control Act of 1944, the Pilot Watershed Program, and the Watershed Protection and Flood Prevention Act of 1954 (Hamilton, 1977; Caldwell, 1999). This infrastructure had an immediate, positive impact on local communities by reducing the ravages of flooding, improving crop yields, increasing municipal water supplies, and enhancing recreational opportunities (Bennett and Dunbar, 2003). These projects provide an estimated $1.5 billion in benefits annually (USDA-NRCS, 2008).

However, flood-control reservoirs in the United States and worldwide currently are under renewed scrutiny. First, there is growing interest in removing or decommissioning some structures due to safety concerns or economic obsolescence, or to improve ecologic indices and recreational opportunities (Pohl, 2002; Bednarek, 2001; Hart et al., 2002; Robbins and Lewis, 2008. Dam removal also reconnects fragmented water courses, the benefits of which extend far beyond the immediate corridor occupied by the structure (Ward and Stanford, 1995; Graf, 1999). Second, thousands of flood-control reservoirs nationwide are rapidly approaching their economic service life. By 2020, more than 85% of the 83,000 dams in the United States (based on the National Inventory of Dams, U.S. Army Corps of Engineers), and nearly 50% of the 11,000 USDA-NRCS structures will be near the end of their design life (Bennett et al., 2002; Doyle et al., 2003). Moreover, aging flood-control dams and reservoirs may require significant rehabilitation for their continued service. For the USDA-NRCS structures, the primary reasons for rehabilitation include replacement of deteriorating components, a change in hazard classification, excessive sedimentation in the reservoir, and failure to meet dam safety regulations or resource needs of the watershed (Caldwell, 2000). This paper addresses excessive sedimentation—one of the reasons for dam rehabilitation.

Reservoirs by design trap incoming sediment and associated chemicals and compounds. Mean annual rates of storage capacity loss for reservoirs typically range from 0.1% to 4% or more worldwide (Dendy, 1968; McHenry, 1974; Crowder, 1987; Mahmood, 1987), and USDA-NRCS reservoirs typically have trap efficiencies of 80%–100% (Dendy, 1974, 1982; DeCoursey, 1975; Dendy and Cooper, 1984). Dendy (1974) noted that average annual loss of reservoir storage capacity is inversely related to the size of the reservoir, which can be attributed to decreased sediment delivery ratios with increasing drainage areas (Walling, 1983). Moreover, as a reservoir fills with sediment over time, its efficiency to trap sediment decreases (Verstraeten and Poesen, 2000). Reservoirs also record variations in sediment loadings and sediment-associated water-quality parameters within the drainage basin, including changes in land use (Van Metre et al., 2004; Van Metre and Mahler, 2004); loadings of heavy metals from natural and anthropogenic sources (Arnason and Fletcher, 2003; Audry et al., 2004), the impacts of agricultural activities (Van Metre et al., 1997, 1998); and the sources, retention, and fluxes of nutrients (Stallard, 1998; Hambright et al., 2004).

Even with modest rates of sedimentation, all reservoirs will reach a stage where flood control is immeasurably compromised, and corrective action may be required. Choosing a path to decommission or rehabilitate an aging flood-control reservoir may depend, in whole or in part, on the disposition of the deposited sediment. That is, the amount of sediment sequestered in the reservoir and its textural characteristics could determine rehabilitation strategies and removal options (Fan and Morris, 1992; Hotchkiss and Huang, 1995), and the quality of the sediment and its pore water could pose serious concerns to the environment, especially with respect to agrichemicals, pollutants, nutrients, and other environmentally important elements (e.g., Bednarek, 2001). These questions were addressed initially in a pilot study conducted by Bennett et al. (2002). Two relatively small reservoirs in Oklahoma were selected for assessment by the USDA-NRCS, wherein a variety of analytic, geochronologic, stratigraphic, and geophysical techniques were employed. These techniques proved to be useful for the rapid, cost-effective assessment of sediment quantity and quality within such reservoirs, which is especially important given that the USDA-NRCS and other federal agencies would need to evaluate thousands of such aging structures in a relatively short time.

Research efforts on sedimentation issues within aging flood-control reservoirs have expanded in both regional and scientific context. While the focus has remained on the characteristics of the sediment, additional research has considered reservoirs in markedly different geographic regions and watershed areas, the analysis of sediment-associated chemical signatures and their temporal and spatial variations, and the geochemical assessment of both the sediment and pore water below the lake bottom using in situ and oxic redox conditions. The overall goal of this collaborative research program is to quantify the characteristics of the sediment deposited in reservoirs so that environmentally sound, economically viable, and socially acceptable rehabilitation strategies can be designed and implemented. The objectives of the current paper are (1) to summarize the assessment technologies and results obtained for flood-control reservoirs from different locations, and having different sizes, focusing on the physical and geochemical characteristics of the deposited sediment, and (2) to discuss the results in the context of aging flood-control reservoirs and general strategies for rehabilitating or decommissioning these structures.

FIELD LOCATIONS

The discussion of sedimentation issues within aging flood-control reservoirs is restricted to those assessed for the USDA-NRCS and the U.S. Army Corps of Engineers (COE). In total, 10 reservoirs in three states were studied: five in Mississippi, three in Oklahoma, and two in Wisconsin (Table 1). The reservoirs in Mississippi include Hubbard-Murphree, Grenada Lake, and three lakes in the Holly Springs National Forest (Lt 14A-4, Drewery Lake, and Denmark Lake), the reservoirs in Wisconsin include White Mound Lake and Twin Valley Lake, and the reservoirs in Oklahoma include Sugar Creek 12 and 14 and Sergeant Major 4. As noted already, these reservoirs were examined at the direct request of federal and state agencies, rather than for scientific purposes, and the extent of deposited sediment analysis was guided by availability of funds and other resources.

METHODS

While various methods are available to secure and characterize the sediment deposited in reservoirs, the focus here is on the techniques successfully employed in these studies, and this should not be viewed as the sole prescription for reservoir assessment. Continuous, undisturbed sediment cores of the deposited sediment can be obtained from the water surface using a vibracoring system. One such system has a 1 horsepower (hp) motor that drives a pair of weights mounted eccentrically on two shafts, housed within a watertight aluminum chamber (Fig. 1; Bennett et al., 2002). The chamber (driver) connects to the top of an aluminum irrigation pipe, 1.5 mm in thickness and 76 mm in diameter, which is cabled to a 4.2-m-high aluminum tripod fitted with a battery-operated winch. The tripod mounts to a portable raft system that is assembled on-site and towed by a johnboat. Once the raft is anchored, the core pipe and chamber are slowly lowered to the lake bottom, the motor is engaged using an electrical generator on the johnboat, and the pipe sinks into the sediment. Once penetration into the sediment ceases, the motor is disengaged, and the core is recovered. Allowing the core to vibrate in place for several minutes before hoisting helps plug and seal the core bottom. All recovered cores then are sealed and returned to the laboratory or opened on-site, their locations are determined

TABLE 1. SUMMARY INFORMATION FOR THE RESERVOIRS EXAMINED, METHODS, AND INVESTIGATION

Reservoir	Constructed by	Year built	Drainage area (km^2)	Land use (%) Agriculture	Land use (%) Pasture, grassland, or rangeland	Land use (%) Forest	Methods Coring	Methods Acoustic survey	Methods Geochronologic analysis	Methods Agrichemical analysis	Methods Sediment chemical analysis
Mississippi											
Hubbard-Murphree	USDA-NRCS	1963	8.2	4	18	78	Y	N	N	Y	Y
Lt 14A-4	USDA-NRCS	1962	2.8			100	Y	N	Y	Y	N
Drewery Lake	USDA-NRCS	1965	1.7			100	Y	N	Y	Y	N
Denmark Lake	USDA-NRCS	1965	2.8			100	Y	N	Y	Y	N
Grenada Lake	USACOE	1954	3379	20	20	55	Y	Y	Y	Y	N
Wisconsin											
White Mound Lake	USDA-NRCS	1970	18.1		NA		N	Y	N	N	N
Twin Lake	USDA-NRCS	1965	32.4		NA		N	Y	N	N	N
Oklahoma											
Sugar Creek 12	USDA-NRCS	1964	8.2	41	50	9	Y	Y	Y	Y	Y
Sugar Creek 14	USDA-NRCS	1962	5.1	Some	NA		Y	Y	Y	Y	Y
Sergeant Major 4	USDA-NRCS	1955	15.1	6	91		Y	Y	Y	Y	Y

Note: NA—not available; Y—yes; N—no; USACOE—U.S. Army Corps of Engineers; USDA-NRCS—U.S. Department of Agriculture–Natural Resources Conservation Service.

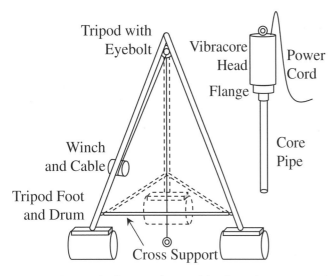

Figure 1. Schematic diagram of a portable vibracoring system and dedicated floating tripod.

using a global positioning system (GPS) receiver, and their water depths are noted. Cores as long as 3 m have been recovered in water depths in excess of 12 m (limited only by length of the electrical connection to the generator). Sediment compaction is a common phenomenon associated with vibracoring reservoir sediment. The vibracorer itself can compact sediment during penetration (Finkl and Khalil, 2005), while the sediment can become differentially compacted due to the mass of the overburden (Evans et al., 2002). Figure 2 shows the variation in sediment porosity with depth below the lake bottom for the 32 cores obtained in Grenada Lake. At this location, sediment typically has a porosity of ~0.6 near the top of the deposited layer, and porosity decreases to ~0.5 and 0.4 at depths of 1 and 2 m below the bottom of the lake. All sediment core data presented herein have been corrected for differential compaction using the method proposed by Evans et al. (2002). In those few cases where sediment porosity data were unavailable, an average correction was used based on the porosity with depth trend line shown in Figure 2 coupled with the method of Evans et al. (2002). This latter approach assumes that all sediment cores obtained using this vibracorer in reservoirs composed of clay- and silt-dominated material would be equally affected by compaction.

The sediment within each core can be sampled for analysis. A circular saw typically is used to cut the core pipe lengthwise, the relative depth below the top of the sediment is noted, and the sediment is sampled incrementally. Bulk density can be determined by weighing oven-dried samples of known core volumes. For particle size analysis, ~10 g samples of sediment are oven dried at 70 °C, crushed, and sieved to less than 2 mm, treated with 30% H_2O_2 for removal of organic matter, and then shaken overnight in 10 ml of 5% sodium hexametaphosphate for complete dispersion. Total clay (<2 μm) is determined by the pipette method (U.S. Department of Agriculture–Natural Resources Conservation Service, 1996). Total sand (2000–64 μm) is determined by wet sieving the remaining sample though a 64 μm sieve and weighing the oven-dried fraction. Total silt (64–2 μm) is calculated by subtracting the sand and clay fractions from the original sample weight.

For bulk chemical analysis, ~5 g samples of oven-dried, crushed sediment obtained from discrete layers within select cores were sent to an accredited environmental laboratory. This sediment was digested using four acids (HF, $HClO_4$, HNO_3, and HCl; a near-total digestion process) and analyzed for as many as 48 elements using an inductively coupled plasma (ICP) spectrometer, whereas cold-vapor atomic absorption was used to measure Hg. It should be noted that discussion of element concentrations, by default, does not consider transport pathways of these materials through the watershed or reservoir, and it does not discriminate between elements adsorbed to clay-sized particles or organic material and the chemical composition of the minerals. Total carbon (C; % by mass) was determined by combusting 2 g sediment samples in a carbon-nitrogen analyzer (errors less than ±3%). The elements, their detection limits, and their analytical error (relative standard deviation) as compared to certified standards and sample duplicates, respectively, are: Al = 0.01% (±95.9%, ±11.4%); As = 0.5 ppm (±2.1%, N.A.); Cr = 2 ppm (±2.8%, N.A.); Cu = 1 ppm (±8.7%, ±5.3%); Hg = 5 ppb (±3.3%, ±7.3%); Pb = 3 ppm (±19.8%, ±14.6%); and Zn = 1 ppm (±4.2%, ±9.9%). Agrichemical analysis of depth-integrated sediment samples from cores (sampling a constant volume of sediment along a discrete core length) was carried out by an accredited laboratory where the concentrations of priority pollutants and polychlorinated biphenyls (PCBs) were determined using standard methods (SW-846 8081a, SW-846 8082; U.S.-EPA, 1997; with detection limits of 1 ppb). Limited

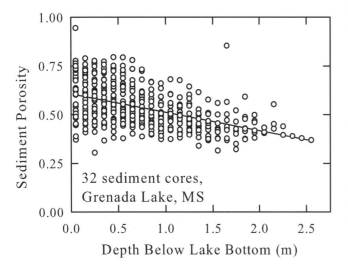

Figure 2. Variation in sediment porosity with depth below the lake bottom for 32 cores obtained in Grenada Lake, Mississippi (MS). These data and trend line are used to correct for differential sediment compaction within the cores.

financial resources precluded higher-resolution agrichemical analysis of the sediment.

Sediment also can be analyzed for radioactive elements to facilitate dating of stratigraphic horizons. Although both ^{14}C and ^{210}Pb have been analyzed in these reservoirs, ^{137}Cs has proven to be a consistent and reliable geochronologic method. In the Northern Hemisphere, atmospheric deposition of ^{137}Cs (30 yr half-life) due to aboveground nuclear testing first occurred in 1954 ± 2, and maximum deposition occurred in 1964 ± 2 (Ritchie and McHenry, 1990). It is generally accepted that the maximum detection levels for ^{137}Cs in reservoir sediment in service since 1960 or longer demarcates this 1964 ± 2 geochronologic horizon. For ^{137}Cs analysis, ~50 g samples of dried, crushed, and sieved (2 mm) sediment samples, taken at discrete intervals (0.1 m) in select cores, were sent to an accredited laboratory. Typical detection limits for a period of 80,000 s using an Ortec GEM High-Purity Germanium Coaxial detector are ~0.5 DPM/g (disintegrations per minute per gram) for a 10 g sample. Unlike other geochronologic techniques, ^{137}Cs analysis only demarcates a single horizon in the sediment core, and it provides no additional information on deposition rates over time. Attempts to use ^{210}Pb to age-date sediment (Appleby and Oldfield, 1992) within two cores from Grenada Lake were not fully successful (Bennett and Rhoton, 2003), and these results are not reported here.

Longitudinal sediment surveys can be coupled with at-a-point core analyses to map the deposited sediment thickness and distribution within the reservoir. Geophysical surveys have been used to conduct this analysis, including conventional seismic methods with a boomer source and multifrequency acoustic systems (Bennett et al., 2002). Because many of these reservoirs have shallow water (~1 m), the close proximity of the seismic source and receiver to the sediment bottom caused much unwanted noise in the recorded signal that could not be sufficiently removed through postprocessing of the data, and these results are not discussed here. The results obtained using the acoustic profiling system, on the other hand, have been much better. This system is composed of acoustic transducers with central frequencies of 200, 48, and 24 kHz, a hydrophone, and an electronics module with built-in computer for controlling data acquisition and GPS navigation (Fig. 3; Dunbar et al., 1999; Bennett et al., 2002; Bennett and Dunbar, 2003). The transducers can be deployed from the side of a johnboat and held at a fixed depth (decimeters) below the water surface. Data are digitally recorded at 16 bit resolution, and all depth-dependent information is normalized relative to daily records of lake level, if applicable. Data postprocessing typically includes assigning a velocity for the acoustic signal (~1500 m/s), scaling each trace by a factor that increases with traveltime (spherical divergence), and tracing the water bottom and base of sediment on each profile, which is often accomplished in concert with the sediment analysis of the core. In shallow-water flood-control reservoirs, reverberation within the water layer tends to mask the true subsurface returns, and these reverberations can be removed using predictive deconvolution (Özdoğan, 1987).

Pore-water geochemical conditions also can be determined for the sediment cores. Freshly obtained and carefully sealed sediment cores can be drilled at discrete intervals (0.1 m) so that electrodes for in situ measurement of pH and redox potential can be deployed. Equilibrium between the probes and the sediment is achieved within 0.5–2 h. This sediment then can be further analyzed through characterization of the pore water and through leaching experiments. For the pore-water characterization, the cores are cut into 0.1 m segments, capped, and transferred to N_2-filled glove bags for removal of the sediment under anoxic conditions. The sediment is placed in polyethylene bottles and immediately centrifuged to separate the solids and pore water. The centrifuged samples are re-introduced to an N_2-filled glove bag, and the pore water is passed through 0.2 μm filter. A portion of the filtered pore water is acidified with high-purity HNO_3

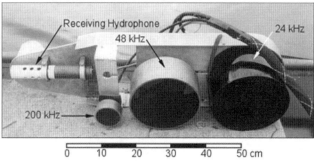

Figure 3. Photographs of a reservoir acoustic profiling system showing the water-resistant control module and differential global positioning system (DGPS) navigation electronics, and the acoustic profiler transducer array. The profiling system produces acoustic signals with three widely separated frequencies (200, 48, and 24 kHz) in rapid succession as the profiles are traversed.

and retained for analysis of 48 elements using ICP. Total and reduced Fe concentrations in the acidified samples are measured by atomic absorption spectroscopy and a colorimetric (ferrozine) technique, respectively. For the leaching experiments, 20 g sediment samples are mixed with 0.8 L of deionized water and kept in suspension by a magnetic stirrer. For leaching under oxidizing conditions, the suspension is left open to the atmosphere. For leaching under reducing conditions, the water is boiled to remove all dissolved gases prior to addition of sediment, and then it is placed into a N_2-filled glove bag with an oxygen absorbent. Both pH and Eh are monitored over time, and filtered aqueous samples are collected over time for ICP analysis.

RESULTS

Discrimination of Pre-Reservoir and Deposited Sediment

A fundamental factor to the assessment of sedimentation issues within flood-control reservoirs is the ability to discriminate between the sediment deposited since the dam was constructed (herein termed deposited sediment or sediment deposit) and the sediment that originally mantled the landscape (herein termed pre-reservoir sediment). This is particularly important in the determination of remaining storage capacity, when as-built construction plans or surveys are unavailable, and in the assessment of sediment quality, because the focus would be on the sequestration and future fate of environmentally important elements and compounds.

Continuous cores obtained from the reservoir can be used to discriminate between the pre-reservoir and deposited sediment as well as to define significant stratigraphic horizons. Under ideal conditions, the geochronologic, textural, and physical characteristics of the sediment should reveal a consistent and concordant stratigraphic picture. Sediment cores from different locations and varying levels of quantification and processing are shown in Figure 4. For Grenada Lake, the peak in normalized ^{137}Cs activity (the activity at a given stratigraphic position divided by the maximum observed in the core), interpreted as the 1964 ± 2 time line, occurs stratigraphically higher than a marked shift in texture from clay-dominated to silt-dominated sediment. This is consistent with the 1954 construction date of the dam. Moreover, sediment bulk density also increases asymptotically with depth toward a much higher value than observed near the core top. The five sediment cores analyzed in Grenada Lake for ^{137}Cs all show similar peak profiles within the upper 0.7 m of sediment, and 15 of the 20 cores analyzed for sediment texture (sand, silt, and clay splits) all display similar clay enrichment and silt depletion near the core tops and a textural crossover to silt enrichment and clay depletion with depth (Bennett et al., 2005). As such, discriminating the pre-reservoir sediment from the deposited sediment in Grenada Lake can be readily accomplished with much confidence.

Not all reservoir surveys can afford such investment in sediment analysis, yet discriminating pre-reservoir sediment from the deposited sediment still can be accomplished. For Drewery

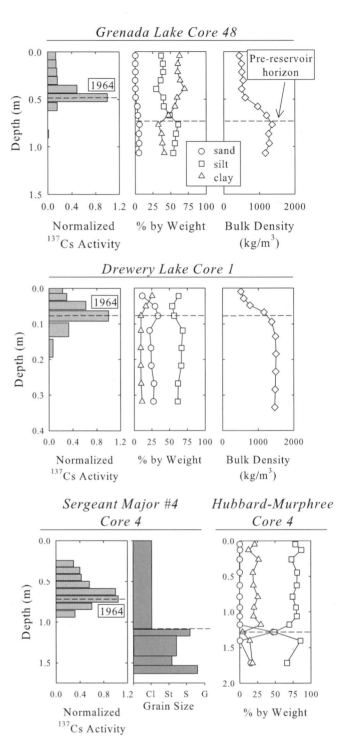

Figure 4. Select sediment cores from various reservoirs showing normalized ^{137}Cs activity, texture, bulk density, and grain size (Cl—clay, St—silt, S—sand, G—gravel) as a function of depth below the lake bottom. The peak in ^{137}Cs activity is interpreted as the 1964 ± 2 time line, the pre-reservoir sediment horizon within the sediment data is identified, and all depths were corrected for differential sediment compaction except the stratigraphic log for Sergeant Major 4.

Lake, the peak in ^{137}Cs occurs less than 0.1 m below the lake bottom, and deviations in both sediment texture and bulk density are observed (Fig. 4). In this case, bulk density is a valuable stratigraphic marker for defining the horizon between these sediment deposits because similar changes to bulk density with depth are observed in two other reservoirs located within the same locality (Wren et al., 2007). For Sergeant Major, no quantitative textural data exist to complement the ^{137}Cs data, except for schematic stratigraphic logs and photographs obtained in the field (Fig. 4). In this reservoir, a relatively coarse-grained deposit found at depth within three of the four sediment cores obtained (ranging from 0.8 m to 1.1 m below the lake bottom) is located stratigraphically below the 1964 ± 2 time line that occurs in the relatively fine-grained sediment. The coarse-grained sediment, stratigraphically pervasive in the deeper portions of the reservoir, is interpreted as the pre-reservoir sediment horizon (Bennett et al., 2002). Finally, only sediment texture with depth is available for cores obtained in Hubbard-Murphree (Fig. 4). Yet this reservoir also displayed a pervasive sand layer at depth within all cores obtained, ranging in depth from 0.8 m to 1.6 m from the lake bottom, and this stratigraphic unit also is interpreted as the pre-reservoir sediment horizon (Bennett, 2001). Pre-reservoir horizons in lake cores often can be marked by a distinctive layer of decaying leaves and sticks (Van Metre et al., 2004).

Discrimination of the deposited sediment within a reservoir can be accomplished using a range of stratigraphic markers and interpretations. Ideally, more than one marker should be used, and spatial corroboration of the marker(s) within the reservoir or locally within nearby reservoirs should be demonstrated. For example, Van Metre et al. (2003) suggested that a minimum of three sediment cores should be secured along the major axis of a reservoir for sediment characterization. Yet when time and resources are limited, it may be possible, under certain circumstances, to obtain a single sediment core from the reservoir pool and to accurately identify the pre-reservoir sediment horizon based on texture or bulk density. Such ability would greatly facilitate rapid assessment of sediment quality and satisfy budgetary constraints. It is noted here that these generalizations should be restricted to those reservoir systems dominated by silt- and clay-sized sediment rather than coarser-grained materials.

Mapping Deposited Sediment Thickness and Assessing Storage Capacity and Sedimentation Rates

Sediment cores and their analysis and interpretation can provide measurements of reservoir sedimentation rates. Annual at-a-point linear sedimentation rates are defined by the thickness of the deposited sediment divided by the time since dam construction (e.g., mm/yr; corrected for differential compaction; Evans et al., 2002). The deposited sediment thickness can be deduced in a number of ways. Annual at-a-point mass sedimentation rates are defined as the accumulated mass per unit area since dam construction (simply the linear sedimentation rate multiplied by bulk density; kg/m^2-yr). These rates then can be used to interpret annual losses of storage capacity by assuming constant rates spatially within the reservoir. It is well known that at-a-point rates of sediment can vary widely, and additional metrics should be employed.

A reservoir survey is one way of determining basinwide rates of sedimentation and loss of storage capacity. Multifrequency acoustic signals that operate in the 10–200 kHz range have proven to be most effective in shallow-water reservoirs. Examples of acoustic survey output are shown in Figure 5. In these examples, recordings of the frequencies are combined to produce a single, false-color cross section through the water column and sediment. Several qualitative observations can be made from these figures. First, the depth of penetration into the subsurface sediment fill is restricted to a few meters. Second, reservoir sediment can display varying reflective intensities due to variations in its physical characteristics, such as bulk density and texture. Third, relict alluvial valleys can be easily identified, and spatial variation in sediment thickness can occur over relatively short distances (Grenada Lake, Fig. 5). Fourth, stratigraphic and sedimentologic relations can be easily discerned, including the pinching-out of acoustic layers and the onlapping of sediment onto the subsurface bathymetry (Twin Valley Lake, Fig. 5).

While much subsurface information can be gathered using an acoustic profiling system, the primary purpose of this activity is to map the thickness of the deposited sediment and to calculate an accurate volume of the remaining storage capacity. This is accomplished in three steps: First, the acoustic profiles must be digitally processed, as noted previously. Second, all interpreted core data must be transferred to the acoustic profiles. That is, the stratigraphic horizon interpreted from the sediment analysis and the pre-reservoir sediment must be identified in the seismographs and verified. Should the core locations coincide directly with an acoustic profile, the process of transferring the core data will be unambiguous. Should the cores not coincide with a specific acoustic line, as these data often are not collected at the same time, then the nearest sediment core is projected onto the acoustic line. Third, this pre-reservoir sediment horizon must be correlated and extended in space so that all acoustic profiles have this surface identified. To date, the validation of the acoustic data by these stratigraphic relationships has been relatively straightforward, and representative examples of these interpreted acoustic signals are shown in Figure 6. Discrepancies between sediment thickness determined from direct sediment sampling (coring) and acoustic surveying are due to projecting the core data onto the seismograph (the cores may be located hundreds of meters away), spatial variations in the lake bathymetry, and the grid resolution of the acoustic data.

The mapping of sediment thickness within a reservoir now can be accomplished. Because all acoustic data are spatially discretized, the difference in the elevations between the reservoir water bottom and the interpreted pre-reservoir sediment horizon, which represents the thickness of the accumulated sediment, can be exported, gridded, and contoured. Select examples of isopach maps rendered using this methodology are shown in Figure 7. For

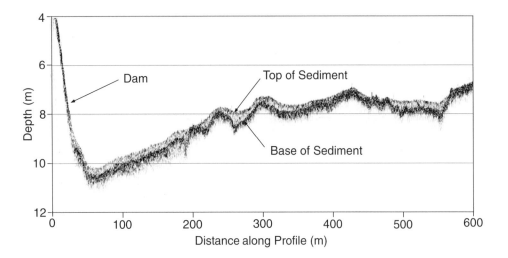
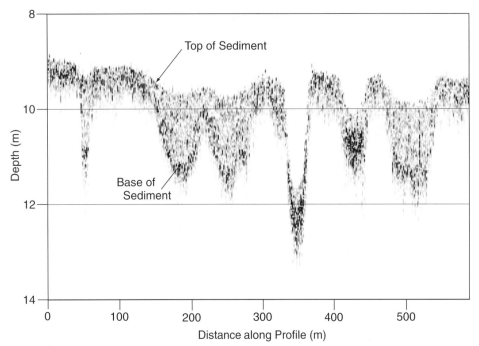

Figure 5. Examples of geophysical survey output from Twin Valley Lake (above) and Grenada Lake (below), with the top and bottom of the deposited sediment identified.

Sugar Creek 12, which has a relatively small reservoir area, 10 sediment cores were used to define the deposited sediment stratigraphy, and three of these were analyzed for ^{137}Cs (Bennett et al., 2002; Bennett and Dunbar, 2003). For Grenada Lake, which has a relatively large reservoir area, 50 sediment cores were used to define the deposited sediment stratigraphy, and five of these were analyzed for ^{137}Cs. The example from White Mound Lake, however, had no core data or ^{137}Cs analysis to facilitate its interpretation. In this case, the identification of the pre-reservoir sediment horizon was based entirely on the interpretation of the acoustic stratigraphy obtained from the acoustic survey (Dunbar et al., 2002). In general, sedimentation thickness tends to be greater near the pools of these reservoirs.

Sedimentation rates and storage capacity assessments based on these coupled coring–acoustic survey data compilations are considered very accurate. Yet these volumetric determinations do not consider all locations of sediment deposition due to the presence of the reservoir. Most coring and acoustic surveying activities generally occur over a few days, especially in relatively small reservoirs, and are conducted at a given reservoir stage below its flood level. As a result, deposited sediment subaerially exposed in the reservoir, such as the flood pool, would not be included in these volume determinations. Moreover, tributary channels feeding into the reservoirs, which would clearly be affected hydraulically by the water level within the reservoir, generally are not surveyed, and these deposits also would be excluded from the volumetric determinations.

Localized rates of reservoir sedimentation also may not reflect regional trends in suspended sediment yields. The reservoirs in Mississippi and Oklahoma are located in ecoregions

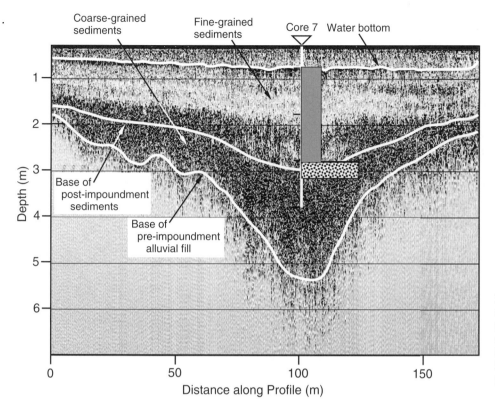

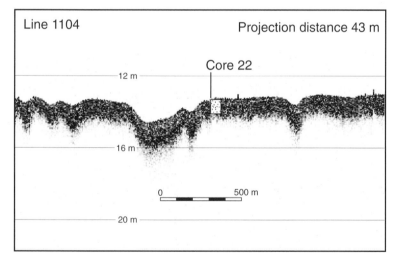

Figure 6. Processed and interpreted acoustic profile lines showing the relative locations of sediment cores for Sugar Creek 12 (above) and Grenada Lake (below; core 22 is projected onto line from a distance of 43 m). Core stratigraphy is depicted schematically.

wherein rivers transport relatively high amounts of suspended sediment (Simon et al., 2004). The reservoirs in Wisconsin, however, are located in an ecoregion wherein sediment yields are less than those in Mississippi and Oklahoma. One might expect higher rates of reservoir sedimentation in watersheds such as Mississippi and Oklahoma, but this is not the case. For the reservoirs shown in Figure 7, approximate losses of storage capacity due to sedimentation, given as a percent of the total storage (and km^3) and determined from sediment coring and/or acoustic surveying, are as follows: 5% (3.60×10^{-5} km^3) for Sugar Creek 12, Oklahoma, 6% (2.03×10^{-5} km^3) for White Mound Lake, Wisconsin, and 3% (4.97×10^{-2} km^3) for Grenada Lake, Mississippi. While the percent losses to storage capacity among these reservoirs are about the same, the total volume of sediment sequestered in Grenada Lake is significantly larger. As noted earlier, sediment delivery ratios decrease with increasing drainage area (Walling, 1983), and this phenomenon greatly conditions the observed sedimentation within Grenada Lake. Here, a simplified sediment balance was constructed for Grenada Lake's southern tributary, the Yalobusha River. It was found that of the sediment eroded from its channel boundaries, determined from historic survey data, ~76% of this remains stored in upstream environments, ~16% is stored

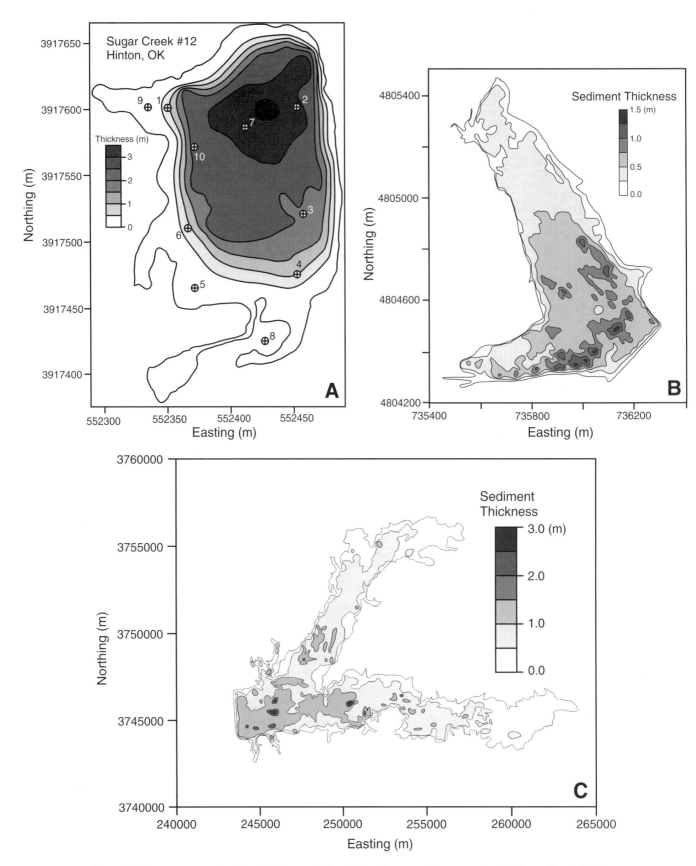

Figure 7. Contour maps of sediment thickness deduced from the acoustic surveys for (A) Sugar Creek 12 (dam located on east side of lake), (B) White Mound Lake (dam located on southeast side of lake), and (C) Grenada Lake (dam located on western side of lake; upper arm is the Skuna River and the lower arm is the Yalobusha River). OK—Oklahoma.

TABLE 2. SUMMARY OF SEDIMENT QUALITY GUIDELINES FOR SELECT ELEMENTS AND COMPOUNDS IN FRESHWATER ECOSYSTEMS THAT REFLECT CONSENSUS-BASED THRESHOLD EFFECT CONCENTRATIONS (TEC) AND PROBABLE EFFECT CONCENTRATIONS (PEC; MACDONALD ET AL., 2000)

Element or compound	Consensus-based TEC (ppm)	Consensus-based PEC (ppm)
Arsenic (As)	9.79	33
Chromium (Cr)	43.4	111
Copper (Cu)	31.6	149
Mercury (Hg)	0.18	1.06
Lead (Pb)	35.8	128
Zinc (Zn)	121	459
DDE	3.16	31.3
DDD	4.88	28
DDT	4.16	62.9
BHC-γ (Lindane)	2.34	4.99
Heptachlor epoxide	2.47	16

Note: DDD—dichlorodiphenyldichloroethane; DDE—dichlorodiphenyldichloroethylene; DDT—dichlorodiphenyltrichloroethane; BHC-alpha—alpha benzene hexachloride.

in the reservoir, and ~8% has exited the lake (Bennett et al., 2005; Bennett and Rhoton, 2009). Thus, it is difficult to make broad generalizations regarding reservoir sedimentation rates without more detailed information.

Agrichemical Concentrations within the Deposited Sediment

Since all of the reservoirs discussed here occur within historically cultivated watersheds, a fundamental factor to the assessment of sedimentation issues is determining the existence of agrichemicals and other possible pollutants within the reservoir and their temporal and spatial variation. MacDonald et al. (2000) summarized consensus-based concentration guidelines for assessing the environmental effects of elements or compounds in freshwater sediment. Two categories are defined therein: (1) a threshold effect concentration (TEC), below which adverse effects on the environment are not expected to occur, and (2) a probable effect concentration (PEC), above which adverse effects are expected to occur more often than not. Table 2 lists these consensus-based TEC and PEC values for elements and compounds examined here. The concentrations of agrichemicals detected in these eight reservoirs, comprising a total of 59 cores and the analysis of more than 60 current-use and residual agrichemicals and compounds, are summarized in Table 3. The results from these analyses show that, in general, sediment quality is good in all of these reservoirs with respect to agrichemicals and priority pollutants. This declaration is based on the limited number of positive detections of specific agrichemicals and compounds, and their relatively low concentrations as compared to the consensus-based effect levels (Table 2). The values reported here also are consistent with agrichemical concentrations found

TABLE 3. SUMMARY OF CONCENTRATIONS OF SELECT AGRICHEMICALS DETECTED IN RESERVOIR SEDIMENT

Compound	DL	Reservoir (number of cores analyzed)							
		SC 12 (12)	SC 14 (2)	SM 4 (7)	Lt 14A-4 (3)	Den. (9)	Drew. (8)	HM (2)	GL (16)
DDD	1	0 to 14	ND	ND	ND	ND	ND	ND	0 to 18
DDE	1	0 to 125	ND	0 to 10	ND	ND	ND	ND	0 to 26
DDT	1	ND	ND	ND	0 to 9	6 to 8	0 to 6	ND	0 to 7
Heptachlor	1	ND	ND	ND	NT	NT	NT	ND	0 to 114
Methyl parathion	0.025	0 to 5	0 to 4	0 to 4	ND	ND	ND	ND	NT
Metolachlor	0.05	0 to 2	ND	ND	0 to 4	0 to 33	ND	ND	NT
BHC-α	1	ND	ND	ND	NT	NT	NT	ND	0 to 37
BHC-β	1	ND	ND	ND	NT	NT	NT	ND	0 to 527
BHC-γ	1	ND	ND	ND	NT	NT	NT	ND	0 to 451

Note: All units are in parts per billion (ppb); ND—not detected; NT—not tested; DL—detection limit; SC 12—Sugar Creek 12; SC 14—Sugar Creek 14; SM 4—Sergeant Major 4; Den.—Denmark; Drew.—Drewery; HM—Hubbard-Murphree; GL—Grenada Lake. The list of potential agrichemicals and pollutants assessed in these reservoirs could include some combination of the following: Herbicides (2,4-D, 2,4,5-T, 2,4,5-TP [Silvex]); organophosphorus pesticides (Alachlor, Atrazine, Azinphos methyl, Bifenthrin, Bolstar, Chlorfenapyr, Chlorpyrifos, Coumaphos, Cyanazine, Alpha-Cyhalothrin, Demeton-S, Diazinon, Dichlorvos, Dimethoate, Disulfoton, EPN, Ethoprop, Ethyl parathion, Fensulfothion, Fenthion, Malathion, Merphos, Methyl parathion, Metolachlor, Mevinphos, Monocrotophos, Naled, Pendimethalin, Phorate, Ronnel, Stirophos, TEPP, Tokuthion, Trichloronate, Trifluralin); priority pollutant pesticides and polychlorinated biphenyls (PCBs; Aldrin, Alpha-BHC, Beta-BHC, Gamma-BHC [Lindane], Delta-BHC, Chlordane, 4,4′-DDT, 4,4′-DDE, 4,4′-DDD, Dieldrin, Endosulfan I, Endosulfan II, Endosulfan sulfate, Endrin, Endrin aldehyde, Heptachlor, Heptachlor epoxide, PCBs [1016, 1221, 1232, 1242, 1248, 1254, 1260], Toxaphene).

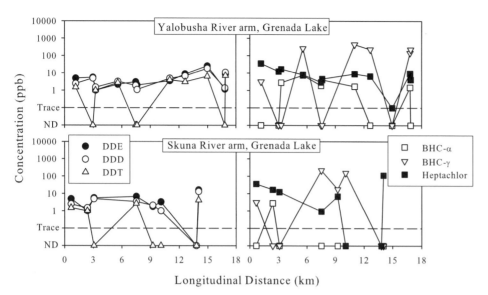

Figure 8. Spatial variation in depth-averaged concentrations of DDD, DDE, DDT, BHC-alpha, BHC-gamma, and heptachlor within the Yalobusha and Skuna River arms of Grenada Lake. Longitudinal distance begins at dam embankment (0 m) and moves upstream (refer to Fig. 7), and all agrichemical core data are projected onto this profile. ND—detected; DDD—dichlorodiphenyldichloroethane; DDE—dichlorodiphenyldichloroethylene; DDT—dichlorodiphenyltrichloroethane; BHC-alpha—alpha benzene hexachloride.

in agricultural watersheds and reservoirs in Mississippi and elsewhere (Cooper et al., 1987, 2002; Cooper, 1991; Knight and Cooper, 1996; Bennett and Rhoton, 2009).

Agrichemical concentrations within depth-averaged sediment samples do not display systematic variations in space. For relatively small reservoirs, where the distance between cores may be tens of meters, the concentrations of agrichemicals do not vary significantly. For relatively larger reservoirs, little spatial variation also has been observed. Figure 8 shows the variation in concentrations of select agrichemicals for longitudinal transects starting at the most upstream core location on the Yalobusha River arm and Skuna River arm of Grenada Lake (11 and 8 cores, respectively), and extending downstream to the pool region of the lake. Along these two transects, compounds such as DDD, DDE, DDT, BHC-alpha[1], and heptachlor, when detected, are found in nearly all sediment samples in comparable concentrations, although some compounds do increase or decrease slightly with distance. This suggests that there is little correlation between agrichemical concentration and longitudinal distance within each arm of the reservoir, and that no marked difference exists between the agrichemical loadings from the Skuna and Yalobusha River basins to Grenada Lake in spite of the large differences in land use and channelization (Bennett and Rhoton, 2009). More importantly, there is no indication of any abnormally high concentrations within the basin that would require special remediation measures. Higher-resolution sampling and analysis of the sediment might show peaks in agrichemical concentrations as a function of depth (time), but the depth-averaging employed here would suppress such temporal variation. For example, Van Metre et al. (1997) showed that concentrations of total DDT (DDT + DDE + DDD) in sediment cores from White Rock Lake, Texas, increased from the mid-1940s to a maximum around 1965, about the same time as DDT usage in the United States peaked.

The sediment contained within reservoirs can be analyzed for environmentally important elements and nutrients. The concentrations of select elements as a function of clay content for four reservoirs are depicted in Figure 9, and these data are restricted to the deposited sediment. As shown, As can range in concentration from 3 to 20 ppm, Pb from 2 to 50 ppm, Hg from 15 to 250 ppb, and C from 0.1% to 3%. All of the trace-element values fall below the consensus-based PEC for freshwater sediment, and all fall below the consensus-based TEC except for As (Table 2). Moreover, these concentrations from the reservoirs in Mississippi are comparable to those reported previously (Cooper et al., 2002), but they are significantly higher than stream channel samples within these same regions (Thompson, 2005). Element concentrations also are strongly correlated with sediment texture. Figure 9 demonstrates that element concentrations can increase by an order of magnitude or more as sediment becomes increasingly rich in clay content. This observation is to be expected, because clay-size aluminosilicate sediment is the most important trace element–bearing phase (Windom et al., 1989; Schropp et al., 1990).

Deviations in element concentration as a function of clay content can be indicative of relative enrichment or depletion of these sediment loadings. For Grenada Lake, Hg concentration shows a strong linear relationship with clay content (Fig. 9), yet Hg concentrations in Hubbard-Murphree, Sergeant Major, and Sugar Creek 12 all show strong deviations from this linear relationship. This variability may be due to differential contributions from organic detritus or atmospheric sources rather than mineralogic sources (Daskalakis and O'Connor, 1995). The variation in carbon content with clay concentration displayed in Figure 9 is more likely due to relative depth below the lake bottom, because near-surface sediment in these reservoirs tends to have higher

---
[1]DDD—dichlorodiphenyldichloroethane; DDE—dichlorodiphenyldichloroethylene; DDT—dichlorodiphenyltrichloroethane; BHC-alpha—alpha benzene hexachloride.

carbon concentrations (2% or 3%) than deeper sediment (<1%; e.g., Bennett and Rhoton, 2009). Carbon content within deposited reservoir sediment can decrease (Dean, 1999; Juracek, 2004) or remain about the same (Van Metre and Callender, 1997) with depth below the lake bottom.

The concentration of elements within reservoirs can be depicted stratigraphically and related to the temporal and spatial variations in chemical and sediment loadings. Extensive sediment coring and sediment analysis were conducted in Grenada Lake, and Figure 10 illustrates the vertical variation in the concentration or weight percent of As, Cr, Cu, Hg, Pb, Zn, and C (select elements), as well as sand, silt, and clay, within select cores with depth below the lake bottom. As noted previously herein, the concentrations of these elements in the deposited sediment of Grenada Lake all fall below the consensus-based PEC for freshwater sediment (Table 2). The 1954 time line of the pre-reservoir sediment horizon also is shown, as deduced from the geochronologic and stratigraphic analyses. From these profiles, it would appear that the sediment deposited in this reservoir contains higher concentrations of environmentally important elements as compared to the pre-reservoir sediment located stratigraphically lower in these same cores. Statistical tests performed on these and other data support this conclusion. All trace-element concentrations are statistically greater in the deposited sediment as compared to the pre-reservoir sediment, and these differences can range from 2 to 10 times larger (Bennett and Rhoton, 2007). Because of this clear dependency of element concentration on sediment texture (Fig. 9), the vertical profiles of trace-element concentration for select cores from Grenada Lake can be normalized by the mass percent of clay contained in the sample (Fig. 11; Windom et al., 1989; Schropp et al., 1990; Summers et al., 1996). Statistical tests performed on these normalized concentration data comparing the pre-reservoir sediment to the deposited sediment, however, produce different results. Using all sediment samples, the trace elements normalized by clay in the deposited sediment are statistically different and lower in magnitude (by about −200%) as compared to the pre-reservoir sediment. Finally, because the 1954 time line is definable in all sediment cores, the vertical variation in trace-element concentration can be used as a surrogate for temporal variation in chemical signatures. Nonparametric statistical analysis was performed on all sediment core data to determine if trace-element concentrations normalized by clay content show any trend with core depth. Of the analyses performed on these normalized trace elements (120 total), only 14% display a statistically significant trend with core depth, and of these, most (82%) show a decrease with depth (i.e., an increase with time before present) of the normalized trace-element concentration (Bennett and Rhoton, 2007). While the sediment sequestered in Grenada Lake since its construction is rich in clay, and it has a higher concentration of environmentally important elements, there is no statistical evidence to support an enrichment of these elements if the data are normalized by clay content, and there appears to be little variation in trace-element concentration vertically, or over time, in the sediment cores. The premise of defining enrichment ratios for select elements is to discriminate temporal variations in anthropogenic loadings that are in excess of natural loadings (e.g., Audry et al., 2004). For Grenada Lake, there is a clear enrichment of clay-sized sediment deposited in the reservoir, but there is no evidence that the signatures of select sediment-associated elements are enriched relative to their background signals.

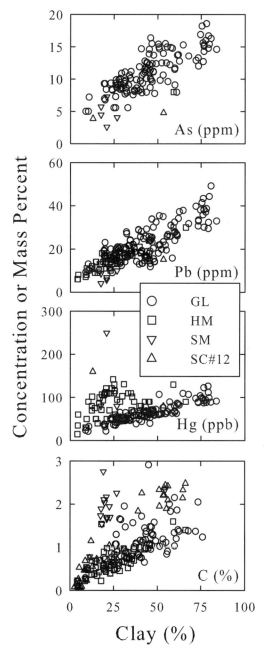

Figure 9. Variation in the concentration or mass percent of As, Pb, Hg, and C as a function of clay content for select sediment samples obtained in the following reservoirs: Grenada Lake (GL), Hubbard-Murphree (HM), Sergeant Major (SM), and Sugar Creek 12 (SC#12).

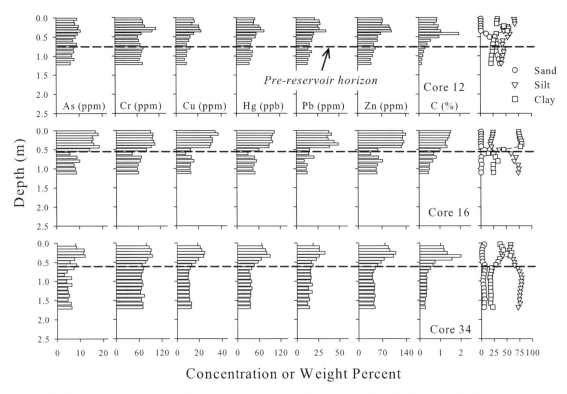

Figure 10. Variation in the concentrations or weight percents of As, Cr, Cu, Hg, Pb, Zn, C, sand, silt, and clay within the sediment of select cores in Grenada Lake with depth below the lake bottom (corrected for differential compaction). Also shown are the interpreted pre-reservoir sediment horizons within each core, and the units.

Pore-Water Conditions with the Deposited Sediment

Dredging or the removal of deposited sediment remains a viable alternative to extend the economic service life of flood-control reservoirs, if the costs to do so are not prohibitively high. Yet the disposition of the dredged material and its potential environmental impact are heavily conditioned on the geochemical composition and characteristics of the sediment. Two separate activities were designed to assess the geochemical conditions of the deposited sediment: in situ determination of pore-water geochemical signatures, and leaching experiments under oxic and anoxic atmospheric conditions.

Pore-water conditions within sealed sediment cores can elucidate the in situ geochemical environment below the lake bottom. Figure 12 shows the depth variation of pH and Eh, pore-water concentrations of As and Fe, and the variation in sediment texture for core 36 from Grenada Lake. The pH of in situ pore water ranges from 5.5 and 6.5, yet these values show no consistent trends with depth. Redox potentials for the same core vary between −70 mV and +140 mV, and the most positive values are associated with sandier strata and the pre-reservoir sediment horizon. In general, the Eh values are negative and consistent with the presence of soluble Fe(II), as confirmed by colorimetric analyses of the pore water and Eh-pH diagrams (McBride, 1994, p. 260–261). Reduced Fe concentrations range between 0 and 100 ppm and are almost identical to total dissolved Fe concentrations. Based on this and other cores analyzed, As is the most significant trace element in this pore water below the bottom of Grenada Lake, with concentrations ranging from 0.04 to 0.10 ppm in the most recent sediment (Fig. 12). Arsenic has been observed in both near-surface pore water and deep groundwater in a wide range of environments; sources can be natural or anthropogenic and modulated by microbial activities (Smedley and Kinniburgh, 2002; Oremland and Stolz, 2003). It is likely that the As in the pore water of Grenada Lake is anthropogenic in origin, because monosodium methanearsonate is a common methylated arsenical herbicide used pervasively in Mississippi on cotton farmlands (Bednar et al., 2002). Based on these limited data, it would appear that potentially poor water-quality indices persist within the bottom sediment of Grenada Lake.

Geochemical conditions within sediment cores from Hubbard-Murphree show similar trends as those observed in the Grenada Lake. The redox potentials in cores collected in May 2001 were mildly reducing, with Eh values ranging from −100 mV to 100 mV, and a few cores showed an increase in Eh with depth (Beard et al., 2003). The increase in Eh suggested that oxidizing groundwater was upwelling through the bottom sediment, and

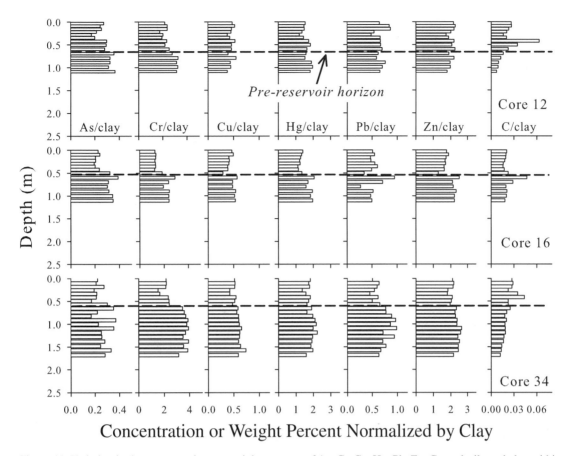

Figure 11. Variation in the concentrations or weight percents of As, Cr, Cu, Hg, Pb, Zn, C, sand, silt, and clay within the sediment of select cores in Grenada Lake with depth below the lake bottom (corrected for differential compaction), as normalized by clay content (units are ppm/% clay content by mass for all elements except Hg, which is ppb/% clay content by mass). Also shown are the interpreted pre-reservoir sediment horizons within each core.

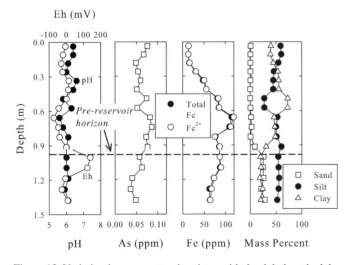

Figure 12. Variation in pore-water chemistry with depth below the lake bottom for core 36 from Grenada Lake (corrected for differential compaction), showing pH, Eh, and As and Fe (total Fe and Fe[II]) concentration. Also shown is the depth variation in sediment texture and the interpreted pre-reservoir sediment horizon.

a piezometer nest installed within the lake verified this hydrologic phenomenon. The redox potentials for cores measured in August 2001, however, were strongly reducing, with Eh values ranging from −200 mV to −400 mV and with no apparent trend with depth (Beard et al., 2003).

Aerating this sediment by dredging the reservoir will increase the redox potential and possibly alter the mobility of some elements. These changes in mobility can be assessed through leaching experiments, which compare element concentrations in water under mildly oxidizing and reducing conditions over a discrete period (in this case, 3–4 d). Figure 13 shows the results of a leaching experiment for one pair of sediment samples obtained from Hubbard-Murphree (Beard et al., 2003; Davidson et al., 2005). The Eh values of slurries open to the atmosphere typically remain near 0 mV initially (up to 2 d), and then gradually climb to near 100 mV, whereas the pH remains relatively constant, fluctuating between 5.5 and 6.0 (Fig. 13). The Eh under the oxygen-free atmosphere exhibited larger variation over time, generally ranging between −150 mV and −350 mV, whereas pH typically started above 8, dropped rapidly over a several-hour

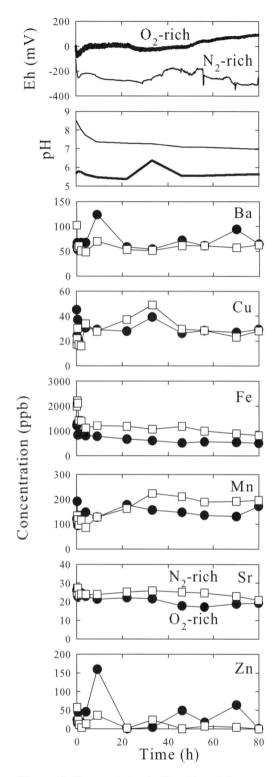

Figure 13. Time variation in Eh, pH, and the concentration of Ba, Cu, Fe, Mn, Sr, and Zn in slurries of water and sediment from a core from Hubbard-Murphree under mildly oxidizing (O_2-rich; thick line, closed symbols) and reducing (N_2-rich; thin line, open symbols) atmospheric conditions. Sharp drops in Eh under reducing conditions resulted when the magnetic stirrer stopped for brief intervals.

period to near neutral values, and then drifted or dropped slowly until the end of the experiment.

The concentration of elements mobilized during leaching experiments depends upon the redox potential. Most of the elements analyzed were found in solution at higher concentrations under reducing conditions as compared to the mildly oxidizing conditions (Ca, Mg, Na, K, Fe, Sr, Al, and Mn; Beard et al., 2003); select examples are shown in Figure 13. A few additional elements appear in higher concentrations under reducing conditions during the first 1–2 d (Ba, Cu and Zn), but they are nearly the same in reducing and oxidizing systems after 3 d. Of the elements found in lower concentrations under oxidizing conditions, only Fe and Mn are redox sensitive (e.g., McBride, 1994). For the non-redox-sensitive elements, the change in pH is likely not the cause for the decrease in concentration relative to the results found under reducing conditions. This decrease is more likely due to precipitation of Fe and Mn oxides, where trace metals and elements from the solution coprecipitate with or adsorb onto these oxides (Jenne, 1968; Du Laing et al., 2009). This hypothesis is supported by a comparison of the filter papers used during each leaching experiment. The filters used under oxidizing conditions were uniformly covered with a rust-colored residue that was absent from the filters used under reducing conditions (Beard et al., 2003). Based on this experiment, it is likely that anoxic lake-bottom sediment exposed to the atmosphere would precipitate Fe and Mn oxides and scavenge environmentally important cations from the pore water of the soil, potentially even the As found in the pore water of Grenada Lake. It is not known if such coprecipitation or adsorption would remove completely all free cations in this process or if other geochemical processes would alter the redox and pH conditions of the sediment and pore water.

DISCUSSION

The collaborative efforts summarized here began in response to the need by federal agencies to assess sedimentation issues within aging flood-control reservoirs nationwide (e.g., Caldwell, 1999). With this in mind, research activities focused on quantifying the characteristics of reservoir sediment so that effective decommissioning and rehabilitation strategies could be designed and implemented. Moreover, the techniques employed had to be fast to implement, provide accurate data, and be relatively inexpensive due to the large number of flood-control structures in need of assessment. The techniques described here, including vibracoring, sediment dating, geophysical surveys, and chemical analysis of the sediment all proved to be satisfactory for these intended purposes. It is envisioned that these technologies and approaches could be adopted by agencies such as the USDA-NRCS and COE as part of their national reservoir management program because of the good results obtained here, the relatively small expense of the equipment, and the ease with which field practitioners can be trained to collect data and conduct surveys.

Several important generalizations can be made regarding sedimentation issues within aging flood-control reservoirs based on observations in 10 reservoirs:

1. Discriminating the pre-reservoir sediment horizon within reservoirs can be accomplished with much confidence using textural, stratigraphic, and geochronologic indices. However, corroboration using more than one approach, and the extension of this important horizon within the reservoir, should be adopted as the norm.
2. Reservoirs in Oklahoma and Mississippi are dominated by silt- and clay-sized sediment. Dendy (1974, 1982) summarized the sedimentation characteristics of 17 reservoirs spanning the southern United States, and he found that fine sediment less than 8 μm in size was fairly evenly distributed both vertically and horizontally within the sediment deposits, whereas coarse sediment 64 μm and larger in size usually was concentrated near the upstream end of the reservoir pool, at or near the permanent pool elevation. Although relict bathymetry can cause local variations in sediment thickness, the deposited sediment examined here tends to blanket the lake bottom rather uniformly. Moreover, limited data also suggest that sediment composition displays both temporal and spatial uniformity in spite of potential changes within the watershed (Bennett and Rhoton, 2007). Given this uniformity in physical and chemical composition, relatively few cores, potentially one to three, may be all that is required to assess, in very broad terms, the thickness and quality of sediment in reservoirs, especially in relatively small reservoirs.
3. Acoustic surveying of reservoirs is shown to be very useful in the accurate determination of sediment volumes and storage capacities even in relatively shallow-water lakes and when as-built construction plans are unavailable. Coupled with a few sediment cores in select locations, this approach to reservoir surveying is considered extremely economical and accurate.
4. The quality of sediment deposited in reservoirs ultimately depends upon land use and anthropogenic loadings to the receiving water. None of the reservoirs examined here is considered atypical for watersheds in the southern United States with historical agricultural activities (Cooper et al., 1987, 2002; Cooper, 1991; Knight and Cooper, 1996). Moreover, while agrichemical and environmentally important elements were found in measurable concentrations within the deposited sediment, none appears to pose any major or significant risk to freshwater ecosystems based on consensus-based thresholds (Tables 2 and 3). This declaration, however, may not be entirely accurate. In Grenada Lake and its tributaries, fish consumption advisories are in place for several species due to bioaccumulated Hg, DDT, and toxaphene (Bennett and Rhoton, 2007). It has been contended that the presence of high concentrations of clay has played a role in these poor ecological indices. That is, the abundant transport and deposition of clay in Grenada Lake, and its adsorptive potential, have markedly facilitated the bioaccumulation of sediment-associated trace elements and compounds within the local fish populations (Mayer et al., 1996; Cope et al., 1999; Bennett and Rhoton, 2007).
5. Pore water below the bottom of lakes in Mississippi tends to be mildly to strongly anoxic and slightly acidic, which allow particular chemical elements to be found as aqueous phases, including As. Yet this same pore water also contains Fe and Mn, and their subsequent oxidation under normal atmospheric conditions can either coprecipitate (scavenge) or adsorb these trace elements, thus potentially isolating them from the environment. Preliminary research has shown that coprecipitation or adsorption can effectively remove select elements from pore water obtained from reservoirs, but additional research is required to determine the efficacy of this geochemical process in removing environmentally important trace elements.

CONCLUSIONS

Flood-control reservoirs designed and built by federal agencies have been extremely effective in reducing the ravages of floods nationwide, as well as providing additional services and benefits to the local communities. However, these same structures are under renewed scrutiny because of growing interest in removing dams for a variety of reasons and the aging of this watershed infrastructure. The disposition of the sediment deposited within flood-control reservoirs, in whole or in part, may play an important role in strategies designed to rehabilitate or decommission these structures. This paper summarizes collaborative research activities to assess sedimentation issues within aging flood-control reservoirs and to provide guidance to federal agencies on such tools and technologies.

Ten flood-control reservoirs located in Oklahoma, Mississippi, and Wisconsin were studied using some combination of vibracoring, stratigraphic, geochronologic, geophysical, chemical, and geochemical technologies and analyses. The following conclusions can be drawn from this work.

1. Identification of the pre-reservoir horizon within the reservoir sediment is of paramount importance, and securing continuous, undisturbed sediment cores greatly facilitates this analysis. This horizon then can be identified by using variations in sediment texture or bulk density with depth below the lake bottom to demarcate this horizon stratigraphically, or by analyzing and interpreting ^{137}Cs concentration to demarcate this horizon geochronologically. While each approach could be applied in isolation, it is recommended that more than one approach be used for corroboration.
2. Multifrequency acoustic surveying of the sediment below a reservoir is shown to effectively discriminate the most

important acoustic horizons within the deposit. When coupled with interpreted core stratigraphy, total sediment thickness in the reservoir and remaining storage capacity can be accurately determined.

3. Reservoirs effectively trap sediment that tends to be dominated by silt- and clay-sized materials, which have a strong association with agrichemicals and environmentally important trace elements. For those reservoirs in Oklahoma and Mississippi, the silt and clay sediment tends to blanket the relict topography uniformly, and tends to be somewhat thicker near the reservoir pools. Agrichemicals common to agricultural watersheds in the southern United States are found in this sediment but not in atypical concentrations, and these agrichemical signatures show no variation in time or space. Moreover, while clay-dominated reservoir sediment has higher concentrations of trace elements, as expected, these concentrations show no enrichment once normalized by clay content, and these elements show little variation in time or space within the reservoir.

4. Analysis of the pore water within two reservoirs in Mississippi shows that in situ conditions are mildly to strongly anoxic and slightly acidic, and elements such as As and Fe(II) are observed in aqueous phases. Yet when select pore water is systematically exposed to oxic conditions, concentrations of Fe and Mn diminish due to oxide precipitation, and other select trace elements also decrease in concentration, either coprecipitated with (scavenged by) or adsorbed onto these oxides.

In summary, the study, carried out to quantify characteristics of reservoir sediment, shows that the techniques employed proved highly satisfactory for the intended purposes. It is envisioned that these techniques and approaches can be adopted by agencies as part of their national reservoir management programs, and therefore can be used in the design of rehabilitation and decommissioning strategies for aging flood-control reservoirs.

ACKNOWLEDGMENTS

Support for this work was provided by the U.S. Department of Agriculture–Agricultural Research Service and the U.S. Army Corps of Engineers. James Evans, Kyle Juracek, and an anonymous referee kindly reviewed this work and made many helpful suggestions.

REFERENCES CITED

Appleby, P.G., and Oldfield, F., 1992, Application of lead-210 to sedimentation studies, *in* Ivanovich, M., and Harmon, S., eds., Uranium-Series Disequilibrium—Applications to Earth, Marine, and Environmental Sciences (2nd ed.): Oxford, Clarendon Press, p. 731–778.

Arnason, J.G., and Fletcher, B.A., 2003, A 40+ year record of Cd, Hg, Pb, and U deposition in sediments of Patroon Reservoir, Albany County, NY, USA: Environmental Pollution, v. 123, p. 383–391, doi:10.1016/S0269-7491(03)00015-0.

Audry, S., Schäfer, J., Blanc, G., and Jouanneau, J.M., 2004, Fifty-year sedimentary record of heavy metal pollution (Cd, Zn, Cu, Pb) in the Lot River reservoirs (France): Environmental Pollution, v. 132, p. 413–426, doi:10.1016/j.envpol.2004.05.025.

Beard, W.C., III, Davidson, G., Bennett, S.J., and Rhoton, F.E., 2003, Evaluation of the Potential for Trace Element Mobilization Resulting from Oxygenating Sediments from a Small Mississippi Reservoir: Oxford, Mississippi, USDA-ARS (U.S. Department of Agriculture–Agricultural Research Service) National Sedimentation Laboratory Research Report 37, 70 p.

Bednar, A.J., Garbarino, J.R., Ranville, J.F., and Wildeman, T.R., 2002, Presence of organoarsenicals used in cotton production in agricultural water and soil of the southern United States: Journal of Agricultural and Food Chemistry, v. 50, p. 7340–7344, doi:10.1021/jf025672i.

Bednarek, A.T., 2001, Undamming rivers: A review of the ecological impacts of dam removal: Environmental Management, v. 27, p. 803–814, doi:10.1007/s002670010189.

Bennett, S.J., 2001, Sediment Quality Issues within a Flood Control Reservoir, Tallahatchie County, MS: Oxford, Mississippi, USDA-ARS (U.S. Department of Agriculture–Agricultural Research Service) National Sedimentation Laboratory Research Report 23, 58 p.

Bennett, S.J., and Dunbar, J.A., 2003, Physical and stratigraphic characteristics of sediments impounded within flood control reservoirs, Oklahoma: Transactions of the American Society of Agricultural Engineers, v. 46, p. 269–277.

Bennett, S.J., and Rhoton, F.E., 2003, Physical and Chemical Characteristics of Sediment Impounded within Grenada Lake, MS: Oxford, Mississippi, USDA-ARS (U.S. Department of Agriculture–Agricultural Research Service) National Sedimentation Laboratory Research Report 36, 161 p.

Bennett, S.J., and Rhoton, F.E., 2007, Reservoir sedimentation and environmental degradation: Assessing trends in sediment-associated trace elements in Grenada Lake, MS: Journal of Environmental Quality, v. 36, p. 815–826, doi:10.2134/jeq2006.0296.

Bennett, S.J., and Rhoton, F.E., 2009, Linking upstream channel instability to downstream degradation: Grenada Lake and the Skuna and Yalobusha River basins, Mississippi: Ecohydrology, v. 2, p. 235–247, doi:10.1002/eco.57.

Bennett, S.J., Cooper, C.M., Ritchie, J.C., Dunbar, J.A., Allen, P.M., Caldwell, L.W., and McGee, T.M., 2002, Assessing sedimentation issues within aging flood control reservoirs in Oklahoma: Journal of the American Water Resources Association, v. 38, p. 1307–1322.

Bennett, S.J., Rhoton, F.E., and Dunbar, J.A., 2005, Texture, spatial distribution, and rate of reservoir sedimentation within a highly erosive, cultivated watershed: Grenada Lake, MS: Water Resources Research, v. 41, p. W01005, doi:10.1029/2004WR003645.

Caldwell, L.W., 1999, Rehabilitating our nation's aging small watershed projects, *in* Soil and Water Conservation Society Annual Conference: Biloxi, Mississippi, 8–11 August 1999: Ankeny, Iowa, Soil and Water Conservation Society.

Caldwell, L.W., 2000, Good for another 100 years: The rehabilitation of Sergeant Major Creek watershed, *in* Proceedings from the Association of State Dam Safety Officials Annual Conference: Lexington, Kentucky, Association of State Dam Safety Officials, p. 194–206.

Cooper, C.M., 1991, Persistent organochlorine and current use insecticide concentrations in major watershed components of Moon Lake, USA: Archiv für Hydrobiologie, v. 121, p. 103–113.

Cooper, C.M., Dendy, F.E., McHenry, J.R., and Ritchie, J.C., 1987, Residual pesticide concentrations in Bear Creek, Mississippi, 1976 to 1979: Journal of Environmental Quality, v. 16, p. 69–72, doi:10.2134/jeq1987.00472425001600010014x.

Cooper, C.M., Smith, S., Jr., Testa, S., III, Ritchie, J.C., and Welch, T., 2002, A Mississippi flood control reservoir: Life expectancy and contamination: International Journal of Ecology and Environmental Sciences, v. 28, p. 151–160.

Cope, W.G., Bartsch, M.R., Rada, R.G., Balogh, S.J., Ruppecht, J.E., Young, R.D., and Johnson, D.K., 1999, Bioassessment of mercury, cadmium, polychlorinated biphenyls, and pesticides in the Upper Mississippi River with zebra mussels (*Dreissena polymorpha*): Environmental Science & Technology, v. 33, p. 4385–4390, doi:10.1021/es9902165.

Crowder, B.M., 1987, Economic costs of reservoir sedimentation: A regional approach to estimating cropland erosion damages: Journal of Soil and Water Conservation, v. 42, p. 194–197.

Daskalakis, K.D., and O'Connor, T.P., 1995, Normalization and elemental sediment contaminations in the coastal United States: Environmental Science & Technology, v. 29, p. 470–477, doi:10.1021/es00002a024.

Davidson, G.R., Bennett, S.J., Beard, W.C., III, and Waldo, P., 2005, Trace elements in sediments of an aging reservoir in rural Mississippi: Potential for mobilization following dredging: Water, Air, and Soil Pollution, v. 163, p. 281–292, doi:10.1007/s11270-005-0731-x.

Dean, W.E., 1999, The carbon cycle and biogeochemical dynamics in lake sediments: Journal of Paleolimnology, v. 21, p. 375–393, doi:10.1023/A:1008066118210.

DeCoursey, D.G., 1975, Implications of floodwater-retarding structures: Transactions of the American Association of Agricultural Engineers, v. 18, p. 897–904.

Dendy, F.E., 1968, Sedimentation in the nation's reservoirs: Journal of Soil and Water Conservation, v. 23, p. 135–137.

Dendy, F.E., 1974, Sediment trap efficiency of small reservoirs: Transactions of the American Society of Agricultural Engineers, v. 17, p. 898–908.

Dendy, F.E., 1982, Distribution of sediment deposits in small reservoirs: Transactions of the American Society of Agricultural Engineers, v. 25, p. 100–104.

Dendy, F.E., and Cooper, C.M., 1984, Sediment trap efficiency of a small reservoir: Journal of Soil and Water Conservation, v. 39, p. 278–280.

Doyle, M.W., Harbor, J.M., and Stanley, E.H., 2003, Toward policies and decision-making for dam removal: Environmental Management, v. 31, p. 453–465, doi:10.1007/s00267-002-2819-z.

Du Laing, G., Rinklebeb, J., Vandecasteelec, B., Meersa, E., and Tacka, F.M.G., 2009, Trace metal behaviour in estuarine and riverine floodplain soils and sediments: A review: The Science of the Total Environment, v. 407, p. 3972–3985, doi:10.1016/j.scitotenv.2008.07.025.

Dunbar, J.A., Allen, P.M., and Higley, P.D., 1999, Multifrequency acoustic profiling for water reservoir sedimentation studies: Journal of Sedimentary Research, v. 69, p. 521–527.

Dunbar, J.A., Higley, P.D., and Bennett, S.J., 2002, Acoustic Imaging of Sediment Impounded within USDA-NRCS Flood Control Dams, Wisconsin: Oxford, Mississippi, USDA-ARS (U.S. Department of Agriculture–Natural Resources Conservation Service) National Sedimentation Laboratory Research Report 30, 34 p.

Evans, J.E., Levine, N.S., Roberts, S.J., Gottgens, J.F., and Newman, D.M., 2002, Assessment using GIS and sediment routing of the proposed removal of Ballville Dam, Sandusky River, Ohio: Journal of the American Water Resources Association, v. 38, p. 1549–1565, doi:10.1111/j.1752-1688.2002.tb04364.x.

Fan, J., and Morris, G., 1992, Reservoir sedimentation: I. Reservoir desiltation and long-term capacity: Journal of Hydraulic Engineering, v. 118, p. 370–384, doi:10.1061/(ASCE)0733-9429(1992)118:3(370).

Finkl, C.W., and Khalil, S.M., 2005, Vibracore, in Schwartz, M.L., ed., Encyclopedia of Coastal Science, Volume 24, Encyclopedia of Earth Sciences Series: Dordrecht, the Netherlands, Springer, p. 1026–1035.

Graf, W.L., 1999, Dam nation: A geographic census of American dams and their large-scale hydrologic impacts: Water Resources Research, v. 35, p. 1305–1311, doi:10.1029/1999WR900016.

Hambright, K.D., Eckert, W., Leavitt, P.R., and Schelske, C.L., 2004, Effects of historical lake level and land use on sediment and phosphorus accumulation rates in Lake Kinneret: Environmental Science & Technology, v. 38, p. 6460–6467, doi:10.1021/es0492992.

Hamilton, T.T., 1977, Building a better small watershed program: Journal of Soil and Water Conservation, v. 32, p. 150–157.

Hart, D.D., Johnson, T.E., Bushaw-Newton, K.L., Horowitz, R.J., Bednarek, A.T., Charles, D.F., Kreeger, D.A., and Velinsky, D.J., 2002, Dam removal and opportunities for ecological research and river restoration: Bioscience, v. 52, p. 669–681, doi:10.1641/0006-3568(2002)052[0669: DRCAOF]2.0.CO;2.

Hotchkiss, R.H., and Huang, X., 1995, Hydrosuction sediment-removal systems (HSRS): Principles and field test: Journal of Hydraulic Engineering, v. 121, p. 479–489, doi:10.1061/(ASCE)0733-9429(1995)121:6(479).

Jenne, E.A., 1968, Controls of Mn, Fe, Co, Ni, Cu, and Zn concentrations in soils and water: The significant role of hydrous Mn and Fe oxides, in Gould, R.F., ed., Trace Inorganics in Water: Advances in Chemistry Series No. 73: Washington, D.C., American Chemical Society, p. 337–387.

Juracek, K.E., 2004, Sedimentation and Occurrence and Trends of Selected Chemical Constituents in Bottom Sediment of 10 Small Reservoirs, Eastern Kansas: U.S. Geological Survey Scientific Investigations Report 2004-5228, 80 p.

Knight, S.S., and Cooper, C.M., 1996, Insecticide and metal contamination of a mixed cover agricultural watershed: Water Science and Technology, v. 33, p. 227–234, doi:10.1016/0273-1223(96)00204-1.

MacDonald, D.D., Ingersoll, C.G., and Berger, T.A., 2000, Development and evaluation of consensus-based sediment quality guidelines for freshwater ecosystems: Archives of Environmental Contamination and Toxicology, v. 39, p. 20–31.

Mahmood, K., 1987, Reservoir Sedimentation: Impact, Extent, and Mitigation: Washington, D.C., The World Bank, World Bank Technical Paper 71, 118 p.

Mayer, L.M., Chen, Z., Findlay, R.H., Fang, J., Sampson, S., Self, R.F.L., Jumars, P.A., Quetla, C., and Donard, O.F.X., 1996, Bioavailability of sedimentary contaminants subject to deposit-feeder digestion: Environmental Science & Technology, v. 30, p. 2641–2645, doi:10.1021/es960110z.

McBride, M.B., 1994, Environmental Chemistry of Soils: New York, Oxford University Press, 406 p.

McHenry, J.R., 1974, Reservoir sedimentation: Water Resources Bulletin, v. 10, p. 329–337, doi:10.1111/j.1752-1688.1974.tb00572.x.

Oremland, R.S., and Stolz, J.F., 2003, The ecology of arsenic: Science, v. 300, p. 939–944, doi:10.1126/science.1081903.

Özdoğan, Y., 1987, Seismic Data Processing: Tulsa, Oklahoma, Society of Exploration Geophysics, 526 p.

Pohl, M.M., 2002, Bring down our dams: Trends in American dam removal rationales: Journal of the American Water Resources Association, v. 38, p. 1511–1519, doi:10.1111/j.1752-1688.2002.tb04361.x.

Ritchie, J.C., and McHenry, J.R., 1990, Application of radioactive fallout cesium-137 for measuring soil erosion and sediment accumulation rates and patterns: A review: Journal of Environmental Quality, v. 19, p. 215–233, doi:10.2134/jeq1990.00472425001900020006x.

Robbins, J.L., and Lewis, L.Y., 2008, Demolish it and they will come: Estimating the economic impacts of restoring a recreational fishery: Journal of the American Water Resources Association, v. 44, p. 1488–1499, doi:10.1111/j.1752-1688.2008.00253.x.

Schropp, S.J., Lewis, F.G., Windom, H.L., and Ryan, J.D., 1990, Interpretation of metal concentrations in estuarine sediments of Florida using aluminum as a reference element: Estuaries, v. 13, p. 227–235, doi:10.2307/1351913.

Simon, A., Dickerson, W., and Heins, A., 2004, Suspended-sediment transport rates at the 1.5-year recurrence interval for ecoregions of the United States: Transport conditions at the bankfull and effective discharge?: Geomorphology, v. 58, p. 243–262, doi:10.1016/j.geomorph.2003.07.003.

Smedley, P.L., and Kinniburgh, D.G., 2002, A review of the source, behaviour, and distribution of arsenic in natural waters: Applied Geochemistry, v. 17, p. 517–568, doi:10.1016/S0883-2927(02)00018-5.

Stallard, R.F., 1998, Terrestrial sedimentation and the carbon cycle: Coupling weathering and erosion to carbon burial: Global Biogeochemical Cycles, v. 12, p. 231–257, doi:10.1029/98GB00741.

Summers, J.K., Wade, T.L., Engle, V.D., and Malaeb, Z.A., 1996, Normalization of metal concentrations in estuarine sediments from the Gulf of Mexico: Estuaries, v. 19, p. 581–594, doi:10.2307/1352519.

Thompson, D.E., 2005, Solid-Phase Geochemical Survey of the State of Mississippi: An Atlas Highlighting the Distribution of As, Cu, Hg, Pb, Se, and Zn in Stream Sediments and Soils: Jackson, Mississippi, Mississippi Department of Environmental Quality, 45 p.

U.S. Department of Agriculture–Natural Resources Conservation Service, 1996, Soil Survey Laboratory Methods Manual: Soil Survey Investigations Report 42, 693 p.

U.S. Department of Agriculture–Natural Resources Conservation Service, 2008, Watershed Rehabilitation Progress Report: http://www.nrcs.usda.gov/programs/wsrehab/ (accessed 6 May 2011).

U.S. Environmental Protection Agency, 1997, Test Methods for Evaluating Solid Waste, Physical/Chemical Methods. Washington, D.C., U.S. Environmental Protection Agency Integrated Manual SW-846.

Van Metre, P.C., and Callender, E., 1997, Water-quality trends in White Rock Creek Basin from 1912–1994 identified using sediment cores from White Rock Lake reservoir, Dallas, Texas: Journal of Paleolimnology, v. 17, p. 239–249, doi:10.1023/A:1007923328851.

Van Metre, P.C., and Mahler, B.J., 2004, Contaminant trends in reservoir sediment cores as records of influent stream quality: Environmental Science & Technology, v. 38, p. 2978–2986, doi:10.1021/es049859x.

Van Metre, P.C., Callender, E., and Fuller, C.C., 1997, Historical trends in organochlorine compounds in river basins identified using sediment cores from reservoirs: Environmental Science & Technology, v. 31, p. 2339–2344, doi:10.1021/es960943p.

Van Metre, P.C., Wilson, J.T., Callender, E., and Fuller, C.C., 1998, Similar rates of decrease of persistent, hydrophobic and particle-reactive contaminants in riverine systems: Environmental Science & Technology, v. 32, p. 3312–3317, doi:10.1021/es9801902.

Van Metre, P.C., Jones, S.A., Moring, J.B., Mahler, B.J., and Wilson, J.T., 2003, Chemical Quality of Water, Sediment, and Fish in Mountain Creek Lake, Dallas, Texas, 1994–97: U.S. Geological Survey Water-Resources Investigations Report 03-4082, 69 p.

Van Metre, P.C., Wilson, J.T., Fuller, C.C., Callender, E., and Mahler, B.J., 2004, Collection, Analysis, and Age-Dating of Sediment Cores from 56 U.S. Lakes and Reservoirs Sampled by the U.S. Geological Survey, 1992–2001: U.S. Geological Survey Scientific Investigations Report 2004-5184, 180 p.

Verstraeten, G., and Poesen, J., 2000, Estimating trap efficiency of small reservoirs and ponds: Methods and implications for the assessment of sediment yield: Progress in Physical Geography, v. 24, p. 219–251.

Walling, D.E., 1983, The sediment delivery problem: Journal of Hydrology (Amsterdam), v. 65, p. 209–237, doi:10.1016/0022-1694(83)90217-2.

Ward, J.V., and Stanford, J.A., 1995, Ecological connectivity in alluvial river ecosystems and its disruption by flow regulation: Regulated Rivers: Research and Management, v. 11, p. 105–119, doi:10.1002/rrr.3450110109.

Windom, H.L., Schropp, S.J., Calder, F.D., Ryan, J.D., Smith, R.G., Jr., Burney, L.C., Lewis, F.G., and Rawlinson, C.H., 1989, Natural trace metal concentrations in estuarine and coastal marine sediments of the southeastern United States: Environmental Science & Technology, v. 23, p. 314–320, doi:10.1021/es00180a008.

Wren, D.G., Wells, R.R., Wilson, C.G., Cooper, C.M., and Smith, S., Jr., 2007, Sedimentation in three small erosion control reservoirs in northern Mississippi: Journal of Soil and Water Conservation, v. 62, p. 137–144.

Manuscript Accepted by the Society 29 June 2012

Using ground-penetrating radar to determine the quantity of sediment stored behind the Merrimack Village Dam, Souhegan River, New Hampshire

David J. Santaniello*,†
Noah P. Snyder*
Department of Earth and Environmental Sciences, Boston College, 140 Commonwealth Avenue, Chestnut Hill, Massachusetts 02467, USA

Allen M. Gontz*
Department of Environmental, Earth, and Ocean Sciences, University of Massachusetts–Boston, 100 Morrissey Boulevard, Boston, Massachusetts 02135, USA

ABSTRACT

We investigated the viability of ground-penetrating radar (GPR) as a method to estimate the quantity of sediment stored behind the Merrimack Village Dam on the Souhegan River in southern New Hampshire. If the predam riverbed can be imaged, the thickness and volume of the reservoir deposit can be calculated without sampling. Such estimates are necessary to plan sediment management after dam deconstruction. In May 2008, we surveyed six cross sections with a Mala Geosciences ProEx 100 MHz GPR. In a related study, topographic surveys were conducted in 2008–2009 to monitor the sediment flux associated with the removal of the Merrimack Village Dam in August 2008. Within a month of the removal, these surveys mapped the predam riverbed in the uppermost cross sections in the former impoundment. We compared these surveys to our interpreted GPR images for one cross section to determine a calibrated velocity for the impounded sand of 0.043 ± 0.020 m/ns. We also estimated the radar velocity of the deposit by analyzing hyperbolic reflections in the GPR images, and found a similar result (0.039 m/ns). Using the calibrated velocity, we estimated a total volume of sediment stored behind the Merrimack Village Dam of 66,900 ± 9900 m^3, which compares well to a previous estimate (62,000 m^3) based on a depth-to-refusal survey. Our findings indicate that GPR is a useful technique for quantifying impounded sediment prior to dam removal in reservoirs containing 1–10 m of sand overlying a coarser predam riverbed, but it may be less effective in settings with finer and/or thicker impounded sediment.

---

*E-mails: david.santaniello.1@bc.edu, noah.snyder@bc.edu, allen.gontz@umb.edu.
†Now at Department of Earth and Planetary Sciences, University of California–Santa Cruz, 1156 High Street, Santa Cruz, California 95064, USA.

Santaniello, D.J., Snyder, N.P., and Gontz, A.M., 2013, Using ground-penetrating radar to determine the quantity of sediment stored behind the Merrimack Village Dam, Souhegan River, New Hampshire, *in* De Graff, J.V., and Evans, J.E., eds., The Challenges of Dam Removal and River Restoration: Geological Society of America Reviews in Engineering Geology, v. XXI, p. 45–57, doi:10.1130/2013.4121(04). For permission to copy, contact editing@geosociety.org. © 2013 The Geological Society of America. All rights reserved.

INTRODUCTION

Numerous dams were constructed on the rivers of the United States during the seventeenth through twentieth centuries. These dams allowed rivers to become more navigable by boats and for log drives, provided a source of power and drinking water, and aided in flood protection. However, dams also altered the natural landscape and disturbed ecosystems, and now many structures have greatly exceeded their design life and have the potential for failure. As environmental protection and public safety have become greater concerns, these dams have started to be removed. The removal process is complex, and many questions must be answered before deconstruction can begin. In particular, knowing the quantity of sediment stored behind a dam is a prerequisite. Here, we study the applicability of ground-penetrating radar (GPR) as a means to rapidly survey the volume of the impounded sediment. To test the method, we compared a GPR survey conducted about 2 mo prior to the removal of the Merrimack Village Dam on the Souhegan River in southern New Hampshire with postremoval exposure of impounded sediment and the predam riverbed.

Our aim is to provide a useful method for future dam removal projects. GPR is a noninvasive procedure, and the field equipment can be operated by a small team in a short time at reasonable cost. Estimates of total volume of impounded sediment calculated from GPR surveys can aid contractors in determining a sediment-management plan after removal. In some cases, dredging may be necessary, and volume estimates provide a means for calculating costs. In other situations, the river may be able to transport the impounded sediment without causing harm to downstream ecosystems and structures (Doyle et al., 2003).

GPR has been used successfully in studying fluvial and lacustrine sediments. For example, Haeni et al. (1987) used GPR to determine the extent and thickness of lake bottom sediments. Lesmes et al. (2002) created a three-dimensional (3-D) model of a sand point bar. Clement et al. (2006) studied interactions between GPR and the water table in a sand-filled channel. The most similar study to this one that we have found is Hunter et al. (2003), where the authors used GPR to analyze the sedimentation of a reservoir in Alaska and to evaluate sediment trap designs. GPR was used in part to determine the thickness of gravel bars upstream of the dam in order to determine the sediment storage and dredging capacity. This project will build upon previous experiments and determine whether GPR can be used successfully and efficiently in determining the volume of sediment stored in a reservoir prior to dam removal.

STUDY AREA

The field site for this study is the Merrimack Village Dam in Merrimack, New Hampshire (Fig. 1). This run-of-the-river dam was located on the Souhegan River, ~500 m upstream of the confluence with the Merrimack River, and was deconstructed in August 2008. Dams have existed at this location since ca. 1734 (NHDHR, 2006). The reach was preferable because of a 6-m-high bedrock falls immediately downstream, so mills could exploit this drop plus that of the dam structure itself (Fig. 2). The modern Merrimack Village Dam (55 m long and 6.25 m high) was built in 1907 and modified in 1934, when the New Hampshire Highway Department constructed a concrete spray skirt and spillway over the dam structure (New Hampshire Public Service Commission, 1934; NHDHR, 2006). When the Merrimack Village Dam was deconstructed, the Souhegan River flowed unimpeded to the Merrimack River for the first time in over 250 yr.

Prior to removal, the Merrimack Village Dam impoundment was filled with sediment, primarily fine to coarse sand (Gomez and Sullivan Engineers, 2006, personal commun.; Pearson et al., 2011). Based on the river morphology upstream of the impoundment, the predam riverbed was believed to be composed of coarser sediment (boulders and cobbles, probably fluvially reworked glacial till) and some bedrock outcrops. This assumption was confirmed by postremoval exposure of the surface. In order to quantify how much sediment had accumulated behind the dam, Gomez and Sullivan Engineers (2006, personal commun.) performed a depth-to-refusal survey. In 2004, the firm installed seven transects across the impoundment and measured the thickness of the deposit by pounding a steel rod to the point of refusal every 3.64 ± 0.99 m along each transect (Fig. 1; Gomez and Sullivan Engineers, 2006, personal commun.). After the surveys were complete, a decision was made to allow the Souhegan River to naturally erode and transport the impounded sediment downstream after the removal. On 28 May 2008, less than 3 mo before the Merrimack Village Dam removal began, we deployed a Mala Geosciences ProEx 100 MHz GPR unit in a canoe and paddled across the impoundment in a series of cross sections (Fig. 1). The goal of performing the GPR survey was to image the predam riverbed surface in order to measure the thickness and volume of the reservoir deposit. Pearson et al. (2011) showed that the Souhegan River prior to dam removal had some capacity to excavate sediment from the impoundment during floods. We anticipated that relatively low summer flows between the time of the GPR survey and the removal would mean little change would occur in the impoundment during that interval. Indeed, the removal began on 6 August 2008 during the highest discharge since modest high flows during the March–April snowmelt season. Immediately upon removal of the Merrimack Village Dam, the Souhegan River incised into the previously impounded sand (Fig. 3), and within 2 mo, it had eroded to the coarser predam riverbed in the upstream parts of the reservoir (Fig. 4; Pearson et al., 2011). This full-scale exposure of the deposit provided us the opportunity to ground truth the GPR interpretations, by comparing topographic surveys performed using a total station by Pearson et al. (2011) with the interpreted predam riverbed on the GPR images. This ability to test the predictions of the total thickness of impounded sediment using GPR allows us to assess whether this is a viable method for quantifying sediment volumes in more general cases.

This site is an ideal test of GPR to image impounded sediment for a variety of reasons. Incision following the dam removal allowed us to observe the former reservoir deposit, which

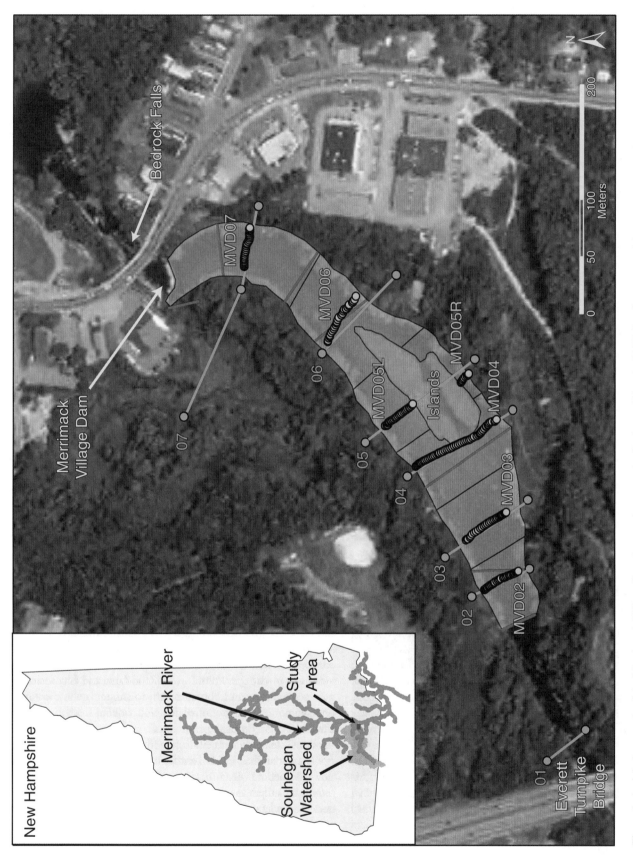

Figure 1. Aerial photograph of the Merrimack Village Dam impoundment on the Souhegan River in 2005 showing topographic survey cross sections (orange; Pearson et al., 2011), GPR cross sections (green) with accompanying areas (blue), and depth-to-refusal transects (pink; Gomez and Sullivan Engineers, 2006, personal commun.). Inset shows the location of the study area in New Hampshire (42°51′27″ N, 71°29′35″ W).

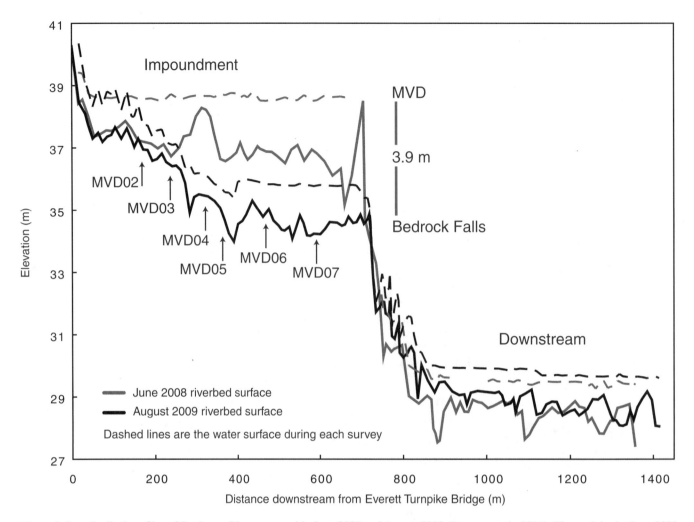

Figure 2. Longitudinal profiles of Souhegan River surveyed in June 2008 and August 2009 (Pearson et al., 2011). The peak in the June 2008 riverbed shows the top of the Merrimack Village Dam (MVD). The August 2009 survey shows a 3.9 m drop in the riverbed elevation associated with the dam removal. Arrows indicate the locations of the topographic survey and GPR cross sections in the impoundment (Fig. 1).

Figure 3. Photograph taken in July 2009 of typical post-dam-removal exposures of impounded sand and thin organic-rich layers. Field book is 12 cm wide.

included layers of fine and organic material interbedded with the sand. These layers were generally less than 0.2 m thick (Fig. 3), and therefore probably too thin to be imaged by GPR, because the observed vertical resolution of the images is 0.21 m. We also found few ferromagnetic materials within the impounded sediment, so magnetic permeability should not affect the imaging. Isolated pieces of buried metallic refuse existed in the sand deposit; however, these small metallic objects should not have affected imaging over the entire area.

Most importantly, soon after the dam deconstruction, the two upstream-most cross sections that were imaged with GPR (MVD02 and MVD03; Fig. 1) were incised to the level of the predam riverbed. At that location, this surface is composed of boulders and cobbles, with some bedrock outcropping, similar to the riverbed upstream of the impoundment (Fig. 4). As of the summer of 2010, the predam riverbed had not been exposed extensively downstream from cross section MVD04, but some

Figure 4. Photographs looking across and upstream from the left (north) bank of the impoundment at the MVD03 cross section. (A) Photograph from April 2008, before the dam removal, showing the sandy bottom and high water level of the impoundment. (B) Photograph from October 2008, after the dam removal, showing the exposed boulders of the predam riverbed.

boulders were cropping out in the thalweg. We assumed the predam riverbed surface would be coarser than the impounded sand that overlies it, although the low gradient of the riverbed upstream of the bedrock-floored former dam site suggests that it may not be as boulder-rich as the steeper reach upstream (Fig. 2).

Another factor that increased the chances of success of this study is the reflection coefficient (R) at the interface of a layer of sand with one of boulders and cobbles. R is directly related to the relative permittivity (ε_r) of the material of both layers (Table 1). The difference between wet granite (7), the approximate composition of the large clasts, and saturated sand (49) is great enough to produce a strong reflection (Table 2). We calculated ε_r values of medium and coarse sand by substituting the estimated velocity of the GPR waves through the impounded sand layer into Equation 1 (Baker et al., 2007; Blindow, 2009):

$$v \approx \frac{c}{\sqrt{\varepsilon_r}}, \qquad (1)$$

where c is the speed of light. The conductivity of the water measured in the field was used to calculate the conductivities of the other media using Archie's law (Archie, 1942; McNeill, 1991). The layer of boulders and cobbles, established as the predam riverbed, was expected to be visible in the GPR record if it was within the range of the electromagnetic (EM) waves. If the predam riverbed had been buried deeply below the surface, the EM waves would attenuate before returning to the receiving antenna, and the layer would not be recorded. Fortunately, the conductivity of freshwater was low enough to allow the EM waves to travel several meters into the subsurface. Furthermore, except at the midchannel islands (Fig. 1), the impounded deposit could be at most 3.9 m thick (including the overlying water), i.e., the difference between the top of the Merrimack Village Dam and the bedrock below (Fig. 2). These factors suggested that the predam riverbed surface could be imaged by the GPR surveys. However, GPR units display imaged layers as a function of time, not depth. Converting the time measurements to depths requires an estimate of the velocity of the EM waves through the media. In the next section, we present two methods we used to estimate the radar velocity of the impounded sand. Selecting the correct calibration

TABLE 1. LIST OF VARIABLES, SYMBOLS, AND UNITS

Variable	Symbol	Units
Impounded sediment thickness	h	m
Calculated sediment thickness based on calibrated velocity	h_{cal}	m
Calculated sediment thickness based on hyperbolic velocity	h_{hyp}	m
Relative permittivity	ε_r	Dimensionless
Reflection coefficient	R	Dimensionless
Two-way traveltime	t_r	ns
Average two-way traveltime	$t_{r,avg}$	ns
Velocity	v	m/ns
Calibrated velocity	v_{cal}	m/ns
Hyperbolic velocity	v_{hyp}	m/ns
Volume	V	m^3
Estimated volume based on calibrated velocity	V_{cal}	m^3
Estimated volume based on hyperbolic velocity	V_{hyp}	m^3

TABLE 2. CONSTITUTIVE PARAMETERS FOR MERRIMACK VILLAGE DAM MEDIA

Medium	Relative permittivity, ε_r	Relative permeability, μ_r[†]	Conductivity, σ (S/m)
Freshwater	81*	1	0.035
Saturated medium sand	49	1	0.009
Saturated coarse sand	49	1	0.013
Wet granite	7*	1	0.005

*Values are from Daniels et al. (1995).
[†]Values are from Baker et al. (2007).

is critical to having a credible method for determining the volume of impounded sediment in future dam removal projects.

ESTIMATES OF GPR VELOCITY IN IMPOUNDED SEDIMENT

Hyperbolic Analysis and GPR Data Interpretation

Radar velocity can be determined through the use of GPR surveying methods and postprocessing algorithms (Annan, 2009). The simplest field method, a common midpoint survey (CMP), requires separate receiving and transmitting antennas that can be moved independently. Federal Communications Commission (FCC) regulations prevent the sale of such systems in the United States and require all GPR systems to encase the transmitter and receiver in a single shielded unit (Breed, 2005). The CMP is also difficult to conduct over water due to the inability to constrain survey geometry and spacing. Based on the configuration of the antennas used with the MALA Geosciences Pro-Ex system, a CMP survey was not possible to directly determine the subsurface velocity.

Presently, the most common assessment of velocity uses postprocessing algorithms to analyze the returns from point-source reflections (Cassidy, 2009). Single-point reflections from large, outsized clasts or objects will produce a reflection on the raw GPR imagery that is hyperbolic in nature. This reflection is the result of the spherical nature of the radar pulse emitted from the GPR transmitter. The radar wave reaches and reflects from points along the wave front equally. When this front interacts with a large clast (e.g., boulder, pipe, drum), the reflection from that object is recorded as if it were directly below the transmitter-receiver unit (Cassidy, 2009; Fig. 5).

Postprocessing algorithms are provided in analytical software for GPR and include graphical systems to match a standard hyperbolic pattern to one observed on the record. The width of the hyperbolic reflection is directly related to the radar velocity in the geologic units that encase or overlie the point source. In general, wider hyperbolic reflections indicate slower velocities.

To analyze the GPR imagery acquired at the Merrimack Village Dam site, we postprocessed the raw GPR data with GPR Slice v. 7.0 developed by the GeoArchaeometric Laboratory. The raw data were lightly processed to enhance the shallow layering through the application of a user-defined gain. GPR surveying over water through a canoe produced a regularly occurring harmonic echo. This signal was reduced through the application of a background removal filter. Data were further enhanced through the application of a band-pass filter with the high pass set at 150 MHz and the low pass set at 75 MHz. The resultant imagery was visually checked for spatial and temporal effects that resulted from nonsynchronous data acquisition and survey initiation. These effects were generally recognized by segments of apparently perfectly parallel reflections at the start or end of a line. Such sections were clipped from the data file.

We examined the processed data file using the migration filter suite in GPR Slice, first indentifying layering and surfaces, and then measuring returns from point reflectors. The tool allows the user to select hyperbolic returns on the data record and match the form of the hyperbolic reflection on the record to a standard hyperbolic pattern (Fig. 6). The software calculates a radar velocity based on the hyperbolic reflection and GPR frequency. We began our analysis on the MVD03 cross section, because the pre-dam riverbed was well exposed by incision after the dam removal

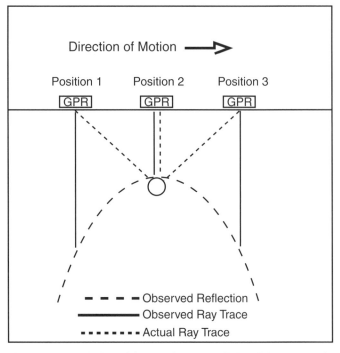

Figure 5. Theoretical model portraying how a GPR unit images an object as a hyperbola.

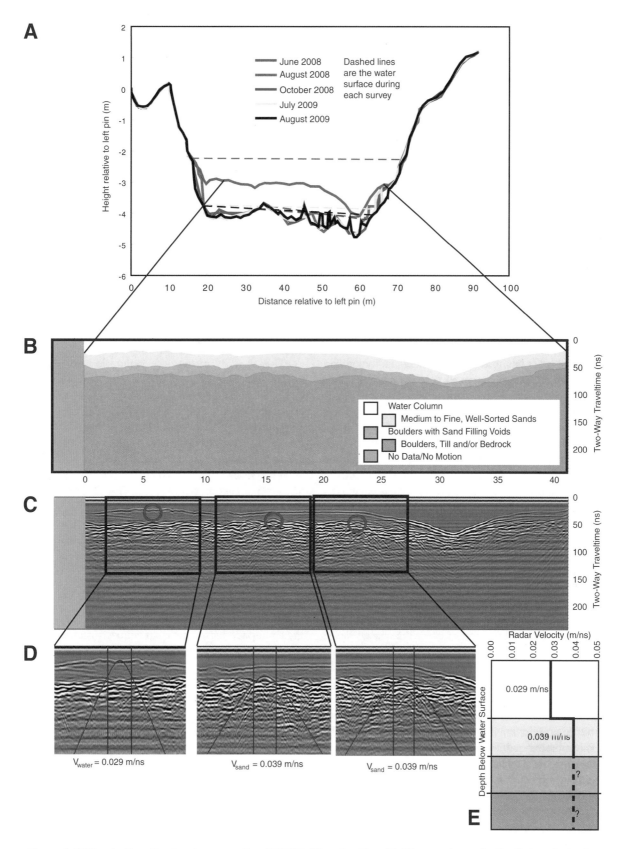

Figure 6. GPR velocity estimates at cross section MVD03. Three boulders (C, D) were chosen for the first method using analysis of hyperbolic reflectors. The second method compared the thickness of the impounded sand from total station surveys (A) with the average traveltime of the interpreted dam deposit (B). The velocity profile of the interpreted units (E) shows an increase between the water column (0.029 m/ns) and the reservoir sediments (0.039 m/ns); however, the velocities of the units buried beneath the impounded sand are unresolved.

(Fig. 4). As described already, the predam riverbed is considerably coarser than the overlying sand. Therefore, we focused our GPR analysis on two characteristic surfaces: (1) the sediment-water interface and (2) the highest surface formed by interlocking hyperbolic point reflections (caused by boulders). We interpreted the material bounded by these surfaces (unit 2) to be the sand deposited after dam construction. At MVD03, the GPR data indicated that there were three geophysical stratigraphic units (Fig. 6). Unit 1 was characterized by an internally transparent character with a parallel upper bounding surface and an intense lower bounding surface. We interpreted unit 1 as the water within the impoundment. Unit 2 was characterized by parallel to subparallel internal reflections that often exhibited cut and fill features. The upper bounding surface was observed as coincident with the lower bounding surface of unit 1. The lower bounding surface was observed as composed of overlapping hyperbolic reflections. We interpreted unit 2 as the sandy sediments that filled the impoundment. Unit 3 was characterized by chaotic internal reflections and numerous hyperbolic reflections. Packages of strongly parallel reflections were observed between hyperbolic returns in the uppermost sections of unit 3. The upper bounding surface was coincident with the lower bounding surface of unit 2. The lower bounding surface of unit 3 was not resolved on the GPR data, most likely due to radar attenuation and scattering indicated by the hyperbolic returns and internally chaotic nature of unit 3. We interpreted unit 3 as glacial till, reworked by predam fluvial processes. We interpreted the bounding surface between units 2 and 3 as the predam riverbed. We interpreted the strongly parallel reflectors in the uppermost sections of unit 3 as sandy sediments filling the interstitial spaces between the large boulders present in the layer.

We selected three hyperbolic reflections from GPR line MVD03 for analysis (Fig. 6). One hyperbolic reflection was chosen on the bounding surface between units 1 and 2 to represent the water column, and two hyperbolic reflections were chosen on the bounding surface between units 2 and 3 to represent the sandy infill sediments. The hyperbolic reflections were chosen based on their degree of definition and the amount of the unit above the bounding surface that surrounded the hyperbolic reflection.

The water-column hyperbolic reflection was observed on the bounding surface between units 1 and 2 (the impoundment floor). This hyperbolic was chosen to provide information on the velocity of the overlying water column. Hyperbolic matching in GPR Slice suggested a radar velocity of 0.029 m/ns. The two hyperbolic reflections on the bounding surface between units 2 and 3 indicated velocities (v_{hyp}) of 0.039 m/ns for the sandy impoundment sediments.

Calibration of GPR Interpretations with Direct Measurements

Another method for determining the radar velocity is to compare interpreted layers on the GPR record with the actual thicknesses of imaged media. The most common method for this analysis is to use boreholes or cores to compare the interpreted layers to measured thicknesses. Here, we ground truthed our GPR interpretations of impounded sediment thickness using direct measurements of the same sedimentary package enabled by subsequent exposure of the predam riverbed (Figs. 4 and 6). Pearson et al. (2011) measured the impounded sediment thickness precisely via topographic (and bathymetric) surveys using a Leica Total Station conducted prior to the dam deconstruction in June 2008 and afterward in August 2008, October 2008, July 2009, and August 2009 (Figs. 6 and 7). We used these cross-section locations to image the impounded sediment with the GPR unit in May 2008 (Fig. 1). In order to track our GPR transects, we placed a Trimble GeoXT global positioning system (GPS) on top of the GPR unit to log its location during imaging. Because the unit was not canoed across the impoundment in a perfectly straight line, the GPR cross sections do not match the topographic cross sections exactly, ranging from 1.3 m to 3.9 m downstream, but these resulting differences in the thickness of the impoundment fill should be minor (Fig. 1).

This velocity calibration method is possible because the predam riverbed level was exposed in the upstream cross sections after dam removal (Fig. 4). Here, we assume that the interpretations at cross section MVD03 (described previously) are correct (Fig. 6). We used MVD03 because the impounded sand deposit was eroded away by August 2008, and the boulders and cobbles at the present bed can be assumed to be the predam riverbed. This riverbed surface was constant through the last four topographic surveys (Fig. 6A). Smaller changes (<0.5 m) in the riverbed height occurred from surveying different points on the same boulders along the cross section. This relief reflects the noise on the surface, and it is a source of uncertainty in our calibration. We did not use cross section MVD02 for the velocity calibration because the variability of the riverbed surface (both in the GPR imagery and field exposure) is greater than at MVD03.

We used the August 2009 topographic survey as the measured interface between the impounded sand and the predam riverbed surface (although any of the four postremoval surveys would yield similar results), and the June 2008 pre-dam-removal survey as the water-sand interface (Fig. 6A). We calculated the thickness of the impounded sand (h; unit 2; Fig. 6) by measuring the difference between the two surveys. This method is not affected by compaction of impounded sediment after the dam removal because it is based on a preremoval survey of the sediment in the reservoir. Because the topographic surveys did not measure exactly the same points across the cross section, we interpolated the data in order to calculate a height difference every half meter relative to the left survey marker (Pearson et al., 2011). Using the GPS data recorded during the GPR imaging, we registered the GPR survey end points to the Pearson et al. (2011) surveys (Fig. 6). We used these two points to constrain the data analysis of the survey heights of the riverbed surfaces in order to include only data points that had corresponding GPR time travel measurements. We calculated the mean (h_{avg}) and standard

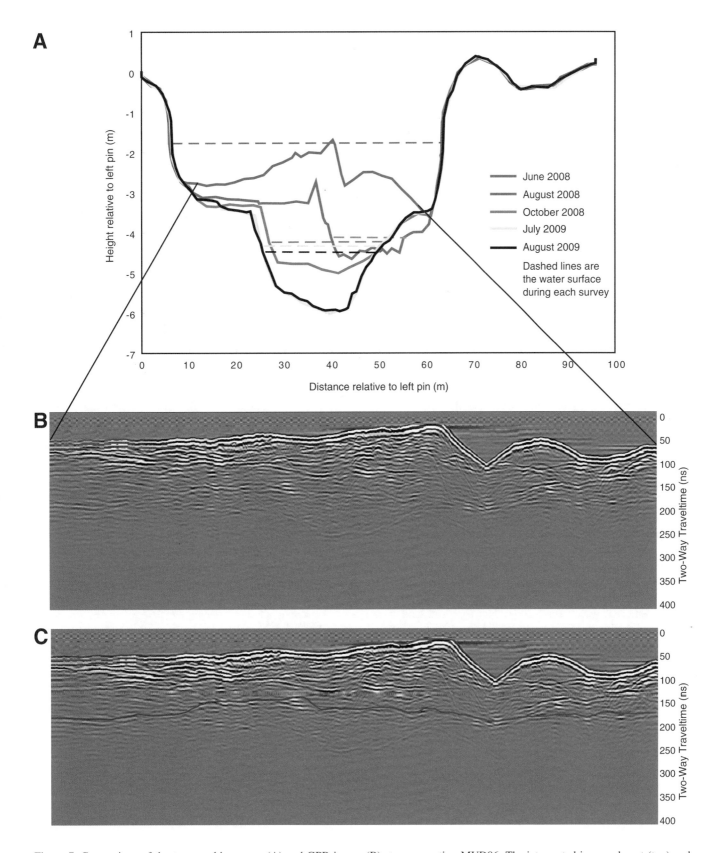

Figure 7. Comparison of the topographic survey (A) and GPR image (B) at cross section MVD06. The interpreted impoundment (top) and predam riverbed (bottom) surfaces are shown in red (C).

deviation of the impounded sand deposit thickness based on the series of height differences.

We then measured the equivalent sediment thickness on the GPR images of the MVD03 cross section. At regular intervals, approximately every 1–2 m along the cross section (Fig. 6), we measured the difference in two-way traveltime (t_r) between the interpreted water-sand interface in the impoundment and interface between the sand and the predam riverbed using the nanosecond scale on the GPR image. By assuming the reservoir bed and predam riverbed are the same on the topographic survey and GPR image, the average thickness, h_{avg}, and average two-way traveltime, $t_{r,avg}$, can be substituted into Equation 2 (Blindow, 2009):

$$h = \frac{t_r}{2} v. \quad (2)$$

By rearranging, it is possible to solve for the calibrated velocity, v_{cal}:

$$v_{cal} = \frac{h_{avg}}{t_{r,avg}/2}. \quad (3)$$

In order to account for the variability in these cross section–based measurements, we calculated standard deviations of h_{avg} and $t_{r,avg}$ at the calibration site and propagated these uncertainties into our estimate of v_{cal}. The value of h_{avg} at the MVD03 cross section is 0.85 ± 0.25 m based on the topographic surveys, and $t_{r,avg}/2$ is 19.9 ± 3.4 ns based on the GPR image interpretation, which yields a v_{cal} of 0.043 ± 0.020 m/ns (Table 3).

Comparison of GPR Velocity Estimation Methods

Our two estimates of GPR velocity of the impounded sand agree well. The hyperbolic analysis gives a velocity (v_{hyp}) of 0.039 m/ns, and the comparison between interpreted and measured stratigraphic thicknesses shows v_{cal} to be 0.043 ± 0.20 m/ns. Both estimates are somewhat slower than the 0.05–0.09 m/ns range of velocities through wet sand cited in the literature (Witten, 2006; Baker et al., 2007). However, wet sand has a relative permittivity (ε_r) between 10 and 30 (Daniels et al., 1995), and freshwater has a ε_r of 81. Therefore, as sand becomes more saturated, the associated ε_r increases. Because the ε_r of saturated sand in the impoundment should be larger than the ε_r of wet sand, the velocity of the EM waves through this sediment should be slower (Eq. 1). Therefore, our velocity estimate for this field site of ~0.4 m/ns appears reasonable. Here, we explore some limitations in each method.

During our hyperbolic analysis, we examined the reflections for signs of symmetry. The morphology of the hyperbolic reflection is strongly related to survey speed, and it was difficult to maintain a constant survey speed using a canoe over water. Any hyperbolic that exhibited asymmetry was removed from the analysis. In general, the hyperbolic reflections in the central portion of all lines exhibited little to no asymmetry, while those close to the beginning and end of the lines tended to exhibit some degree of asymmetry.

In order to calibrate the velocity by comparison with the topographic surveys of impounded sand thickness, we assumed that (1) the interpretations of the impoundment surface and predam riverbed at MVD03 were correct; (2) they were equivalent to the topographic survey lines of June 2008 and August 2009, respectively (Fig. 6); and (3) the topographic and GPR surveys measured the same fill material (Fig. 1). Here, we evaluate each of these assumptions.

The first assumption, that the GPR interpretations are correct, is critical to the calibration. If the wrong surface is interpreted as the pre-am riverbed, the traveltime (t_r) measurements will be incorrect, and an invalid velocity will be calculated. An incorrect velocity will drastically change the volume estimate. However, there is a strong layer of hyperbolic reflections in the GPR record of MVD03 (Fig. 6). Hyperbolic reflections are interpreted as individual reflectors such as buried objects, pipes, or boulders. In a fluvial setting, interpreting hyperbolic reflections as boulders is a valid assumption. Also, the photographic and topographic evidence after the dam removal shows large boulders throughout MVD03 that were previously overlaid with sand (Figs. 4 and 6A).

The second and third assumptions are relatively straightforward. The June 2008 topographic survey was conducted only

TABLE 3. VOLUME CALCULATIONS

Area	Interpreted sediment thickness (h_{cal}) (m)	Interpreted sediment thickness (h_{hyp}) (m)	Area (m²)	Interpreted volume (V_{cal}) (m³)	Interpreted volume (V_{hyp}) (m³)
MVD02	0.81	0.74	2929	2400	2200
MVD03	0.85	0.78	4335	3700	3400
MVD04	2.35	2.14	4719	11,100	10,100
MVD05L	3.11	2.83	2537	3600	3300
MVD05R	2.37	2.16	1150	6000	5500
Islands	3.41	3.16	4503	15,400	14,200
MVD06	2.67	2.43	5745	15,300	14,000
MVD07	1.49	1.36	6294	9400	8600
Total	N.A.	N.A.	32,212	66,900 ± 9900*	61,100

Note: N.A.—not applicable.
*Columns may not sum exactly due to rounding.

6–8 d after the May 2008 GPR survey, during a period of low discharge on the Souhegan River, so the water-sand interface probably did not change due to erosion or deposition within the impounded (Pearson et al., 2011). The close correspondence with the postremoval surveys at MVD03 indicates that the exposed boulders represent the in-place predam riverbed (Figs. 4 and 6A). Finally, because the thickness of the exposed impounded sand did not vary widely over short longitudinal distances, the minor downstream offsets between the GPR canoe-based transect and the topographic cross section are probably not a significant source of uncertainty in the calibration (Fig. 1).

Although this calibration comparing topographic surveys of stratigraphic surfaces to the GPR interpretations is clearly an excellent way to estimate GPR velocity, it has an obvious flaw: It is only possible after the dam has been removed. The hyperbolic analysis provides one means by which to calibrate velocity, but this requires large objects (e.g., boulders) within the deposit. In general, future studies may wish to include a few cores, boreholes, or simply depth-to-refusal surveys within the impoundment to calibrate GPR velocity. Establishing a set of topographic cross sections to be surveyed prior to and after dam removal is an excellent practice to both calibrate GPR surveys after the removal and monitor the excavation of impounded sediment (Collins et al., 2007; Pearson et al., 2011).

IMPOUNDED SEDIMENT VOLUME ESTIMATE

The goal of this study is to produce a credible volume estimate of sediment stored behind the Merrimack Village Dam prior to removal. If this number can be produced, the evidence for the application of GPR in future dam removals will be strengthened.

Impounded Sediment Thickness Estimates

We interpreted our GPR images at each of the study cross sections (Fig. 1) using the methods described herein. At each section, we identified a reflector in the subsurface that we interpreted to be the predam riverbed surface. We then used our GPR velocity estimates (v_{cal} and v_{hyp}) to estimate impounded sediment thicknesses, and calculate a single average value (h_{cal} and h_{hyp}) for each cross section. Because the two velocity estimates are so similar, here we focus on the calculations using v_{cal}, which also include error bounds based on the variability of the sediment thickness at the MVD03 calibration site (v_{cal} = 0.043 ± 0.020 m/ns).

We made our interpretations of the impounded sand layer using some knowledge of the impoundment geometry, for instance, the sediment thickness should be greatest, but probably not more than 3.9 m (including the water in the impoundment), near the dam site (MVD06 and MVD07 cross sections) and thin to nothing upstream of MVD02 (Figs. 1 and 2). With this knowledge, we identified a continuous predam riverbed surface at each cross section, similar to that done at MVD03 (Fig. 6). Unlike MVD03, we could not ground truth our interpretations as extensively because (as of the summer of 2010) the river had not incised to and exposed the predam riverbed across the impoundment downstream of that cross section, although it had clearly reached base level defined by the bedrock around and just upstream of the former dam site, as evidenced by the relatively flat thalweg longitudinal profile in the downstream half of the impoundment (including cross sections MVD04–07; Figs. 1 and 2) and field observations (Pearson et al., 2011). As described already, in this part of the former impoundment, a few boulders were exposed in the deep thalweg, but we did not see a clear, coarse predam riverbed surface like that exposed upstream (Fig. 4). The thicker impoundment deposits downstream of MVD03 include more structure visible in the GPR imagery. For example, MVD06 shows a continuous reflection with varying strength between 150 and 200 ns across the entire cross section, which we interpret as the predam riverbed, and several other reflectors above it that probably represent cutting and filling events within the impoundment (Fig. 7; Pearson et al., 2011). The average predicted thickness of the sediment (h_{cal}) at this location is 2.67 m based on the GPR interpretations, which is reasonable based on the observed base-level change (Fig. 2), and the postremoval incision (Fig. 7A).

Impounded Sediment Volume Estimates

In order to calculate the volume (V) of the impounded sediment based on our GPR images, we divided the impoundment into eight areas, including one per GPR cross section and an additional area including the midchannel islands between transect MVD05L and MVD05R, and calculated eight individual volumes (Fig. 1; Table 3), which is similar to the approach used by Snyder et al. (2004). The seven areas associated with GPR images were defined as spanning from halfway to the downstream cross section to halfway to the upstream cross section. We estimated impounded sediment volume by multiplying h_{cal} or h_{hyp} by the area (Table 3). To propagate the errors, we also made the same calculations based on the upper and lower bounds on v_{cal}. In order to calculate the volume of the island area (Fig. 1), we extrapolated the interpreted predam riverbed surface from the GPR surveys at MVD05L and MVD05R through the islands, and used the June 2008 topographic survey as the top of the impounded sediment.

The volume estimate for the total amount of sediment trapped behind the Merrimack Village Dam is the sum of the eight individual volume estimates. To estimate error bounds, we calculated every possible combination of the total volume based on the minimum, calibrated, and maximum volume for each of the eight areas. We then took the mean and standard deviation of these 6561 (3^8) possible volume estimates, which yields 66,900 ± 9900 m^3 (V_{cal}) based on the calibrated velocity (v_{cal}). Simply using v_{hyp}, we estimate the volume of impounded sediment (V_{hyp}) to be 61,100 m^3.

The most definitive test in determining if GPR is a viable method for studying the quantity of impounded sediment stored behind dams is to compare the volume calculation to a credible, independent estimate. Gomez and Sullivan Engineers (2006, personal commun.) performed a depth-to-refusal survey in 2004 using a similar definition of impoundment area to ours, and they estimated the volume of sediment stored in the impoundment at that time to be 62,000 m^3 (Table 3). All three volume estimates agree within <10%, which provides sufficient evidence to claim that GPR is a viable method for estimating the quantity of sediment stored behind a dam and should be considered for use during subsequent dam removals.

CONCLUSIONS

As the dams built during the seventeenth through twentieth centuries have aged, safety and environmental concerns about these structures have been raised by the government and advocacy groups, and dam removal is an increasingly viable solution. Regardless of the reason for the removal, knowing the volume of sediment stored behind the dam is a necessary prerequisite for planning sediment management during a dam removal project. In some cases, dredging will be necessary; in others, the decision may be made to allow the river to transport the impounded sediment downstream. GPR can be used to obtain this estimate and is a beneficial tool for future dam removals. The technique is noninvasive and is capable of being performed by a small team in a short time span at a reasonable cost.

The goal of this study was to determine whether GPR could be used to estimate the total volume of sediment behind the Merrimack Village Dam. The GPR surveys imaged the entire impounded sand deposit, and our two estimates of the volume agree within 10% with a previous survey of the impoundment (Gomez and Sullivan Engineers, 2006, personal commun.). These results demonstrate that GPR is a viable option for studying the volume of sediment stored behind dams. We suggest that GPR coupled with topographic (and bathymetric) surveys of fixed cross sections in the impoundment provides an excellent means to understand sediment transport behind dams, before, during, and after removal. Our work indicates that using hyperbolic reflections in the GPR imagery is a viable means by which to estimate the radar velocity of impounded sand; however, one could also use cores, boreholes, or depth-to-refusal surveys for this purpose. GPR surveys are also useful for identifying the location of the former channel and any preserved floodplain deposits that might host cultural heritage sites that were drowned during the creation of the impoundment.

GPR is only a viable alternative to other methods if it can image the entire impoundment deposit. Several factors determine whether this imaging is possible. The conductivity of the media must be low, and the reservoir cannot be too deep. Conductivity largely impacts the attenuation rate, and the largest rates that occur during imaging are when the EM wave travels through the water column. The maximum water depth that can be imaged depends on the conductivity and can range from 7 m to 20 m (Sellmann et al., 1992; Moorman and Michel, 1997; Hunter et al., 2003). Similarly, clay also has a high conductivity, and the presence of thick clay layers in the deposit will greatly reduce the penetration depth of the EM waves. Imaging is also aided by having a high contrast in grain size between the impoundment sediments and predam riverbed, such as that seen at our field site. Therefore, GPR is a viable technique for many dams in lower-relief regions like the northeastern United States, but it may not be capable of imaging thicker impoundment deposits behind larger dams in mountainous regions.

ACKNOWLEDGMENTS

This work was supported by American Chemical Society Petroleum Research Fund grant 47858-GB8 to Snyder. Acknowledgment is made to the donors of the American Chemical Society Petroleum Research Fund for support of this research. We thank Adam Pearson (Boston College, now at University of Delaware) and Mathias Collins (National Oceanic and Atmospheric Administration Restoration Center, Gloucester, Massachusetts) for many discussions about the project and sharing topographic data; and Billy Armstrong (Boston College), Ilya Buynevich (Woods Hole Oceanographic Institution, now at Temple University), and Chris Stillman (University of Massachusetts at Boston) for helping with the GPR data collection. Reviews by Jim Pizzuto, Andrew Reeve, and volume editor James Evans greatly improved this contribution.

REFERENCES CITED

Annan, A.P., 2009, Electromagnetic principles of ground penetrating radar, in Jol, H.M., ed., Ground Penetrating Radar: Theory and Applications: Amsterdam, Elsevier, p. 3–40.

Archie, G.E., 1942, The electrical resistivity log as an aid in determining some reservoir characteristics: Transactions of the American Institute of Mining, Metallurgical, and Petroleum Engineers, v. 146, p. 54–62.

Baker, G.S., Jordan, T.E., and Pardy, J., 2007, An introduction to ground penetrating radar (GPR), in Baker, G.S., and Jol, H.M., eds., Stratigraphic Analyses Using GPR: Geological Society of America Special Paper 432, p. 1–18, doi:10.1130/2007.2432(01).

Blindow, N., 2009, Ground penetrating radar, in Kirsch, R., ed., Groundwater Geophysics: A Tool for Hydrogeology, Second Edition: Berlin, Springer-Verlag, p. 227–252, doi:10.1007/978-3-540-88405-7.

Breed, G., 2005, A summary of FCC rules for ultra wideband communications: High Frequency Electronics, January, p. 42–44.

Cassidy, N.J., 2009, Ground penetrating radar data processing, modeling and analysis, in Jol, H.M., ed., Ground Penetrating Radar: Theory and Applications: Amsterdam, Elsevier Science, p. 141–176.

Clement, W.P., Barrash, W., and Knoll, M.D., 2006, Reflectivity modeling of a ground-penetrating-radar profile of a saturated fluvial formation: Geophysics, v. 71, p. K59–K66.

Collins, M., Lucey, K., Lambert, B., Kachmar, J., Turek, J., Hutchins, E.W., Purinton, T., and Neils, D.E., 2007, Stream Barrier Removal Monitoring Guide: Gulf of Maine Council on the Marine Environment, 85 p., http://gulfofmaine.org/streambarrierremoval/ (accessed 10 December 2010).

Daniels, J.J., Roberts, R., and Vendl, M., 1995, Ground penetrating radar for the detection of liquid contaminants: Journal of Applied Geophysics, v. 33, p. 195–207.

Doyle, M.W., Stanley, E.H., Harbor, J.M., and Grant, G.S., 2003, Dam removal in the United States: Emerging needs for science and policy: Eos (Transactions, American Geophysical Union), v. 84, no. 4, p. 29, 32–33, doi:10.1029/2003EO040001.

Haeni, F.P., McKeegan, D.K., and Capron, D.R., 1987, Ground-Penetrating Radar Study of the Thickness and Extent of Sediments beneath Silver

Lake, Berlin and Meriden, Connecticut: U.S. Geological Survey Water Resources Investigations Report 85-4108, 19 p.

Hunter, L.E., Ferrick, M.G., and Collins, C.M., 2003, Monitoring sediment infilling at the Ship Creek Reservoir, Fort Richardson, Alaska, using GPR, in Bristow, C.S., and Jol, H.M., eds., Ground Penetrating Radar in Sediments: Geological Society of London Special Publication 211, p. 199–206.

Lesmes, D.P., Decker, S.M., and Roy, D.C., 2002, A multiscale radar-stratigraphic analysis of fluvial aquifer heterogeneity: Geophysics, v. 67, p. 1452–1464.

McNeill, J.D., 1991, Use of electromagnetic methods for groundwater studies: Geotechnical and Environmental Geophysics, v. 1, p. 191–218.

Moorman, B.J., and Michel, F.A., 1997, Bathymetric mapping and sub-bottom profiling through lake ice with ground penetrating radar: Journal of Paleolimnology, v. 18, p. 61–73, doi:10.1023/A:1007920816271.

New Hampshire Division of Historical Resources (NHDHR), 2006, Individual Inventory Form for Merrimack Village Dam (MER0018): Concord, New Hampshire Department of Cultural Resources.

New Hampshire Public Service Commission, 1934, Questionnaire—Statement Concerning Mills and Their Repairs, Dams and Flowage: Concord, New Hampshire Public Service Commission.

Pearson, A.J., Snyder, N.P., and Collins, M.J., 2011, Rates and processes of channel response to dam removal with a sand-filled impoundment: Water Resources Research, v. 47, W08504, doi:10.1029/2010WR009733.

Sellmann, P.V., Delaney, A.J., and Arcone, S.A., 1992, Sub-Bottom Surveying in Lakes with Ground Penetrating Radar: Hanover, New Hampshire, U.S. Army Cold Regions Research and Engineering Laboratory CRREL Report 92-8, 24 p.

Snyder, N.P., Rubin, D.M., Alpers, C.N., Childs, J.R., Curtis, J.A., Flint, L.E., and Wright, S.A., 2004, Estimating rates and properties of sediment accumulation behind a dam: Englebright Lake, Yuba River, northern California: Water Resources Research, v. 40, W11301, doi:10.1029/2004WR003279.

Witten, A.J., 2006, Handbook of Geophysics and Archaeology: London, Equinox Publishing Limited, 329 p.

MANUSCRIPT ACCEPTED BY THE SOCIETY 29 JUNE 2012

Prediction of sediment erosion after dam removal using a one-dimensional model

Blair Greimann

U.S. Bureau of Reclamation, Technical Service Center, Denver Federal Center, Denver, Colorado 80225, USA

ABSTRACT

The accurate prediction of sediment erosion after dam removal is critical to quantifying the impact of dam removal on the reservoir and downstream environment. A variety of methods can be used to estimate this impact, and one of the most common is to use a one-dimensional mobile bed sediment transport model. I describe a one-dimensional sediment transport model (SRH-1D) and use it to simulate a laboratory experiment of incision through a reservoir delta deposit. The model allows the user to specify the erosion width through the deposit as a function of the flow rate. The model is shown to predict the vertical incision and downstream sediment load with reasonable accuracy if the erosion width is specified. Sensitivity tests to the transport equation parameters, erosion width, and angle of repose are conducted. The sediment loads exiting the dam are shown to be sensitive to the critical shear stress, but they are relatively insensitive to changes to the erosion width and angle of repose. One-dimensional models are shown to require the specification of the erosion width, but the results are not considered to be extremely sensitive to its value, so long as it is approximately equal to the observed river width under the same flow conditions. Further work on modeling of bank erosion is necessary to more accurately predict the long-term evolution of reservoir deposits.

INTRODUCTION

When a reservoir's water level is lowered, the drop in water-surface elevation creates an increase in energy slope through the reservoir delta. The increase in energy slope then causes erosion of reservoir sediment. In many cases, the reservoir is much wider than the river, and the erosion of the reservoir sediment delta is often not uniform across the reservoir delta, but instead is rather focused in a portion of the reservoir width, similar to the width of the river (Bromley and Thorne, 2005; Cui et al., 2006a; Doyle et al., 2003).

Numerical models are often needed to simulate these processes, and while there are several general-application sediment models that allow simulation of sediment transport in alluvial channels of known width, not all models will be able to simulate the correct width of the erosion through the reservoir delta. Two basic categories of sediment models are one-dimensional (1-D) and two-dimensional (2-D) models. The 2-D models compute depth-averaged quantities within a grid defined by the user and can represent local variation in hydraulic variables, but present 2-D models that are coupled to sediment transport processes are computationally expensive. In addition, most 2-D models do not simulate bank erosion processes, though recently this has become an area of more active research (Duan et al., 2001; Dulal et al., 2010; Jang and Shimizu, 2005; Rinaldi et al., 2008; Wang et al., 2010). One example of a 2-D model applied to dam removal is

Greimann, B., 2013, Prediction of sediment erosion after dam removal using a one-dimensional model, *in* De Graff, J.V., and Evans, J.E., eds., The Challenges of Dam Removal and River Restoration: Geological Society of America Reviews in Engineering Geology, v. XXI, p. 59–66, doi:10.1130/2013.4121(05). For permission to copy, contact editing@geosociety.org. © 2013 The Geological Society of America. All rights reserved.

documented in Reclamation (2011), where they applied a 2-D model called SRH-2D (Lai, 2008) to simulate erosion following dam removal on the Klamath River. However, this reservoir did not have a significant delta, and channel incision through delta deposits was not a dominant process.

Because of their simplicity and more widespread availability, 1-D models are most often applied in practical situations (Cui et al., 2006b; Wells et al., 2007; Cui and Wilcox, 2008). 1-D models assume that the water surface, velocity, and other hydraulic variables are constant across the cross section. Because of the averaging, 1-D models cannot represent the transverse variation in shear stress that occurs across a cross section. The 1-D model will underpredict the rates of channel incision because it assumes that the erosion occurs uniformly across the delta deposit. To circumvent these problems, the erosion width can be specified as a constant, as in DREAM (Cui et al., 2006a), or as a function of flow rate, as in HEC-6T (MBH Software, 2001):

$$W = aQ^b, \quad (1)$$

where W is erosion width, Q is flow rate, and a and b are empirical constants. The empirical constants a and b can be unique to each river. MBH assumes $b = 0.5$ and allows the user to enter a.

Cui et al. (2006a, 2006b) performed tests of the DREAM model against laboratory and field data. Cantelli et al. (2007) developed a method to predict the erosional narrowing that occurs during rapid drawdown. In this paper, I use experimental data to test the validity of a 1-D model called SRH-1D (Huang and Greimann, 2010) for predicting the erosion of reservoir deposits following dam removal. I test the validity of the erosion width assumption and calculate the sensitivity of model results to various parameters.

EXPERIMENTAL DESCRIPTION

I used experimental data to test the validity of SRH-1D (Huang and Greimann, 2010) in predicting the erosion of reservoir deposits following dam removal or during reservoir sluicing. Cantelli et al. (2004) performed experiments at the University of Minnesota to simulate the removal of dam. A unique feature of these experiments was the precise measurement of the bed profile and cross-section width upstream of the dam while the flow eroded the deposit. A flume with a slope of 0.018 and a width of 0.061 m was filled with sediment to replicate the sediment deposit behind a dam. The sediment was uniform with a d_{50} of 0.8 mm, which is a coarse sand. The maximum depth of the sediment deposit was ~0.12 m. A channel, 1 cm deep and 27.5 wide, was cut in the middle of the deposit to ensure that the erosion occurred in the middle of the channel. A flow of 0.3 L/s was allowed to erode the deposit. The sediment feed continued at a rate of 0.002 kg/s during the experiment. This was calculated to be an equilibrium supply rate at the given slope and width.

MODEL INPUT

SRH-1D was used to simulate run 6, which is described in detail in Cantelli et al. (2004). The input data for SRH-1D consist of data geometry, flow data, sediment boundary conditions, and sediment transport parameters. The geometry was taken from the description of the experimental setup. Cross sections were spaced 10 cm apart in the numerical model. Each cross section was represented by points spaced 1 cm transverse to the flow. The Manning's roughness coefficient was assumed to be 0.02. A single sediment size of 0.8 mm was assumed. The flow rate and sediment feed rate were taken from the experiment values, which were 0.0003 m³/s and 0.002 kg/s, respectively.

After a test of several transport equations, Parker's (1990) surface-based transport bed-load formula was used to predict sediment transport capacity. Even though the Parker equation was developed for gravel-bed streams and the sediment in the experiment was in the sand size range, the sediment transport in the experiment was dominated by bed load, and it is not surprising that a bed-load formula was most appropriate. In addition, the purpose of this paper is not to recommend the best sediment transport formula for dam removal conditions, but rather to evaluate the erosion width assumption and other model parameters associated with dam removal modeling. Parker (1990) developed an empirical bed-load transport function based on the equal mobility concept and field data. Parker's dimensionless bed-load transport parameter, W_i^*, was assumed to be a single valued function of the dimensionless shear stress parameter, ϕ_i, or,

$$W_i^* = 11.9 f(\phi_i), \quad (2)$$

where the two parameters are defined as

$$W_i^* = \frac{q_{bi} g(s-1)}{p_i (\tau_b/\rho)^{1.5}}, \quad (3)$$

and

$$\phi_i = \frac{\theta_i}{\theta_c \xi_i}, \quad (4)$$

where q_s is volumetric sediment transport rate per unit width; τ_b is total bed shear stress, d_{50} is the median diameter; g is acceleration of gravity; γ is specific weight of water; and s is relative specific density of sediment (ρ_s/ρ). Also, θ_c is the critical Shield's parameter, and θ_i is Shield's parameter of the sediment size class i computed as:

$$\theta_i = \tau_b/(\gamma(s-1)d_i). \quad (5)$$

The parameter ξ_i is the exposure factor, which accounts for the reduction in the critical shear stress for relatively large particles and the increase in the critical shear stress for relatively small particles:

$$\xi_i = (d_i/d_{50})^{-\alpha}, \quad (6)$$

where α is a constant usually fitted to data. The function in Equation 2 was fitted to field data from Oak Creek and is:

$$f(\phi) = \begin{cases} 1 - 0.853/\phi & , \phi > 1.59 \\ 0.000183\exp[14.2(\phi - 1) - 9.28(\phi - 1)^2] & , 1 < \phi \leq 1.59 \\ 0.000183\phi^{14.2} & , \phi \leq 1 \end{cases} \quad (7)$$

Two parameters can be defined to calibrate the Parker's equation to a particular site: θ_c and α. Ideally, these values should be fit to data of the stream being simulated. However, in the absence of data, several references provide guidance, such as Buffington and Montgomery (1997) and Andrews (2000). In the simulations performed here, a value of 0.0386 was used for θ_c, which was the value recommended by Parker (1990). As a sensitivity test, values of 0.05 and 0.025 were also simulated for the value of θ_c. The value of α is not important for the particular experiment being simulated here because only a single size class is being simulated. In field conditions, the hiding coefficient, α, can be very important to defining the behavior of a gravel-bed system.

The erosion width is an important parameter in estimating the delta erosion. Because 1-D models do not have a shear stress that varies across a cross section, it is difficult to estimate the nonuniform erosion that occurs during incision. Cantelli et al. (2004) observed a rapid narrowing of the channel followed by a gradual widening. The narrowing was caused because the vertical erosion rate in the middle of the channel was faster than additional sediment supply from the banks. The highest shear stresses were in the middle of the channel, and therefore the highest erosion rates occurred there. The banks initially did not supply sufficient sediment to maintain a wide section, and the section narrowed. Eventually, the rapid incision slowed, and the bank erosion continued, and the section started to widen. Cantelli et al. (2007) modeled these processes with an extended 1-D model. However, the method of Cantelli et al. (2007) is not straightforward to implement in a general-purpose 1-D numerical model that represents natural cross sections.

The SRH-1D model does not directly simulate lateral transport of sediment because it is a 1-D model; however, it empirically accounts for the processes involved by using a relationship between erosion width and flow rate and an angle of repose condition for bank stability. Equation 1 is used in SRH-1D to determine the erosion width. The boundaries of the erosion width are determined by first finding the centroid of the cross section, and then assuming that W_e is apportioned equally on either side. In a rectangular or trapezoidal cross section, the erosion width would be the bottom width of the channel that defines the area that can be eroded vertically downward. The erosion width is a function of discharge, and as the discharge increases, the erosion width can increase. The experiment simulated in this study was a steady-flow experiment, so the erosion width is constant throughout the simulation.

The initial channel width for the experiments of Cantelli et al. (2004) began at ~26 cm. The width decreased rapidly to less than 5 cm near the dam face, but at ~40 cm upstream of the dam face, the width was not less than 17 cm. After 2.5 min, the width throughout the entire flume was greater than 17 cm and then gradually increased to ~24 cm on average near the end of the 2 h experiment. In the simulation, the erosion width was set to 24 cm for the flow rate of 0.0003 ft^3/s ($a = 14$, $b = 0.5$) to represent the ultimate erosion width of the channel. The value of b was assumed to be 0.5 as in HEC-6T, and the value of a was chosen such that the average erosion width near the end of the experiment was reproduced. As a sensitivity test, an erosion width of 20 cm was also simulated. I did not attempt to simulate the changing width during the experiment.

Bank failure is simulated using an angle of repose condition. SRH-1D requires an angle of repose below and above water. The angle of repose above water was set to 70° because the banks were nearly vertical during the experiment. Because the sand was saturated before the experiment and the water surface was only a few centimeters below the banks, capillary forces were significant between the sand particles and enabled the banks to remain almost vertical. Based upon the video footage from the experiment, the bank collapse was seen to occur as the banks were undercut. In field situations, the bank height is much larger, and the bank angle of repose may change with height above the water. Langendoen and Alonso (2008) and Langendoen and Simon (2008) have developed more detailed treatment of bank failure and implemented it into the CONCEPTS model.

The angle of repose underwater is more difficult to determine. The underwater angle of repose was set to 30°. To test the sensitivity of the model to the angle of repose, simulations were also run for an angle of repose equal to 35° above water.

A summary of the input parameters used in the simulations is given in Table 1. The simulations are referred to simulations 1 through 5, with simulation 1 being referred to as the base run.

TABLE 1. SUMMARY OF INPUT PARAMETERS USED IN SRH-1D SIMULATIONS

Parameter (units)	Simulation				
	1	2	3	4	5
Critical shear stress (no units)	0.0386	**0.05**	**0.025**	0.0386	0.0386
Erosion width (cm)	24	24	24	**20**	24
Angle of repose above water (degrees)	70	70	70	70	**35**
Note: Values changed from base run (simulation 1) appear in boldface italics.					

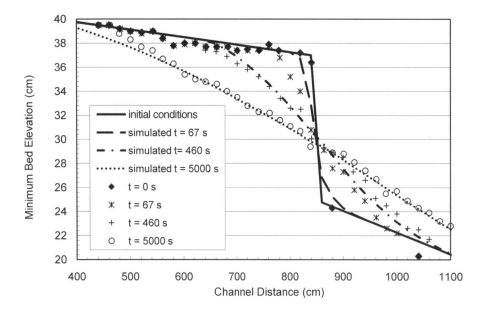

Figure 1. Comparison between Cantelli et al. (2004) and SRH-1D for the minimum bed elevation, base run (run 1).

MODEL RESULTS

The simulated results from SRH-1D were compared against the experimental data of run 6 of Cantelli et al. (2004). The simulated and measured bed elevations are shown in Figure 1. Overall, the agreement between the measured and simulated is satisfactory. The only significant disagreement between measured and predicted values is in the initial stages of channel formation, where the initial incision rates were underpredicted. Because the model does not simulate all processes involved in the erosional narrowing observed in the experiments, the simulated initial channel widths were larger than the measured ones. Therefore, the simulated bed elevations decreased more slowly than the measured ones.

The sediment loads are shown in Figure 2. Initially, the simulated sediment loads fit the measured loads relatively well. However, the simulated sediment loads at the dam face did not decrease as rapidly as the measured loads. The measured sediment loads decreased to approximately the feed rate values after ~600 s. This would indicate that after 600 s, there was no net erosion or deposition between the location of the feed and the dam face. That conclusion is in contradiction to the bed profile and top width evolution. Both the bed profile and the top width indicate that erosion continued throughout the course of the experiment. Therefore, one would expect the sediment discharge at the dam face to be higher than the feed rate. It is expected that there is a bias in the measured sediment discharges at the dam face.

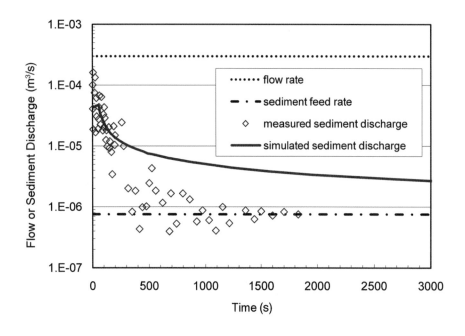

Figure 2. Comparison between Cantelli et al. (2004) and SRH-1D for sediment discharge at the dam face, base run (run 1).

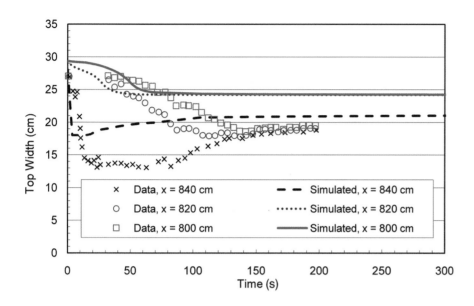

Figure 3. Comparison between Cantelli et al. (2004) and SRH-1D for wetted top width, base run (run 1).

The evolution of the wetted top width is shown in Figure 3. The 1-D model does not predict erosional narrowing accurately because no transverse sediment transport is modeled. The cross-section evolution is shown in Figure 4. The erosion is primarily vertical, with only a minor widening, because the angle of repose above water was assumed to be relatively large. There were no cross-sectional data against which to compare the model.

Sensitivity to Critical Shear Stress

The critical shear stress was increased to 0.05 (simulation 1) and decreased to 0.025 (simulation 2). A comparison of simulations 1, 2, and 3 is given for the bed profile (Fig. 5) and for the sediment discharge at the dam face (Fig. 6).

Because the critical shear stress enters directly into the sediment transport formula, Equation 2, raising the critical shear stress will decrease the simulated sediment transport rates. The decrease in the sediment transport rates causes sediment to erode more slowly. Therefore, the bed elevations of simulation 2 are higher than for simulation 1 upstream of the dam. In addition, the predicted sediment loads at the dam face are lower in simulation 2 than for simulation 1.

Sensitivity to Erosion Width

The erosion width was decreased to 20 cm, and the simulation results from this simulation (simulation 4) were compared against the base run, simulation 1. Decreasing the erosion width causes

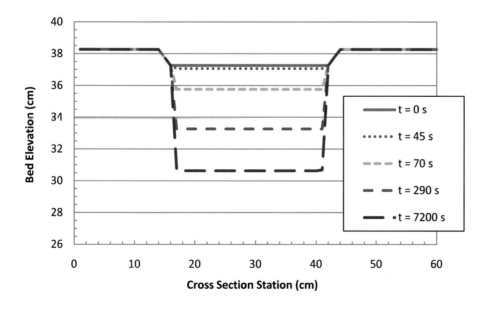

Figure 4. Cross section evolution at $x = 8.0$ m, 40 cm upstream of the dam, base run (run 1).

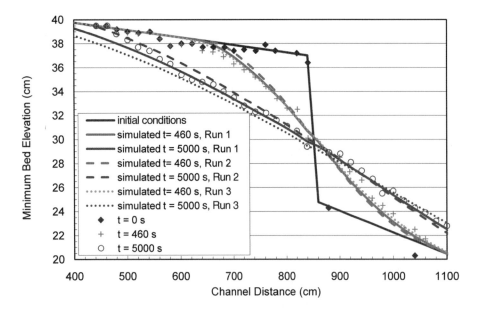

Figure 5. Sensitivity of bed profile to increasing and decreasing critical shear stress (runs 2 and 3).

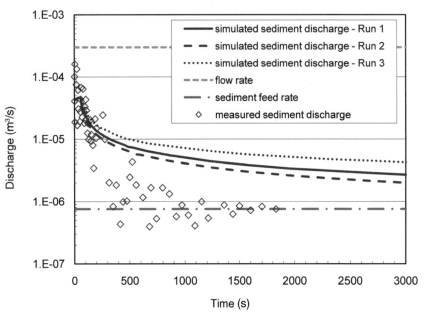

Figure 6. Sensitivity of sediment discharge at dam face to increasing and decreasing critical shear stress (runs 2 and 3).

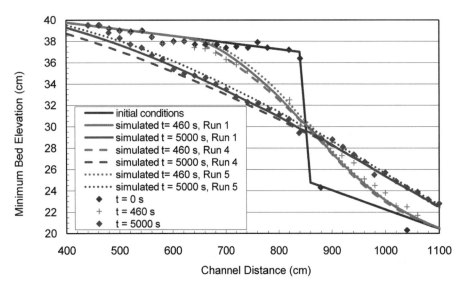

Figure 7. Sensitivity of bed elevation to decreasing erosion width (run 4) and decreasing angle of repose (run 5).

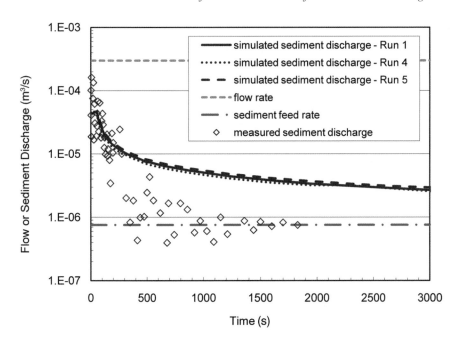

Figure 8. Sensitivity of sediment load to decreasing erosion width (run 4) and decreasing angle of repose (run 5).

the vertical incision to occur more rapidly and progresses further upstream (Fig. 7). However, the sediment loads at the dam face are relatively unaffected by the decrease in erosion width (Fig. 8). The sediment loads did not change significantly compared to simulation 1 because, even though the vertical incision occurred more quickly, the width of erosion was less, and therefore approximately the same volume of sediment was removed (Fig. 9).

Sensitivity to Angle of Repose

The above-water angle of repose was set to 35°. This simulation is termed simulation 5 and was compared against simulation 1. The difference in bed elevations was relatively small, 3 mm or less (Fig. 7), but the erosion depth in simulation 5 was consistently less than the base run. However, the sediment loads at the dam face were relatively unaffected (Fig. 8). The sediment loads were relatively unaffected because, even though the vertical incision was less in simulation 5, the horizontal widening was greater because of the smaller angle of repose and therefore approximately the same volume of sediment was removed (Fig. 9).

DISCUSSION

There are several important issues that need to be addressed when applying a 1-D model to simulate dam removal. Many sediment processes are ignored in a 1-D model because it does not directly simulate transverse sediment movement, and it calculates a single average shear stress for a cross section. To account

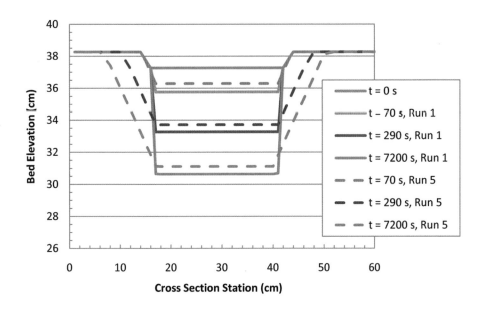

Figure 9. Sensitivity of cross-section shape to decreasing angle of repose (run 5).

empirically for these factors, an appropriate erosion width and angle of repose should be specified. In SRH-1D, the erosion width is specified as a function of flow rate. HEC-6T (MBH Software, 2001) also requires a similar specification, and DREAM (Cui et al., 2006a) requires a fixed base width. Often the erosion width is determined based upon the upstream and/or downstream river width. As a first estimate, Equation 1 could be fit to the upstream and downstream river channels. If the fit coefficients of Equation 1 for the upstream and downstream river channels are significantly different, some judgment may be necessary to determine the most appropriate values. The values of a and b are likely not universal and may need to be adjusted for different situations. However, the erosion rates and downstream transport were not sensitive to small changes in the erosion width, which is similar to the results of Cui et al. (2006b).

The angle of repose can be specified based upon sediment type. The angle of repose for noncohesive sediments is relatively well constrained. However, for cohesive sediments, the angle of repose may be difficult to determine, or it may not be a constant value. It is likely that the stable bank angle for cohesive soils will be a function of the height above the water table and the groundwater gradients in the sediments. Field specific data are necessary to evaluate the angle of repose for removal of a particular dam.

Even though the transverse variation of shear stress and bed load are ignored, in most cases the loss of these details does not significantly alter the uncertainty of the total erosion estimates. The greatest uncertainty may still reside in the calculation of the streamwise sediment transport.

Additional research is needed to develop appropriate models of long-term bank erosion. This is important in determining the concentrations downstream of the dam. Because the reservoir deposit is usually much finer than the riverbed sediment, the reservoir may potentially act as a source of fine sediment for many years as the banks slowly erode. As shown in the experiments of Cantelli et al. (2004), bank caving occurs after the initial channel incision as the flowing water undermines the steep banks. This process is difficult to model with a 1-D model, and no other readily available model currently exists. In most cases, it is assumed that bank erosion after the first few years will be limited to large storm events. However, there is often no quantitative method applied to determining which events will cause additional bank erosion. It may be feasible to use a method similar to that employed in CONCEPTS (Langendoen and Alonso, 2008), where the geotechnical strength of the bank is evaluated as well as the shear stress applied to the bank.

REFERENCES CITED

Andrews, E.D., 2000, Bed material transport in the Virgin River, Utah: Water Resources Research, v. 36, no. 2, p. 585–596, doi:10.1029/1999WR900257.

Bromley, C., and Thorne, C.R., 2005, A scaled physical modeling investigation of the potential response of the Lake Mills delta to different magnitudes and rates of removal of Glines Canyon Dam from the Elwha River, WA, in Moglen, G.E., ed., Proceedings of the 2005 Watershed Management Conference, July 19–22, 2005, Williamsburg, Virginia: Reston, Virginia, ASCE.

Buffington, J.M., and Montgomery, D.R., 1997, A systematic analysis of eight decades of incipient motion studies, with special reference to gravel-bedded rivers: Water Resources Research, v. 33, no. 8, p. 1993–2029, doi:10.1029/96WR03190.

Cantelli, A., Paola, C., and Parker, G., 2004, Experiments on upstream-migrating erosional narrowing and widening of an incisional channel caused by dam removal: Water Resources Research, v. 40, W03304, 12 p.

Cantelli, A., Wong, M., Parker, G., and Paola, C., 2007, Numerical model linking bed and bank evolution of incisional channel created by dam removal: Water Resources Research, v. 43, W07436, doi:10.1029/2006WR005621.

Cui, Y., and Wilcox, A., 2008, Development and application of numerical models of sediment transport associated with dam removal, in Garcia, M.H., ed., Sedimentation Engineering: American Society of Civil Engineers (ASCE) Manuals and Reports on Engineering Practice 110, p. 995–1020.

Cui, Y., Braudrick, C., Dietrich, W.E., Cluer, B., and Parker, G., 2006a, Dam Removal Express Assessment Models (DREAM). Part 1: Model development and validation: Journal of Hydraulic Research, v. 44, no. 3, p. 291–307, doi:10.1080/00221686.2006.9521683.

Cui, Y., Braudrick, C., Dietrich, W.E., Cluer, B., and Parker, G., 2006b, Dam Removal Express Assessment Models (DREAM). Part 2: Sample runs/sensitivity tests: Journal of Hydraulic Research, v. 44, no. 3, p. 308–323, doi:10.1080/00221686.2006.9521684.

Doyle, M., Stanley, E.H., and Harbor, J.M., 2003, Channel adjustments following two dam removals in Wisconsin: Water Resources Research, v. 39, no. 1, 1011, 15 p., doi:10.1029/2002WR001714.

Duan, J.G., Wang, S.S.Y., and Jia, Y., 2001, The applications of the enhanced CCHE2D model to study the alluvial channel migration processes: Journal of Hydraulic Research, v. 39, no. 5, p. 1–12, doi:10.1080/00221686.2001.9628272.

Dulal, K.P., Kobayashi, K., Shimizu, Y., and Parker, G., 2010, Numerical computation of free meandering channels with application of slump blocks on the outer bends: Journal of Hydro-environment Research, v. 3, p. 239–246, doi:10.1016/j.jher.2009.10.012.

Huang, J., and Greimann, B., 2010, User's Manual for SRH-1D, Sedimentation and River Hydraulics—One Dimension Version 2.6: Denver, Colorado, Technical Service Center, U.S. Bureau of Reclamation Technical Report SRH-2010-25, 225 p.

Jang, C., and Shimizu, Y., 2005, Numerical simulation of relatively wide, shallow channels with erodible banks: Journal of Hydraulic Engineering, v. 131, no. 7, p. 565–575, doi:10.1061/(ASCE)0733-9429(2005)131:7(565).

Lai, Y., 2008, SRH-2D Version 2: Theory and User's Manual, Sedimentation and River Hydraulics—Two-Dimensional River Flow Modeling: Denver, Colorado, Technical Service Center, U.S. Bureau of Reclamation, 109 p.

Langendoen, E.J., and Alonso, C.V., 2008, Modeling the evolution of incised streams: I. Model formulation and validation of flow and streambed evolution components: Journal of Hydrologic Engineering, v. 134, no. 6, p. 749–762, doi:10.1061/(ASCE)0733-9429(2008)134:6(749).

Langendoen, E.J., and Simon, A., 2008, Modeling the evolution of incised streams: II. Streambank erosion: Journal of Hydrologic Engineering, v. 134, no. 7, p. 905–915, doi:10.1061/(ASCE)0733-9429(2008)134:7(905).

Mobile Boundary Hydraulics Software, 2001, Sedimentation in Stream Networks (HEC-6T): User's Manual, Mobile Boundary Hydraulics: http://www.mbh2o.com/index.html (accessed September 2012).

Parker, G., 1990, Surface based bedload transport relationship for gravel rivers: Journal of Hydraulic Research, v. 28, no. 4, p. 417–436, doi:10.1080/00221689009499058.

Reclamation, 2011, Hydrology, Hydraulics and Sediment Transport Studies for the Secretary's Determination on Klamath River Dam Removal and Basin Restoration: Denver, Colorado, Technical Service Center, U.S. Bureau of Reclamation Technical Report SRH-2011-02, prepared for Mid-Pacific Region.

Rinaldi, M., Mengoni, B., Luppi, L., Darby, S.E., and Mosselman, E., 2008, Numerical simulation of hydrodynamics and bank erosion in a river bend: Water Resources Research, v. 44, W09428, doi:10.1029/2008WR007008.

Wang, H., Zhou, G., and Shao, X., 2010, Numerical simulation of channel pattern changes: Part I. Mathematical model: International Journal of Sediment Research, v. 25, p. 366–379, doi:10.1016/S1001-6279(11)60004-8.

Wells, R.R., Langendoen, E.J., and Simon, A., 2007, Modeling pre- and post-dam removal sediment dynamics: The Kalamazoo River, Michigan: Journal of the American Water Resources Association, v. 43, no. 3, p. 773–785, doi:10.1111/j.1752-1688.2007.00062.x.

Manuscript Accepted by the Society 29 June 2012

Sediment management at small dam removal sites

James G. MacBroom
Roy Schiff
Milone & MacBroom, Inc., 99 Realty Drive, Cheshire, Connecticut 06410, USA

ABSTRACT

The removal of obsolete and unsafe dams for safety, environmental, or economic purposes frequently involves the exploration of sediments trapped within the impoundment and the subsequent assessment of sediment management needs and techniques. Sediment management planning requires a thorough understanding of the watershed's surficial geology, topography, land cover, land use, and hydrology. The behavior of sediments is influenced by their age, consolidation, and stratigraphy. All watersheds have a history that helps forecast sediment loads, quality, gradation, and stratigraphy. Impounded sediment deposits may include coarse deltas and foreset slopes, fine or coarse bottom deposits, cohesive or organic matter, and wedge deposits immediately behind the dam. Some watersheds have anthropogenic pollutants from agricultural activities, mining, industries, or urban runoff.

The volume and rate of sediment release during and after small dam removal can be limited by active management plans to reduce potential downstream impacts. Management strategies include natural erosion, phased breaches and drawdowns, natural revegetation of sediment surfaces, pre-excavation of an upstream channel, hazardous waste removal or containment, flow bypass plans, and sediment dredging.

INTRODUCTION

This paper presents a practical procedure to evaluate and select sediment management strategies at small dam removal projects based upon direct experience with over 20 dam projects over the past decade (Table 1) and literature research. Examples of sediment management strategies selected for a variety of projects are presented here. Sediment management typically includes four steps: (1) initial assessment, which includes estimating sediment characteristics and volumes, (2) estimation of the potential erosion of impounded sediment, (3) evaluation of the potential downstream sediment impacts on natural and anthropogenic systems, and (4) assessment of alternative management techniques to minimize excessive sediment release or to confirm that a large sediment transport event is tolerable (Fig. 1).

As the number of dam removals has grown and the science and practice have advanced, sediment management has emerged as a design issue whether the dam is removed because it is obsolete, unsafe, ecologically harmful, or simply unwanted. Sediment quantity and quality often determine if removal of a dam is ultimately feasible. If dam removal is possible, sediment management typically dictates the removal design, including proposed channel conditions and construction sequence.

Both active and passive sediment management techniques have been used at dams. Active sediment management has historically been most common, especially along streams where

TABLE 1. EXAMPLE DAM REMOVAL PROJECTS

Dam	Sediment type	Dam management	Sediment management	Dam height (FT)	Dam length (FT)	Status	Decision factors*
Norwalk Mill Pond, Connecticut	Silt & clay	Vertical removal	Dredge contaminants	6	200	Removed 1998	9
Anaconda, Connecticut	Sand & gravel	Horizontal partial removal	Partial containment	11	327	Removed 1999	1, 7
Freight, Connecticut	Gravel	Full removal	Natural erosion	4	158	Removed 1999	2
Platts Mill, Connecticut	Sandy gravel	Horizontal partial removal	Dredge hotspots	10	231	Removed 1999	1, 7, 9
Union City, Connecticut	Sandy gravel	Full removal	Partial dredge	7	190	Removed 1999	1, 7, 9
Edwards Dam, Maine	Sand	Full removal	Natural erosion	25	917	Removed 1999	2
Billington St., Massachusetts	Silty sand	Partial removal	Dredge sediment	9	100	Removed 2002	1, 4, 7, 9
Chase Brass Dam, Connecticut	Gravel	Full removal	Partial dredge	6	180	Removed 2004	1, 2, 7, 9
Cuddebackville, New York	Sand	Partial removal	Natural erosion	10	125	Removed 2004	2, 10
Heishman Mill, Pennsylvania	Silt & sand	Fish bypass channel	Containment	8	162	Constructed 2004	4, 7
Carbonton Dam, North Carolina	Silt & sand	Full removal	Partial dredge, natural erosion	20	300	Removed 2005	1, 3, 7
Lowell Dam, North Carolina	Silt & sand	Full removal	Partial relocation	9	210	Removed 2005	1, 3, 7
South Batavia, Illinois	Silt & clay	Horizontal removal	Containment	10	700	Removed 2005	1, 6, 11
Ballou, Massachusetts	Sand & gravel	Full dam removal	Dredge sediment	8	60	Removed 2006	1, 3
Zemko Dam, Connecticut	Silt & clay	Partial vertical removal	Containment	10	150	Removed 2007	1, 3, 5, 7
Rice Creek, Michigan	Silt & clay	Relocate river around dam	Containment	11	180	Bypassed 2008	1, 3, 7
Fort Covington, New York	Sand	Full removal	Natural erosion	10	200	Removed 2009	2
Briggsville, Massachusetts	Coarse gravel	Full removal	Partial dredge	15	145	Removed 2010	1, 7, 11
Veazie Dam, Maine	Sand	Full removal, breach legacy dam	Natural erosion	32	850	Pending 2013	2
Howland, Maine	Trace gravel	Bypass channel	Containment	19	757	Pending 2014	8
Dufresne Dam, Vermont	Silt & clay	Partial removal	Containment	9	300	In progress	1, 5, 7
Great Works, Maine	Trace	Full removal, breach legacy dam	Natural erosion	19	1020	Removed 2012	2, 6

*Decision factors: 1—regulatory requirements; 2—low sediment volume; 3—high sediment volume; 4—historic site; 5—downstream habitat; 6—downstream water intake; 7—water quality; 8—recreation; 9—sediment quality; 10—endangered species; 11—infrastructure.

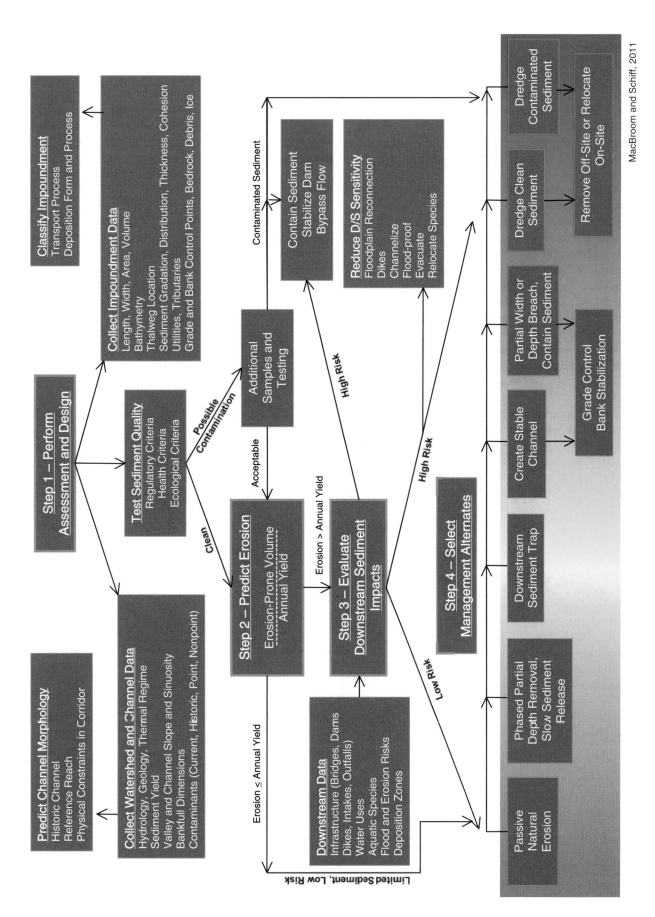

Figure 1. Sediment management approaches for dam removal.

environmental and social risks are high (Table 2). Active approaches such as natural channel design (Rosgen and Silvey, 1996; Schiff et al., 2007) reduce the risk of a sediment transport event that is excessively large relative to the predicted annual watershed yield. Active sediment management also has the advantage of accelerating the formation of a stable upstream channel across the former pool area that often leads to quicker establishment of improved fish passage, floodplain connectivity, and instream habitat (Table 2).

The fact that active sediment management approaches can often work against natural river processes by limiting channel migration coupled with the large initial disturbance due to channel construction and high construction cost has led to frequent consideration and use of passive sediment management. Passive sediment management allows natural processes to partially or completely adjust the sediment deposit in the impoundment and form a new channel with dimensions that are proportional to water discharge and sediment load (Lane, 1955). Recent large dam (Marmot Dam, Condet Dam, Elwha Dams) removals have illustrated a tolerance for temporary increases in sediment load at rural locations that subsequently flush through the channel over time once it is verified that future channel size and location are acceptable.

INITIAL ASSESSMENT AND DESIGN

Engineered dam removal projects include the following steps: assessing existing data and plans, conducting a dam inspection, reviewing dam safety reports, performing a watershed characterization, reviewing or studying river ecology, identifying sediment characteristics, estimating sediment volume, and assessing potential downstream impacts. At most sites, sediment and water-quality sampling and testing are performed to determine if pollutants exist. The presence of polluted sediment has a strong influence on sediment management options, with some level of containment and disposal likely being required.

Dam removal projects may be subject to local, state, and federal regulatory programs that can dictate some of the assessment and design elements. Applicable permits depend on site-specific conditions such as size of the impoundment, amount of existing sediment, and presence of a Federal Emergency Management Agency (FEMA) regulatory floodway. Permits are often required from the municipal conservation commission, the municipal zoning board of adjustment, the municipal planning board, the FEMA National Flood Insurance Program, the U.S. Environmental Protection Agency (EPA) Clean Water Act Section 401 Water Quality Certificates, and the U.S. Army Corps of Engineers Section 404 Discharge. The regulatory programs may require public notice and participation, historic and archaeological studies, and environmental impact assessments. Federal funded projects are generally required to comply with the National Environmental Policy Act.

Design documents need to address construction methods, procedures, and the proposed work. Typical plans include access roads, staging areas, water drawdown procedures, dam details, demolition sequence, river flow control methods and sequence, utility relocation, powerhouse decommissioning, and sediment management and containment. Construction plans may also show downstream measures to accommodate increases in sediment loads and restoration of the impoundment area after dam removal.

Multidiscipline teams are formed for planning and design of dam removal projects. Professional team members usually include land surveyors, hydrologists, hydraulic engineers, dam safety engineers, fluvial geomorphologists, ecologists, landscape architects, and construction managers. Supplemental staff may include historic preservation specialists, archaeologists, geologists, boring contractors, land-use planners, and mechanical-electrical engineers. Many states require that dam removal projects be led by a professional licensed engineer registered in that state.

Preconstruction and postconstruction monitoring is an important task to evaluate project performance, document results, and facilitate adaptive management if needed. For larger projects, monitoring should include physical, chemical, and biologic metrics. Smaller projects or those with low ecological risk may be monitored with photo-documentation.

Sediment Characterization

The physical and chemical characteristics of impounded sediments must be fully explored and tested to forecast potential dam removal impacts and develop appropriate management strategies. Initial assessment may find limited existing sediment in the impoundment, which is contrary to popular perception that all dams have a large amount of trapped sediment upstream. Many run-of-the-river dams with short flow retention times do not contain measurable sediments.

TABLE 2. SCENARIOS FOR ACTIVE AND PASSIVE SEDIMENT MANAGEMENT

Active	Passive
Large sediment volume in impoundment	Small sediment volume in impoundment
Contaminated sediment	Clean sediment
Rare and endangered species	Common and resilient species
Water-supply sources	No threat to source water
Significant aquatic habitats	Resilient aquatic habitat
Flood and erosion hazards	No increase to flood and erosion hazards
Recreation sensitive to sedimentation	Robust recreation

Multiple types of sediment deposits have been documented in impoundments; they are a function of inflow rates, sediment gradation, impoundment shape and size, impoundment flow velocities, and retention times. Classic delta and bottomset deposits are found in some large impoundments. Small impoundments and run-of-the-river low dams can have wedge-shaped deposits against the back of the dam with uniform bottom deposits through the entire pool area without a delta. Field identification and sampling of sediment deposits are two of the most important initial tasks during dam removal studies. Fieldwork must be carefully planned with clear goals and objectives, such as gaining enough information to understand the type of sediment deposit and preparation of a bathymetric map depicting the bottom contours. The historic channel should be highlighted from physical surveys or historic evidence from predam photographs or maps. Fathometers mounted on small boats are invaluable for rapid field identification of former channels, sediment bars, boulders, log debris, and previous-generation dams under water in the impoundment.

Experience has shown that sediment probing and testing should be performed in at least two phases so that the preliminary results can be used to guide more detailed additional phases of assessment. Initial project planning for assessing the sediment includes how and where to access the pool, the size and types of boats required for safely making observations, identification of proper survey and other equipment, and safety needs. The initial field reconnaissance should probe sediment thickness, locate fine- and coarse-grained deposits, and explore for debris and foreign matter. Surficial spot samples can be obtained with buckets, hand ponar dredges, simple tube samplers, or core samples to visually classify deposits. Sediment thicknesses, density, and composition can be explored with metal rods, plastic pipes, or wood rods. Steel rods with threaded joints or pipes with couplings are used to access deeper water.

The reconnaissance survey results are used to define areas for detailed study and formal sampling for laboratory analysis. Common physical tests to explore sediment erodibility and handling methods include sieve tests, shear tests, void ratio, density, and moisture content. Wolman (1954) pebble counts have been used to assess coarse delta and downstream channel bed material.

Sediment Quality

Many rivers in urban, industrial, or agricultural watersheds convey waste products as well as water, sediment, and organic matter. Wastes are transported to impoundments where low water velocity leads to increased sediment deposition and localized buildup of pollutants. Many pollutants, such as excess nutrients and heavy metals, preferentially adhere to fine-grained sediment such as silt and clay, so grain size is a crude way to initially predict where pollutants may be found. Many of the obsolete small dams being considered for removal are in old industrial complexes with historic pollution sources or are located in areas that have grown to be urban centers where the likelihood of historical exposure to point source pollution discharges is high. Investigating sediment quality has become an essential part of dam removal feasibility studies for all but the smallest and most isolated structures.

Sediment quality assessments can begin by contacting local, state, and federal environmental agencies to collect existing data on known sediment and river-water pollutants linked to local and watershed land uses. Massachusetts Department of Environmental Protection permitting guidelines list a typical due-diligence process as reviewing existing and historic on-site land uses, watershed land uses, review of state and federal databases, and checking of health department files. Many states have geographic information system (GIS) records of pollution sources that can expedite the initial research on the project site.

The initial sediment testing program is usually a survey of sediment types with one to ten samples of the fine-grained material for laboratory analysis. Areas of high concentration above local and state thresholds are typically then subject to follow-up chemical and toxicity testing at a sampling density of one sample per 1000 cubic yards (760 cubic meters) of sediment. Chemical tests for contamination need to reflect watershed conditions and potential discharges, and commonly include heavy metals, polychlorinated biphenyls (PCBs), polycyclic aromatic hydrocarbons (PAHs), and hydrocarbons in developed areas. Follow-up samples should be obtained from low-velocity zones near shore, in coves or embayments, and in deep water to expand the investigation of fine-grained sediments. Hand core and bucket grab samples are collected in shallow water and soft sediments, while mechanical sampling methods are needed in deep water or firm sediments.

Sediment Volumes

Dam removal feasibility studies should explore and estimate the volume of impounded sediment. Manual probes with thin steel rods or augers can easily penetrate 10 ft (3 m) of loose fine-grained sediment but seldom penetrate more than 3 ft (~1 m) of dense or coarse-grained material. Mechanical augers or vibrating systems are needed for larger deposits and dense materials. The sediment thickness may also be estimated by comparing the predam ground elevations in past survey or as-built plans with the existing sediment surface elevations. Sediment probing at the Great Works and Veazie Dams on the Penobscot River in Maine revealed only traces of sediment due to the sediment trapping at upstream dams and high annual flood velocities during the spring snowmelt flood that scour and remove sediment. The impoundment upstream of Dufresne Pond Dam in Vermont, on the contrary, was filled with 6 ft (1.83 m) of silty organic material. Probing in the impoundment revealed the historic cobble channel bed under the deposited fine sediment.

The position of sediment within an impoundment helps indicate whether the material is likely to erode following dam removal. Sediments against the back of the dam, along the face of the delta, and in the thalweg are most likely to erode. Sediment along the sides of wide pools and in embayments is less likely to erode. Some of the sediments deposited in impoundments may be situated at the upstream end of the impoundment and could rest above the spillway elevation in delta formations or in the backwater zone along inflowing channels. If these sediments rise above the water surface for the majority of flows, they can begin to be vegetated, which increases their resistance to erosion. For example, at the Zemko Dam in Connecticut and the Lowell Dam in North Carolina, portions of the impounded sediment were located above the water surface during normal flows and had become vegetated despite consisting of very loose erodible material. At the Cuddebackville Dam in New York, the existing sediment delta was so high that it became a well-vegetated island that bifurcated the upstream channel and was not prone to erosion.

The ratio between the volume of erosion-prone impounded sediments to the watershed's annual sediment yield is an important indicator of the risk of a large-scale sediment transport event following dam removal. Releasing the equivalent of multiple years of sediment yield could overwhelm sediment transport, leading to long-term smothering or channel instability. The volume of erosion-prone sediment should also be compared to the downstream channel's annual sediment transport capacity from modeling or empirical bankfull flood transport capacity (e.g., Simon et al., 2004). The transport capacity of the upstream channel is also useful, as it may limit sediment delivery to the impoundment.

At several sites where small dams have been removed, the maximum possible volume of impounded sediment was equivalent to approximately one year's sediment yield from the watershed, limiting the likelihood of long-term downstream impacts following dam removal. In addition, downstream transport capacity remained relatively high, so the long-term deposition of increased sediment loads was not anticipated. A comparison of erosion-prone impounded sediment volume to annual yield and downstream transport capacity is the first of many considerations to explore passive approaches to channel restoration following dam removal.

Estimating watershed sediment yields can be difficult due to the lack of data. Among the techniques that can be used are direct suspended and bed-load measurements, U.S. Geological Survey gauge data, regional sediment yield equations, and erosion models. Unfortunately, sediment yield and transport estimates are not precise and have a wide range of variability. The American Society of Civil Engineers (Vanoni, 1975) provides helpful guidance on regional sediment yields and forecasting methods.

During analysis of the South Batavia Dam in Illinois, sediment probes and a cross-section survey of the pool led to an estimate of 17,700 tons of sediment in the impoundment. This large volume of impounded sediment created initial concern about interfering with the operation of a downstream water-supply intake should a large sediment transport event take place. Data from the Illinois Environmental Protection Agency led to an estimated annual sediment load of 46,400 tons per year—the impounded sediment volume was 38% of the annual load. The assessment indicated that release of a percentage of the total annual sediment load would have a limited impact on the downstream water-supply intake and that the sediment would likely move through the system with several floods. The dam was successfully removed after an initial partial breach and drawdown, as neither excessive channel deposition nor malfunction at the filtration equipment was reported. Observations one year after dam removal showed that the majority of sediment in the impoundment remained in place.

Impoundment Drawdown Tests

Small dams often provide an opportunity to draw down water levels for a close inspection of the impounded sediments. Drawdown tests require an operable low-level outlet gate with adequate capacity that exceeds the inflow rate during low flows. Granular sediments that dry and compact (i.e., consolidate) rapidly gain strength and allow foot access to observe their gradation and enable probing or test boring. Having the water drawn down allows easier access to search for the historic channel bottom and predict where the new channel will form. In some cases, the drawdown will change the understanding of the historic channel. For example, at Dufresne Pond Dam in Manchester, Vermont, the historic channel originally appeared to flow at the existing spillway, yet, after drawdown, the deepest part of the impoundment and the historic cobble bed were located at the middle of the embankment away from the spillway. This finding suggests that the spillway was constructed outside of the original channel, and then the channel was moved.

Prolonged or full drawdown tests may result in sediment erosion that provides initial insight into sediment transport and channel evolution. Drawdown tests also enable inspection of the normally submerge dam elements. The drawdown process at Carbonton Dam in North Carolina was used to confirm construction joint locations and locate unknown low-level outlets that were eventually used for water control during the dam removal. Extensive submerged log jams were also found in the pool and on bridge piers, allowing for their removal before they could wash downstream in an uncontrolled manner. The drawdown at Lowell Dam along the Deep River in North Carolina provided an opportunity to inspect the dam and the upstream bridge foundations. Natural partial dam breaches occurring during floods at the South Batavia and Anaconda Dams allowed further evaluation of both the dams' structures and the impounded sediments during dam removal studies.

The use of low-level controls is not always feasible for drawdown as gates are often inoperable at old dams that have not been maintained. At Carbonton Dam in North Carolina, the gates were inoperable and had to first undergo extensive repairs. Water control can be established at existing uncontrolled openings during dewatering and construction with the use of metal plates or rock cofferdams in front of openings. The low-level outlet at Zemko Dam was operable but was not located at the base of the dam, so water and sediment remained in the impoundment. A partial drawdown exposed the sediment surface and allowed the loose silt and clay to undergo revegetation.

Legacy Dams

Many favorable dam construction sites have been utilized in the past by earlier dams that predate current structures and impoundments. It is not unusual to find partial or complete nineteenth- and early twentieth-century dams submerged in the impoundments created by modern-era dams. The presence of legacy dams can have a significant impact upon the dam removal process and postdam conditions. The presence of legacy dams and timber cribs can alter the anticipated river flow pattern, block fish passage, create a secondary pool that remains after the modern dam is removed, and create navigation and recreation hazards. Legacy dams may also have trapped sediment in their pool prior to being submerged, in addition to accumulating sediment after their submergence.

The presence of legacy dams may go undetected unless detailed historical studies are conducted and thorough site inspections of the modern impoundment are performed. Legacy dams were identified on the Penobscot River that were submerged over 100 years ago. Historic records of log driving and lumbering mention their existence. The legacy dams were confirmed first by a boat-mounted sonar unit and hand probes with a steel rod, and later by detailed bathymetric survey.

Our dam removal projects suggest that legacy dams are common in tandem with more current dams. A complete 30-ft-high (9-m-high) timber crib dam that was full of sediment was located on the Pootatuck River in Connecticut. A mostly complete nineteenth-century 1200-ft-long (366-m-long) masonry and timber crib dam was found submerged upstream of the Great Works Dam in the Penobscot River in Old Town, Maine. A 20-ft-high (6-m-high) dam was found upstream of the present (1912) Veazie Dam on the Penobscot River. The hydraulic modeling that was conducted for the removal of the Great Works and Veazie Dams indicated that the legacy dams would restrict fish passage and alter flow conditions after the modern dams were removed. Plans to remove the Great Works and Veazie Dams also included partial removal of the legacy dams (Milone and MacBroom, Inc., 2011).

ESTIMATE POTENTIAL EROSION OF IMPOUNDED SEDIMENTS

A conceptual model of the ways in which upstream channels behave after dam removal has been developed by Mac-Broom (MacBroom, 2005). This empirical procedure is used to forecast how the postdam channel may evolve, and thus if sediments are likely to erode. The model is based on observations showing that there are a wide variety of postdam channel processes, and previous evolution models are not universal. Several channel and impoundment characteristics are used as indicators of future processes, including the type and location of deposits, ratio of the impoundment width to channel width, and presence or absence of the original channel thalweg. Channel evolution in reservoirs with cohesive sediment is different than those with coarse sediment. Secondary metrics include the predam channel slope, sediment surface slope, lateral channels, and potential revegetation coverage.

There is a wide variation in the percentage of the impounded sediment that is eroded after small dam removals. Dams on steep river channels with granular sediment would have high erosion rates in the former impoundment once the dam was removed. In narrow impoundments, such as the Freight Street Dam in Connecticut and Cuddebackville Dam in New York, most of the sediment eroded in just a few months, without stepped headcuts. In contrast, wide, shallow impoundments, such as at South Batavia Dam in Illinois and Zemko Dam in Connecticut, have very little sediment loss (less than 20% of the initially impounded sediment). Doyle et al. (2003) found similar results with sediment erosion rates of 7.8% and 15.5% in the first year at the LaValle and Rockdale Dams on low-gradient rivers in Wisconsin. A recent review of 12 dam removal projects found that pools that were 2.5 times wider than the channel lost less than 15% of their sediment volume (Sawaske and Freyberg, 2012). Only a portion of sediment in wide impoundments is likely to erode (Randle, 2003).

For initial estimates of erosion-prone sediment, assume that the postdam channel quickly erodes down to the predam channel bed. For narrow pools (i.e., less than twice the bankfull width of the upstream channel or a regime channel), assume all sediment erodes. For wider pools, assume that the newly eroded channel will be as wide as the upstream or equivalent regime channel, with side slopes at the angle of repose to form a channel cross section. The minimum erosion-prone volume is then the channel cross-sectional area times the pool length.

EVALUATE DOWNSTREAM SEDIMENT IMPACTS AND RESILIENCE

Dam removal often results in erosion and downstream transport of previously impounded sediments. The type and extent of sediment management during and after dam removal are influenced by the river's resilience to additional sediment loading, potential adverse impacts, regulatory requirements on sediment release, and the social acceptance of having a river with a temporarily increased sediment load. Regulated impacts can fall under protection of endangered species, keeping channels navigable, and specified sediment thresholds for habitat protection. The reduction of recreation quality and aesthetics are often socially unacceptable and can influence sediment management approaches, particularly on projects that are publicly funded and flow through a population center.

Downstream impacts (Table 3) of individual dam removal projects are site specific. The dam removal literature tends to focus on uncontrolled dam removals and sediment release more than engineered dam removals with implemented sediment management plans. A study on the removal of small dams in the

TABLE 3. POTENTIAL DOWNSTREAM IMPACTS OF EXCESSIVE SEDIMENT RELEASE	
Channel aggradation	Culvert and bridge blockages
Channel widening	Channel avulsions
Bank erosion	Excessive floodplain deposition
Higher flood levels	Habitat damage
Intake and outfall obstruction	Reduced quality of recreation
Water-quality reduction	Limited navigation

Mid-Atlantic states concluded that the final downstream channel configuration after dam removal will eventually be similar to predam condition (Skalak and Pizzuto, 2005). Kibler et al. (2011) described increased habitat heterogeneity near a removed dam in Oregon that had formerly impounded coarse sediments, where the increased sediment load allowed bar formation and the change from a plane bed channel to a riffle-pool channel. Doyle et al. (2003) observed widespread fine sediment erosion several kilometers downstream and varying levels of upstream erosion (i.e., headcutting) into the impoundment that appeared to be a function of grain size in two Wisconsin dam removals.

The sediment concentration and volumes of sediment associated with uncontrolled dam removal may have physical, chemical, and biological impacts, and may negatively influence infrastructure and human water uses. If the volume of impounded sediment is equivalent to many years of the watershed's sediment yield, significant impacts are possible. Unfortunately, a universal technique to reliably predict erosion, transport, and deposition rates or impacts does not yet exist. Current prediction techniques range from empirical observations to simple steady-state models to complex one-dimensional and two-dimensional hydraulic models. Sediment quantification and impact prediction methods have uncertainty associated with changing climate, surface and groundwater flow rates, sediment source, and erosion and deposition rates.

Field observations (see Table 1) and sediment transport analysis indicate that loose fine-grained sediments (i.e., silt and fine sand) usually erode rapidly and can travel far downstream in suspension. In contrast, consolidated cohesive clay erodes very slowly, creating rectangular channels in the impoundment with vertical banks prone to slow episodic mass failures. Coarse sediment (i.e., gravel and coarse sand) usually moves as bed load and often is deposited in low-velocity areas near the dam. In low-gradient channels such as the Fort Covington Dam in New York, sand moved as bed-load dunes, initially settled within a few hundred yards of the dam, and took three years to dissipate. The combination of a large mass failure upstream, unknowingly introducing excessive amounts of sand into the system prior to dam removal, and the low sediment transport capacity on the flat channels approaching the St. Lawrence Seaway led to large initial deposits of sand and passive re-formation of the new channel dimensions in the vicinity of the dam. On the contrary, cohesive sediment in low-gradient pools at Zemko Dam and Norwalk Mill Pond Dams in Connecticut had minimal erosion.

The morphology of the downstream valley and channel affects sediment transport and potential sediment impacts. Channel slope and dimensions influence sediment transport, and the flow rates due to watershed hydrology dictate whether transport rates exceed the expected sediment loads. Channels with natural or manmade levees have increased transport capacities due to higher water velocities associated with confined flow. Increased transport could lead to higher downstream channel aggradation, which could lead to channel instabilities and avulsion due to sudden aggradation as floods recede. In developed areas, bed-load sediment increases and aggradation could lead to problems at bridges, culverts, dams, and water intakes.

Incised channels downstream of dams are disconnected from their floodplains and usually transport increased local bed load downstream. Bed-load increases are often not problematic on transport-dominated reaches (Schumm et al., 1984). Channel aggradation following dam removal and increased sediment release deposition can increase the risk of flood and erosion hazards. In reaches with relatively low banks and active floodplains, the frequency and duration of overbank flows on the floodplain can increase. Large sediment deposits and building of single or multiple bars can increase the chances of channel widening, lateral migration, braiding, and avulsion. Excessive bar building typically drives lateral channel migration (FHWA, 2012).

Increased sediment transport can lead to water-quality and ecological impacts. Sedimentation often impacts turbidity, color, taste, odor, and biological oxygen demand. Increased sediment loading can mean increased contaminants, such as heavy metals or nonpoint pollutants, accumulated in the impoundment in association with the deposited sediment. The reduction in water quality can reduce habitat quality and impair biological communities. Although dam removal tends to improve biological conditions (e.g., restore fish passage and naturalize temperature regime), there is potential for ecological risk (e.g., Palmer et al., 2005) that must be properly addressed with a sediment management plan. Sporadic smothering of habitats, smoothing of bed features, and burying of riparian wetland features in the floodplain are all possible following dam removal associated with large amounts of impounded sediment.

EVALUATE SEDIMENT MANAGEMENT METHODS

Many techniques are now available to manage sediment deposits at dam removal projects (Table 4). Selection of the best approach is a function of sediment characteristics, potential downstream impacts, objectives for the new channel across the former impoundment, cost, and regulatory requirements. Sediment management techniques include natural (i.e., passive) erosion, full or partial sediment dredging and removal, sediment relocation, sediment containment or capping, and various combinations of these methods (MacBroom, 2005).

Natural Erosion of Impounded Sediment

Allowing natural river processes to erode and relocate sediment with little or no intervention during and after dam removal

TABLE 4. SEDIMENT MANAGEMENT ALTERNATIVES

Method	Advantages	Disadvantages
Natural erosion (passive)	Lowest cost Restores sediment continuity Bar replenishment Creates bed forms Natural channel dimensions	Downstream aggradation Flood and erosion risks Temporary habitat burial Water-quality impacts Reduced aesthetics
Downstream sediment trap	Low impact Low cost	Limited to small volumes Primarily for bed load
Create stable channel	Moderate cost Minimizes erosion Set desired channel dimension	Flood-related risks Limited to smaller dams Hard to construct
Partial breach, contain sediment	Reduces downstream impact Flexible construction Phased release	Long-term stability Potential fish blocks Safety of dam remnants
Dredge hotspots only	Protects water quality and habitat Limits cost Lower health risk	Must address clean sediment Requires extensive testing Costly contaminant disposal
Full dredge	Reduces downstream impact Removes all sediment Low risk	High cost Contaminant disposal Slow process Suspended sediment releases
Retain/repair dam	Water supply Economic benefits No sediment cost Recreation enhancements	Fails to achieve ecologic goal Dam safety concerns Continued sediment trapping Dam maintenance costly

is an appealing low-cost alternative applicable where small sediment volumes exist or ecological risk is limited. The passive approach to sediment management where natural patterns of erosion and deposition distribute the formerly impounded material during the range of flows is often used when a small risk exists for impairing habitat, water quality, or human uses. Passive sediment management has been employed even when large sediment volumes exist when the sediment pulse of a predicted size and longevity following dam removal was determined to be acceptable and the costs of more controlled interventions precluded other alternatives. For this approach to be employed, an attempt to understand the size and duration of the sediment pulse should be made. The natural erosion process may take several years with large floods before an equilibrium channel evolves with stable bed forms and habitat. In extreme cases, natural erosion may temporarily prevent fish passage, reduce recreation, and impair aesthetics, yet these impacts often self-heal as the river flushes sediment through the channel. Natural erosion is not feasible where sediments are contaminated.

Historical evidence suggests that thousands of mill dams were once scattered over streams in the eastern United States that are now breached or removed, and, thus, channels have been naturally eroding built-up sediments over the long term and may deviate from their true reference stream type (Walter and Merritts, 2008). Channels become incised into the "legacy sediments," and the remaining flat reservoir bottom appears to be a terrace. In a study of 10 Colonial-era low mill pond dams, eight of the dam sites were estimated to have over half of their original reservoir sediment still in place (Merritts et al., 2010).

Natural channel erosion of legacy sediments occurred during the Cuddebackville Dam removal in New York due to a major flood at the end of construction. Lateral sediment deposits were successfully retained in place due to strategic removal of only one of two spillways as the distribution of modern and old sediment deposits was known. Due to the ecological risk associated with the large sediment deposits, a population of the endangered dwarf wedge mussel in the downstream channel was manually relocated away from possible sediment deposition areas prior to dam removal.

The amount and rate of natural sediment erosion and, thus, potential impact can be controlled by construction approach with an understanding of sediment distribution. For example, notching the dam or removing portions of the structure from the top down incrementally over a period of months or years can reduce the likelihood of a large single release and allow the channel more time for sediment transport as it works closer to the balance of sediment and water. The stepped approach to dam removal to control sediment transport can only be performed when enough of the dam is structurally sound that it remains safe during a phased removal.

Sediment Excavation

The partial or full removal of impounded sediments has been practiced at many small dams where downstream sediment releases need to be prevented due to potential adverse impacts. At the Norwalk Mill Pond and Platts Mill Dams in Connecticut, heavy metal contamination in the sediment required excavation to prevent toxic effects. Contaminated sediments were dredged from the shallow impoundments after partial water drawdowns and disposed of at sanitary landfills. Full or partial sediment removal typically has high initial costs but improves environmental quality and minimizes future downstream adverse impacts and risks. A history of industrial uses in the watershed and fine sediment buildup in the impoundment are good indicators that costly excavation, hauling, and disposal of contaminated sediment represent a likely sediment management alternative.

Sediment removal has been accomplished with conventional earthwork equipment such as crane-mounted buckets or backhoes. Backhoes can operate from shore or from created work surfaces such as temporary construction roads made of rock extending into the impoundment or timber pontoons to hold the tracks up in fine-grained soft sediment. Work surface materials are typically removed once excavation is complete. In larger impoundments where dewatering can be controlled, sediment may also be removed by barge-mounted hydraulic dredges. Hydraulic dredging is difficult where large woody debris or cobbles are mixed in sediments such as the material found at the Carbonton Dam removal project in North Carolina.

Partial sediment removal is a common management approach to strike a balance between reducing the risk of environmental harm and project cost. At the Union City and Chase Brass Dams along the Naugatuck River in Connecticut, contaminated sediments were dredged while clean sediment largely remained in place, left to naturally erode. At the Billington Street Dam, the portion of the clean sediment located where the proposed channel was to be located was removed to seed the channel evolution process following dam removal and reduce downstream sediment loading. Sediments at the proposed channel margins and floodplain were left to naturally erode. At the Lowell Dam in North Carolina, the channel across the former reservoir was widened to the predicted bankfull dimensions to initially establish a channel near its stable equilibrium (Lane, 1955) and reduce sediment releases.

Stable Channel Creation

Sediment removal may be limited to excavating and creating a new channel between the impoundment's inlet channel and dam, because this is the material that is typically most vulnerable to immediate erosions following dam removal. Channel creation speeds the trajectory to a stable channel planform and dimensions—one that is not rapidly adjusting due to the new site hydraulics, yet one that is fine-tuning the initial dimensions set during design and construction. It is common to still see changes in slope, planform, and cross-sectional dimensions in postdam created channels, especially when the design allows for restoration of natural processes such as channel migration if no physical site constraints exist (Schiff et al., 2007).

Several design methods are available for determining stable channel shape, form, and dimensions, such as regime theory, natural channel design, and analytical sediment transport techniques (reviewed in MacBroom, 2006). Active channel creation minimizes sediment erosion, accelerates recovery of fish passage, and increases habitat development. The main design task with channel design is to reliably predict the appropriate channel type and dimension. Channel type would ideally return to the reference condition for the valley setting yet may require a less natural, more constrained channel where infrastructure exists in the river corridor. Channel slope and resulting sinuosity for the given sediment size and load are important design elements to understand if ample space exists for the channel, if infrastructure in the corridor is in danger of being undermined, and if structural grade control is required. A detailed sediment transport analysis can reveal enough information on sediment budget, and if the channel is likely to aggrade or degrade over the long term, so that reliance on structural grade can be reduced and bank controls that can ultimately be detrimental over the long term are avoided (e.g., Thompson, 2002).

Impounded sediment thicknesses in excess of the proposed bankfull channel depth should raise concern that the resulting channel is likely to become incised in these modern-day sediment deposits without active sediment management such as excavation. Incised channels in former impoundments increase the chance of continuing excessive sediment transport for longer periods of time following dam removal. Flood flows contained to the channel (i.e., without floodplain access) increase the velocity and erosive power of the water. Ideally, both channel and floodplain are re-established in the former impoundment. Channel and floodplain creation following dam removal can be combined with full or partial sediment removal and sediment relocation.

Channel creation via partial sediment removal was used at the Briggsville, Ballou, and the Billington Street Dam removals in Massachusetts. Each of these impoundments was dominated by coarse (i.e., gravel and cobble) bed-load sediments that would have taken years to naturally erode before new channels self-formed. The high incident bed load and slow response pointed to the need for sediment excavation. Stable channel dimensions that included a compound cross section and flood benches in these confined settings were used to achieve the estimated reference condition for sediment erosion and deposition. Channel habitat features such as pools, riffles, and large woody debris were installed after sediment removal to seed the recovery of predicted bed forms and cover types. Plantings and bioengineering practices were used to stabilize the site in the short term following dam removal and establishment of vegetation.

Sediment Relocation

Drawing down the impoundment for dam removal exposes moderate to large areas of the former bed, providing areas suitable for relocating sediments from other portions of the pool. Redistributing sediment is useful for establishing compound channel cross sections in formerly overwide impoundments. Sediment relocation within the pool area reduces disturbance of upland areas and minimizes haul distances and project cost.

Sediment relocation can be used to cap contaminated sediments with cleaner material making them less vulnerable to long-term erosion. On-site relocation and containment of contaminated sediments, where allowed by regulation, saves expensive off-site disposal fees and can be the single factor making a dam removal project feasible. Partial sediment relocation was conducted at the Lowell Dam in North Carolina and the Dufresne Pond Dam in Vermont to create the desired cross sections and minimize hauling costs.

Partial Dam Removal for Sediment Containment and Channel Stability

Many small dam removals actually consist of partial breaches primarily for dam safety to drain the impoundment or restore habitat connectivity. For the partial removals, much of the original impounded sediment mass remains in place, and sediment continues to be trapped by the dam. Partial dam removals consist of either the full-depth removal of a limited horizontal section of the dam's length or a partial depth removal that leaves a portion of the dam's lower elements in place for grade control. In both of these practices, the dam itself may be used to help provide long-term sediment containment that minimizes sediment releases.

Full-Depth, Partial-Width Dam Removals

Full-depth dam removals are preferred because they allow the riverbed to return to an equilibrium slope and do not create vertical barriers to fish passage or sediment movement. Full-depth removals are appropriate when the total length of the dam is much greater than the bankfull width of the downstream channel and when the dam remnants can be used to contain sediment. The breach width should always be greater than the anticipated bankfull channel width, and the resulting breachway velocities need to be appropriate for the anticipated sediment load, safe for recreational boating, and suitable for fish movement.

Partial removal of the South Batavia Dam on the Fox River in Illinois utilized a full-depth, partial-width removal. The 650-ft-long (198-m-long) low dam had two individual concrete ogee crest spillways separated by an earthen embankment. The predesign feasibility study found the pool contained up to 4 ft (1.2 m) of cohesive sediment in an asymmetric deposit toward the left bank. No sediment was present in line with the right spillway. Due to concerns about excessive sedimentation at downstream water-supply intakes, only the right spillway was removed, and the majority of impounded sediment was retained.

Even when a site appears to have had a full dam removal, small elements of an existing dam may remain in place for sediment control and channel stability. These remaining elements may be only a small portion of the structure and are typically buried in the channel and banks. At the Briggsville Dam in Massachusetts, a full-depth dam removal was performed. The majority of the dam was removed, but the abutments at the edges of the structure were left in place to minimize bank erosion and to narrow the channel cross section at the former dam site where the channel was overwidened. The outlet works for the structure, now located at the top of the riverbank, were left in place to make a pocket park to view the river.

Partial-Depth Dam Removal

Sediments located in impoundments that are partially full of material can be managed by initially removing only the portion of the dam above the proposed top elevation of the sediment and then subsequently removing lower horizontal portions to gradually release sediment and lower the dam. A partial-depth dam removal ultimately allows the pool to be drained and for the re-establishment of an open channel with riverine habitat across the former impoundment at a set elevation. The bottom half of the dam that is left in place controls the local grade. The partial-depth dam removal sediment management option is commonly applied progressively when periodic (e.g., every other year) sediment releases of a predetermined amount are acceptable and determined to be most appropriate for a given site.

The Zemko Dam on Eight Mile Brook in Connecticut was subjected to a partial vertical removal to contain sediment. Up to 4 ft (1.2 m) of very loose silt and clay were observed in the impoundment located on a popular trout and salmon fishery. A primary design objective was to control the fine-grained sediment that could bury spawning gravel beds. The low-level outlet was opened one year before the planned dam removal, enabling a dense emergent wetland cover to form on the exposed sediment surface and initial consolidation of soft sediments. The vegetated cover limited surface erosion. The partial-depth dam removal elevation was carefully selected to minimize the potential for headcuts, channel incision, and a single large-scale sediment transport event. The new channel was lined with rock fill, forming a fish passable roughened ramp from the downstream tailwater to the new upstream channel elevation set by the grade control at the remaining portion of the dam.

The Lake Switzerland Dam in Delaware County, New York, was partially removed and sediment left in place. A 100 year frequency flood occurred in 2011, and the sediment remains in place.

Even for full-dam removal projects, a version of partial-depth removals can be used where a small portion of the bottom of the structure is left in place to maintain desired grade control. For example, the base of the dam approximately 1 ft (.3 m) under the proposed streambed elevation was left in place at the Fort Covington Dam in New York to establish grade control and prevent headcutting in the vicinity of an upstream nearby bridge. The remaining portion of the dam is not visible, so there is no aesthetic impact, and leaving the portion behind reduced construction costs.

Sediment Traps

Sediment traps (e.g., Carter, 1997) used during dam removal are large pits excavated in channels downstream of dams to trap sediment. They are only effective where limited volumes of coarse sediment are present. Fine-grained sediments typically pass over sediment traps in suspensions unless channel slope and dimensions slow water velocity enough. Sediment traps have to be dredged several times during the duration of a dam removal project, so ample storage remains to trap material. The Anaconda Dam partial failure and subsequent removal in 1999 on the Naugatuck River in Connecticut represent a coincidental but successful use of a sediment trap. The emergency removal of the spillway occurred after a flood partially breached the dam, and much of the eroded material settled in a 20-ft-deep (6-m-deep) former gravel extraction site downstream of the dam.

Leave Dam in Place

The "do nothing" alternative of leaving the dam in place can be utilized when it is more important to leave sediment in place than to accomplish dam removal objectives. The presence of highly contaminated sediment, very large quantities of clean sediment, endangered species, and habitats sensitive to disturbance can deter dam removal and river restoration projects. Leaving the dam in place without action does not improve dam safety, and, thus, structures should be repaired and maintained in compliance with dam safety requirements. Sediment management requirements are thus put off to a later date.

Some dams may be left in place due to their historic or social value, in which case fish passage past the barrier should be considered. Fish-passage techniques include conventional fish ladders, bypass channels, and ramps that use fill material to raise the tailwater channel invert up to the dam crest. The Peconic Dam in Long Island is an example of a scenic and historic dam being left in place despite ecological impacts, and a fish ramp was proposed to provide passage. The Heishman Mill Dam in Pennsylvania is a scenic historic dam that blocked anadromous fish passage yet was left in place. A 500-ft-long (152-m-long) bypass channel was constructed around the dam. Similarly, the 19-ft-high (5.8-m-high) Howland Dam will remain in place for recreational use, but its fish ladder will be abandoned and replaced with a long naturalistic fish bypass channel (MacBroom and Bonin, 2004).

Bypass Channels

Bypass channels are mostly applicable on small dams where a channel is built around a dam left in place. Bypass channels have recently been used at the Heishman Dam in Pennsylvania and the Cannondale Dam in Connecticut to divert a portion of the flow around the structure to enable fish passage. The dams remain, leaving water and sediment storage largely in the same state as before the bypass channel was constructed. The continued use of the dam spillway is only suitable if the dam is in good condition and is properly maintained.

A full bypass channel was used at the Rice Creek Dam in Michigan to divert all flow around an unsafe dam that remains in place dry with its upstream stored sediment.

Sediment Disposal

Dam owners and members of dam removal project teams are generally not responsible for removal of sediments from within existing impoundments. Many impoundments upstream of dams have only a small percentage of their original storage capacity due to lack of maintenance. Regular dredging takes place when economic incentives exist, such as with hydroelectric operations where more water storage volume allows for more power generation.

Dam owners and members of dam removal project teams become responsible for the release of sediments following dam removal. The mobilization of impounded sediment is a regulated discharge that must be controlled or disposed of legally. A Regulatory Guidance Letter from the U.S. Army Corps of Engineers (USACE, 2005) addresses federal jurisdiction over sediment discharges from dams. There is concern that fine sediments can degrade aquatic habitat, smother spawning beds, or fill in niches in the streambed important for invertebrates. In general, the discharge of materials that could change the bottom elevation of any portion of water, or the discharge of substantial quantities of bottom sediment, requires a Section 404 permit.

The disposal of clean sediments may be done on-site or integrated with construction projects as general fill. Sediments from upstream of the Carbonton Dam in North Carolina were clean and disposed of on an abutting property. Small quantities of lightly contaminated sediment from the Chase Brass and Union City Dam removal projects in Connecticut were brought to approved municipal sanitary landfills, while larger volumes of mercury-contaminated fine-grained sediments from the Norwalk Mill Pond project in Connecticut were used as the initial impermeable layer at a landfill closure cap. In contrast, the PCB-contaminated sediments already released downstream of the Fort Edward Dam in New York are to be dredged, dewatered, and shipped to an approved hazardous waste disposal site. The sediments located behind the T&H Dam on the Neponset River in Massachusetts have a PCB concentration exceeding 50 ppm and are subject to the Federal Toxic Substance Control Act.

SUMMARY

There is a common perception that all dams have trapped sediments in their impoundments and that dam removals will release sediments that create downstream channel impacts. Experience with low dams confirms that some dams with short retention times do not contain measurable sediments, and an assessment is required to determine the risks due to sediment transport. If sediment is present, planning and design of low dam removal projects should include sediment management plans that evaluate whether or not materials are erodible and whether they would pose substantial ecological and infrastructure risk downstream. The primary elements of creating a sound sediment management plan as part of a dam removal project include (1) initial assessment and design, where sediment quantity and quality are determined, and the impoundment, if possible, is drawn down; (2) estimation of erosion-prone impounded sediment to understand the anticipated risk of dam removal due to sedimentation; (3) investigation of downstream sediment impacts and the ways in which the channel may respond to increased sediment loading over some predicted duration; and (4) analysis of alternatives for managing sediment. Many techniques are available to minimize release of sediments and control the ecological risk associated with dam removal at some locations. The ultimate goal of sediment management is to address the accumulated sediment in the impoundment and move the system back to a natural sediment regime with the least impacts.

REFERENCES CITED

Carter, E.F., 1997, Guidelines for Retirement of Dams and Hydroelectric Facilities: New York, Committee on Guidelines for Retirement of Dams and Hydroelectric Facilities of the Hydropower Committee of the Energy Division of the American Society of Civil Engineers, 222 p.

Doyle, M.W., Stanley, E.H., and Harbor, J.M., 2003, Channel adjustments following two dam removals in Wisconsin: Water Resources Research, v. 39, no. 1011, doi:10.1029/2002WR001714.

Federal Highway Administration (FHWA), 2012, Stream Stability at Highway Structures (Hydraulic Engineering Circular No. 20): Federal Highway Administration, U.S. Department of Transportation, FHWA-HIF-12-004, 328 p.

Kibler, K., Tullos, D., and Kondolf, M., 2011, Evolving expectations of dam removal outcomes: Downstream geomorphic effects following removal of a small, gravel-filled dam: Journal of the American Water Resources Association, v. 47, no. 2, p. 408–423, doi:10.1111/j.1752-1688.2011.00523.x.

Lane, E.W., 1955, The importance of fluvial morphology in hydraulic engineering, *in* Proceedings of the American Society of Civil Engineering: Journal of the Hydraulics Division, v. 81, paper 745, p. 1–17.

MacBroom, J.G., 2005, Evolution of channels upstream of dam removal sites, *in* Proceedings of the American Society of Civil Engineering Conference on Managing Watershed for Human and Natural Impacts: Engineering, Ecological, and Economic Challenges: Reston, Virginia, American Society of Civil Engineers (ASCE), 11 p.

MacBroom, J.G., 2006, River Restoration Post Dam Removal: Short Course Notes: University of Wisconsin Continuing Education Program at University of Massachusetts, 22 p.

MacBroom, J.G., and Bonin, J.A., 2004, Fish Bypass Evaluation, Howland Dam, Maine: Cheshire, Connecticut, Milone and MacBroom, Inc., 7 p.

Merritts, D., Walter, R., and Rahnis, M., 2010, Sediment and Nutrient Loads from Stream Corridor Erosion along Breached Mill Ponds: Lancaster, Pennsylvania, Franklin and Marshall College, 145 p.

Milone and MacBroom, Inc., 2011, Great Works Dam Removal, Alternative Analysis: Cheshire, Connecticut.

Palmer, M.A., Bernhardt, E.S., Allan, J.D., Lake, P.S., Alexander, G., Brooks, S., Carr, J., Clayton, S., Dahm, C.N., Shah, J.F., Galat, D.L., Loss, S.G., Goodwin, P., Hart, D.D., Hassett, B., Jenkinson, R., Kondolf, G.M., Lave, R., Meyer, J.L., O'Donnell, T.K., Pagano, L., and Sudduth, E., 2005, Standards for ecologically successful river restoration: Journal of Applied Ecology, v. 42, no. 2, p. 208–217, doi:10.1111/j.1365-2664.2005.01004.x.

Randle, T.J., 2003, Dam removal and sediment management, *in* Proceedings of the Heinz Center's Dam Removal Research Workshop: Washington, D.C., H. John Heinz Center for Science, Economics, and the Environment, p. 81–103.

Rosgen, D., and Silvey, L., 1996, Applied River Morphology: Pagosa Springs, Colorado, Wildland Hydrology, 343 p.

Sawaske, S.R., and Freyberg, D.L., 2012, A comparison of small dam removals in highly sediment impacted systems in the U.S.: Geomorphology, v. 10, 1016, doi:10.1016/j.geomorph.2012.01.013.

Schiff, R., MacBroom, J.G., and Armstrong Bonin, J., 2007, Guidelines for Naturalized River Channel Design and Bank Stabilization: Concord, New Hampshire, prepared by Milone & MacBroom, Inc. for the New Hampshire Department of Environmental Services and the New Hampshire Department of Transportation, NHDES-R-WD-06-37, 279 p.

Schumm, S.A., Harvey, M.D., and Watson, C., 1984, Incised Channels: Morphology, Dynamics and Control: Littleton, Colorado, Water Resources Publications, 200 p.

Simon, A., Dickerson, W., and Heins, A., 2004, Suspended-sediment transport rates at the 1.5-year recurrence interval for ecoregions of the United States: Transport conditions at the bankfull and effective discharge?: Geomorphology, v. 58, no. 1–4, p. 243–262, doi:10.1016/j.geomorph.2003.07.003.

Skalak, K., and Pizzuto, J., 2005, The geomorphic effects of existing dams and historic dam removals in the Mid-Atlantic region, USA, *in* Proceedings of the 2005 Watershed Management Conference: Managing Watersheds for Human and Natural Impacts: Engineering, Ecological, and Economic Challenges: Williamsburg, Virginia, Environmental and Water Resources Institute (EWRI) of the American Society of Civil Engineers, 12 p.

Thompson, D.M., 2002, Long-term effect of instream habitat-improvement structures on channel morphology along the Blackledge and Salmon Rivers, Connecticut, USA: Environmental Management, v. 29, no. 2, p. 250–265, doi:10.1007/s00267-001-0069-0.

U.S. Army Corps of Engineers (USACE), 2005, Guidance on the Discharge of Sediments from or through a Dam and the Breaching of Dams, for Purposes of Section 404 of the Clean Water Act and Section 10 of the Rivers and Harbors Act of 1899: U.S. Army Corps of Engineers Report 05-04, 5 p.

Vanoni, V., 1975, Sediment Engineering (ASCE Manual No. 54): Reston, Virginia, American Society of Civil Engineers, 745 p.

Walter, R.C., and Merritts, D.J., 2008, Natural streams and the legacy of water-powered mills: Science, v. 319, no. 5861, p. 299–304, doi:10.1126/science.1151716.

Wolman, M.G., 1954, A method of sampling course river-bed material: Transactions, American Geophysical Union, v. 35, p. 951–956.

MANUSCRIPT ACCEPTED BY THE SOCIETY 29 JUNE 2012

Multiyear assessment of the sedimentological impacts of the removal of the Munroe Falls Dam on the middle Cuyahoga River, Ohio

John A. Peck
Nicholas R. Kasper
Office for Terrestrial Records of Environmental Change, Department of Geology and Environmental Science, University of Akron, Akron, Ohio 44325-4101, USA

ABSTRACT

The 2005 removal of the 3.6-m-high Munroe Falls Dam from the middle Cuyahoga River, Ohio, provided an opportunity to assess dam removal channel-evolution models and to anticipate impacts from additional dam removals on the Cuyahoga River. Preremoval geomorphic and sedimentologic conditions were characterized. Monitoring of the river response to dam removal has continued for 5 yr. The dam removal lowered base level and increased flow velocity upstream of the former dam site. Postremoval, the initial channel response was rapid incision to the predam substrate, followed by rapid lateral erosion of the exposed impoundment fill. Four to nine months after removal, dewatering and vegetation of the exposed impoundment fill greatly reduced the rate of lateral erosion. For 2.5 yr postremoval, sandy bar forms were present upstream of the former dam, and sand was transported under all flow conditions of the year. Subsequently, the bed has become armored with gravel. Downstream of the former dam site, the channel aggraded with sand, causing flow to occupy meander bend chutes that had formerly only been active during high flow. A sandy deltaic feature has accumulated 3.3 km downstream in the impoundment created by the Le Fever Dam. The impacts of the Munroe Falls Dam removal are generally well described by published channel-evolution models with minor exceptions due to local geology and hydrology. The similarities between the Munroe Falls and Le Fever Dam impoundments suggest that this study can aid in understanding the impacts of the possible future removal of the Le Fever Dam.

INTRODUCTION

Construction and removal of dams can cause fundamental hydrologic, geomorphic, and ecologic changes to the fluvial environment (Pizzuto, 2002; Doyle et al., 2003; Walter and Merritts, 2008). Dams decrease a river's slope and increase the channel cross-section area within the impoundment. As a result of these changes, both bed shear stress and stream power decrease within the impoundment, and the ability of flow to transport sediment is diminished. Thus, sedimentation increases within the dam reservoir. Dams also affect the downstream fluvial environment. Downstream of the dam, river flow can still transport sediment, but without a sediment source, the riverbed may become armored in coarse-grain sizes (Kondolf, 1997). Compared to preindustrial times, it is estimated that sedimentation within large and small dam reservoirs has reduced the modern global sediment flux to the world's oceans by 20% and 6%, respectively (Syvitski et al., 2005). Sediment retention by numerous small dams has fundamentally changed the geomorphology of streams and floodplains of many eastern U.S. watersheds. Walter and Merritts (2008) found that the modern environment of incised channels and impoundment-fill terraces is largely an inherited artifact of dam activity. The modern environment in many eastern U.S. watersheds is unlike the extensive wetlands and anabranching channels that characterized the fluvial environment prior to European settlement and dam construction (Walter and Merritts, 2008).

Dams can provide many benefits to society, including hydroelectric power and flood protection; however, they can also have adverse impacts on the fluvial environment (Hunt, 1988; Stanley and Doyle, 2003). Although dams have highly variable impacts, there is a growing trend in removing dams in the United States (Grant, 2001). The motivation for removing dams is varied and includes the removal of aging dams with compromised structural integrity (Heinz Center, 2002). In addition, dams are removed to improve the water quality of streams and rivers. Algae blooms occur within the slow-velocity dam pool, where increased nitrate and phosphate levels are common due to runoff from agricultural fertilizers or wastewater treatment facilities (Brainwood et al., 2004). Dissolved-oxygen levels in the dam pool decrease when the algae die and are decomposed. After the dam is removed, increased flow velocity helps to oxygenate the water, thus improving water quality (Tuckerman and Zawiski, 2007). In some cases, little or no improvement is observed following dam removal (Stanley et al., 2002). Adverse impacts can result from dam removal, such as the release of sediment, changing downstream habitat, which can result in large-scale fish kills (Rathburn and Wohl, 2001). Owing to the varied and complex impacts associated with dam removal, there is a need for field studies that document the range of variability associated with removals (Doyle et al., 2002).

This study reports on our continuing efforts to document multiyear changes to the fluvial environment following the removal of the low-head Munroe Falls Dam on the middle reach of the Cuyahoga River, Ohio. Our earlier studies examined natural and anthropogenic watershed activities as preserved in the Munroe Falls impoundment fill (Peck et al., 2007), as well as the short-term changes that occurred within 1.5 yr of the dam removal (Rumschlag and Peck, 2007). Now that 5 yr have passed since the dam was removed, we are able to examine the longer-term changes. With the benefit of a longer time perspective and extending the study reach downstream to the next dam pool, this study assesses conceptual geomorphic models of channel evolution following dam removal.

STUDY AREA

The Cuyahoga River is located in northeast Ohio (Fig. 1). The river erodes into Pleistocene glacial sediments and Pennsylvanian- through Devonian-aged sedimentary rocks as it flows to Lake Erie. The study area is located in the middle Cuyahoga River for a distance of 5 km upstream and 6 km downstream of the former Munroe Falls Dam (Fig. 2). Within the study reach, the glacial till and outwash deposits vary in thickness and are underlain by the Sharon Formation, an interbedded quartzose conglomerate and quartz arenite sandstone (Mrakovich and Coogan, 1974; Foos, 2003). Where the surficial glacial sediments have been eroded, the river flows on bedrock of the Sharon Formation. Covering 840 km^2, the middle Cuyahoga River watershed drains suburban, urban, agricultural, forested, and wetland land uses (Finkbeiner, Pettis, and Strout, Inc., 2002).

Flow of the middle Cuyahoga River is highly regulated by water released from Lake Rockwell, a dammed part of the upper

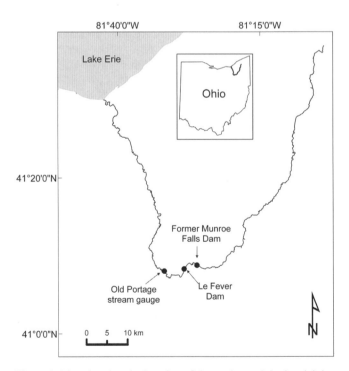

Figure 1. Map showing the location of the study reach in the vicinity of the former Munroe Falls Dam and the Le Fever Dam; and the Old Portage stream gauge. Inset map shows the location of the Cuyahoga River in northeast Ohio.

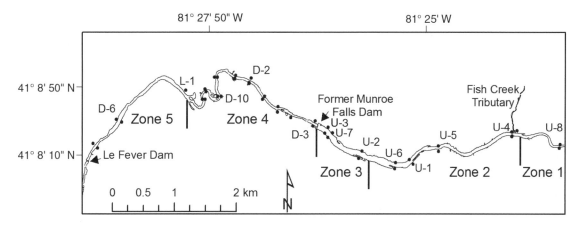

Figure 2. Map showing the location of the study reach in the middle Cuyahoga River. River flow is from east to west. All geomorphic profile locations are indicated by black dots. The study reach has been segmented into the five zones based upon the river's sedimentologic and geomorphologic response to the Munroe Falls Dam removal.

Cuyahoga River, which is used as the public water supply for the city of Akron (OEPA, 2000). The U.S. Geological Survey (USGS) maintains stream gauging stations on the Cuyahoga River, including a gauge at Old Portage, located 15.1 river kilometer (RK) downstream of the former Munroe Falls Dam (Fig. 1). Based upon multidecadal records, the largest average monthly river discharge occurs from December to May, whereas the largest average monthly precipitation occurs from May to September (Kasper, 2010). The mismatch between the maximum precipitation and maximum discharge months is due to enhanced runoff during the winter months when the ground is saturated and frozen. In addition, interception by vegetation limits precipitation from entering the river during the summer months. In the 85 yr Old Portage record, the largest flood on the middle Cuyahoga River occurred 22 May 2004, and the ninth largest flood (6 February 2008) occurred two and a half years after the removal of the Munroe Falls Dam (NOAA, 2011).

The Munroe Falls Dam was located on the middle reach of the Cuyahoga River at RK 80.3 in the city of Munroe Falls, Ohio (Fig. 2). In 1818, Wolcott and Griswold established a saw mill at Munroe Falls (Bronson, 1866a). At a similar but unspecified time, Francis Kelsey also constructed a grist mill at Munroe Falls (Bronson, 1866b). A map of Munroe Falls dated 1838, and reproduced in Lown (2000), shows a dam, four different mill buildings, and raceway. In response to industrial needs, a 3.6-m-tall, arch-shaped, stone weir dam replaced the earlier log dam in 1902 (Finkbeiner, Pettis, and Strout, Inc., 2002). This dam created a slack-water impoundment that extended 6.6 km upstream (Finkbeiner, Pettis, and Strout, Inc., 2002). Owing to its age and lack of maintenance, the dam was found to be in poor condition in the 1990s (ODNR, 1996). A total maximum daily load (TMDL) study found the Munroe Falls impoundment in violation of water-quality standards due to low dissolved oxygen and poor habitat (OEPA, 2000; Finkbeiner, Pettis, and Strout, Inc., 2002; Tuckerman and Zawiski, 2007). After examining alternative methods of improving water quality, the decision was made to remove the Munroe Falls Dam to increase stream flow, water quality, and health of biologic communities in the middle Cuyahoga River (Finkbeiner, Pettis, and Strout, Inc., 2002; Tuckerman and Zawiski, 2007). The dam was completely removed between 15 August and 31 October 2005.

Today, there are three run-of-the-river dams on the Cuyahoga River in the city of Cuyahoga Falls, between 8.4 and 5.8 RK downstream from the former Munroe Falls Dam. Farthest downstream is the 17.4-m-tall Ohio Edison Dam, constructed in 1912 for both hydroelectric power and cooling-water storage (Seguin and Seguin, 2000). Next upstream, is the 3.4-m-tall Sheraton Dam (also called the Mill Dam). The Le Fever Dam (also called the Powerhouse Dam) is 4.1 m tall, 30.5 m wide, and located closest downstream to the former Munroe Falls Dam by 5.8 RK (Fig. 2). In 1810, near where present-day Bailey Road crosses the Cuyahoga River in Cuyahoga Falls, Wilcox and Kelsey built the first grist mill within the study reach (Bronson, 1866b). The mill burned down soon after it was built. The mill was rebuilt in 1811, but later it was flooded, causing Wilcox to quit the milling business and Kelsey to relocate a grist mill to Munroe Falls (Bronson, 1866b). Throughout the nineteenth century, several dams were constructed in the city of Cuyahoga Falls to power a variety of industries (Raub, 1984; Hannibal and Foos, 2003). The present-day Le Fever Dam was built between 1913 and 1915 to replace a dam at that site that was damaged in the flood of 1913 (Raub, 1984). The Le Fever Dam impoundment extends upstream ~2.4 km. The Ohio-Edison, Sheraton, and Le Fever Dams are presently being considered for possible removal.

METHODS

Geomorphic profiling, field observations, photography, sediment sampling, and flow measurements have been used pre- and postremoval to quantify changes to the middle Cuyahoga

River. The details of these methods have been described previously (Peck et al., 2007; Rumschlag, 2007; Rumschlag and Peck, 2007; Kasper, 2010). Briefly, seven river transects were established prior to dam removal, six additional transects were added 1–2 months after removal, and ten transects were added in 2009 to extend the study downstream to the Le Fever Dam (Fig. 2). Additionally, measurement of a longitudinal profile monitored the migration of a midchannel, transverse bar (deltaic feature) at the upstream end of the Le Fever Dam pool (Fig. 2). The geomorphic profiles were most recently surveyed in 2010. Fluvial sediment thickness and type of substrate underlying the fluvial sediment were estimated by hand-pushing an 8-mm-diameter, steel rod into the sediment. Throughout the study reach, the surficial channel-bed sediments were sampled in 2004, 2006, and 2009. Thus, sedimentologic monitoring extends 4 yr following dam removal, whereas geomorphic profile monitoring extends 5 yr postremoval. Only those sediment samples obtained from deep-water areas of the Munroe Falls impoundment in 2004 can be compared to postremoval samples because shallow-water areas within the impoundment became subaerially exposed postremoval. Ten modified-Livingstone piston cores were collected from the Munroe Falls Dam pool, and 11 cores were collected from the Le Fever Dam pool. Detailed information about each sample site can be found in Rumschlag (2007) and Kasper (2010). The sediment samples were measured for wet and dry bulk density and organic content following the method of Dean (1974). Grain-size analysis was performed following the sieve and pipette methods of Folk (1980). Combining the sieve and pipette measurements allowed the percentages of gravel (>2 mm), sand (2 mm to 0.063 mm), and mud (<0.063 mm) to be calculated, as well as the samples' mean and median grain size (d_{50}), sorting, and skewness (Friedman and Sanders, 1978). Down-core concentration profiles of Pb, Cu, Cr, and Zn were obtained following the total digestion methods of Lacey et al. (2001) and Mecray et al. (2001) and using a Perkin-Elmer 7000 atomic absorption spectrophotometer (AAS) having a detection limit of 0.1 mg L^{-1}. Measurement of the National Institute of Standards and Technology (NIST) Buffalo River Sediment Reference Material yielded trace-metal recoveries between 96% and 110% of the NIST reported values, indicating excellent efficiency of the digestion method. Acid blanks were below the detection limit of the AAS, indicating the absence of contamination during the digestion process. Measurement of duplicate sediment samples showed that variability within each sample was much less than down-core variability.

RESULTS

Sediment and Geomorphology Variations

Based upon sedimentologic and geomorphic variations, we had previously divided the study reach into four zones (Peck et al., 2007). We have since added a fifth zone by extending the study farther downstream to the Le Fever Dam (Fig. 2). The zones occur at 3700 m to 5000 m (zone 1), 1000 m to 3700 m (zone 2), 0 m to 1000 m (zone 3), −3300 m to 0 m (zone 4), and −5800 to −3300 m (zone 5) relative to the former Munroe Falls Dam site. In this paper, all positive distances are measured

TABLE 1. AVERAGE SURFICIAL SEDIMENT CHARACTERISTICS WITHIN EACH ZONE OF THE DEEP-WATER CHANNEL ENVIRONMENT

Sample set	Zone	n	Water depth (m)	Organic content (%)	Mean (Φ)	Sorting (Φ)	Gravel (%)	Sand (%)	Mud (%)
Preremoval 2004	1	5	1.91	4	−1.40	0.73	68	31	1
	2	21	2.03	6	1.05	0.80	8	86	6
	3	14	2.70	13	2.63	1.66	8	70	22
	4	13	1.54	4	−1.16	0.68	74	24	2
	5	0	N.D.	N.D.	N.D.	N.D.	N.D.	N.D.	N.D.
Postremoval 2006	1	9	0.97	2	−1.31	1.64	52	48	0
	2	34	0.76	4	0.08	1.09	23	75	2
	3	14	0.87	7	0.03	0.68	18	76	6
	4	15	1.14	3	0.67	0.77	19	79	2
	5	0	N.D.	N.D.	N.D.	N.D.	N.D.	N.D.	N.D.
Postremoval 2009	1	3	0.66	3	−2.29	1.22	59	41	0
	2	19	0.67	3	−2.80	1.56	74	26	0
	3	8	0.74	2	−2.12	1.61	64	36	0
	4	15	1.09	2	−1.78	1.47	53	46	1
	5	28	1.99	10	1.92	1.81	11	72	17

Note: N.D.—no data.

upstream of the former Munroe Falls dam, whereas all negative distances are downstream.

In 2004, when the Munroe Falls Dam was in place, and today near the Le Fever Dam, water depth increased approaching the dam on the upstream side (Table 1; Fig. 3). Both dams created deep-water impoundments, widened the river, decreased slope, and slowed stream velocity. Allowing for varying discharges on the different survey dates, water depth decreased an average of 1.34 m in zones 1, 2, and 3 following the removal of the Munroe Falls Dam and loss of the impoundment. Downstream in zone 4, water depth also decreased by an average of 0.40 m due to deposition of sediment eroded from the former impoundment (Table 1). The following description of the changes that accompanied the Munroe Falls Dam removal begins upstream and proceeds downstream.

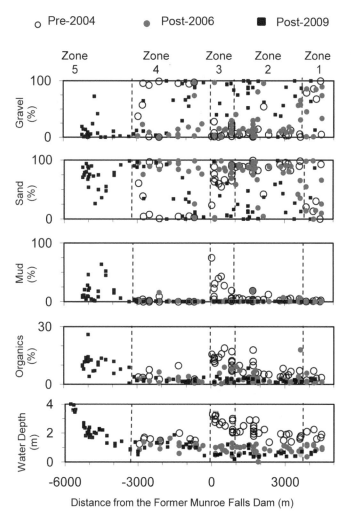

Figure 3. Channel sediment characteristics before (open circles), 1 yr after (gray dots), and 4 yrs after (black squares) the removal of the Munroe Falls Dam. Vertical dashed lines mark zone boundaries. The former Munroe Falls Dam was located at 0 m, and the existing Le Fever Dam is at −5820 m.

Upstream Changes

In zone 1, near the upstream end of the Munroe Falls Dam impoundment, the channel-floor sediment was gravel and sand with low organic and mud content prior to dam removal. In the years following dam removal, the sediment became more poorly sorted but showed little change in size and organic content (Table 1; Fig. 3). Downcutting and lateral erosion were minimal (<40 cm) and occurred within months after removal to expose predam sand and gravel.

Zone 2 displayed pronounce changes in bed sediment and geomorphology over the study period. The channel floor sediment changed from moderately sorted, medium-grained sand to poorly sorted, coarse-grained sand within a year of the removal (Table 1; Fig. 3). For 2.5 yr after removal, numerous linear sand bars were present along the channel margins and in the lee of large woody debris. In addition, small linguoid dunes having eroded crests and superimposed three-dimensional (3-D) ripples covered the bed. Four years after removal, the bars and dunes had largely been transported out of zone 2, and the channel floor was predominantly gravel with low organic content (Table 1; Fig. 3). The major difference between zone 2 and zone 3 was that sediments were finer grained and had significantly higher organic and mud content in zone 3 prior to dam removal (Table 1; Fig. 3). After the removal of the Munroe Falls Dam, the sediment changes previously described for zone 2 also occurred in zone 1 (Table 1; Fig. 3).

The former Munroe Falls impoundment was significantly wider from the middle of zone 2 downstream through zone 3 to the dam (0–2300 m upstream; Peck et al., 2007). Transect U-2 is representative of the postremoval morphologic change that occurred in this reach (Fig. 4). Here, extensive low-velocity shallow-water margins, containing up to 3 m of organic mud, flanked a deeper-water channel (Fig. 4, profile U-2). Postremoval, the channel incised to the predam substrate within 1 mo close to the dam and within 6 mo farther upstream (Rumschlag and Peck, 2007). Once incision reached the predam substrate, further downcutting was negligible. However, lateral erosion into exposed, fine-grained impoundment sediment widened the channel. Lateral erosion resulted in the U-2 right bank having a near-vertical scarp on all surveys from 2006 to 2010 (Fig. 4). Lateral erosion occurred episodically as blocks of the cohesive fine-grained sediment slumped into the channel, generally during winter high-flow conditions. Four months following removal, 6 m of lateral erosion had occurred at profile U-2. During the next 5 mo in winter–spring 2006, a total of 2.5 m of lateral erosion was measured. Over the next 4 yr, measured lateral erosion diminished to 2.1, 1.3, 1.5, and 1 m between annual surveys. This reduction in lateral erosion was evident by the decline in back-stepping of the scarp feature on the U-2 profile (Fig. 4). Because our survey interval was not evenly spaced in time, we have normalized the measured lateral erosion by the time between surveys. To convert the number of days between surveys into months, one month is assigned a value of 31 days. At U-2, and five other transects in the former impoundment, the rate of lateral erosion was highest in the first 9 mo following removal, and

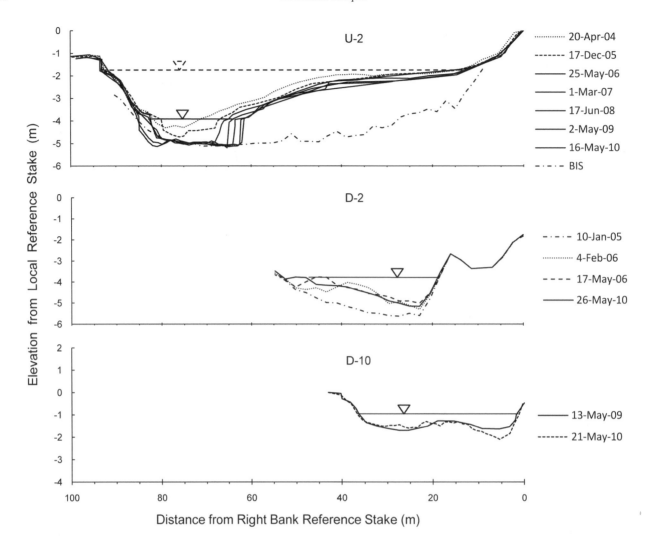

Figure 4. Representative channel cross sections plotted with a 5:1 vertical exaggeration. Refer to Figure 2 for profile locations. Dashed horizontal line represents water level of the impoundment prior to dam removal. Solid horizontal lines represent water levels of the Cuyahoga River on May 2010 surveys. BIS—base of the impoundment sediment fill. Although multiple solid lines are plotted for the U-2 profile, the more recent surveys show continuous right-bank scarp erosion.

generally decreased with time (Fig. 5). In the year following dam removal, when lateral erosion was at its highest rate, mean daily discharge of the Cuyahoga River as measured at Old Portage, did not exceed 55 m³/s (USGS, 2010). The winters of 2007, 2008, and 2009 all had comparable high flows; however, the rate of later erosion declined during this period (Fig. 5). Therefore, factors other than the stream's ability to erode the banks during high flow must play a role in the increased bank stability over time. The decline in lateral erosion is partly due to the exposed impoundment fill undergoing dewatering and compaction (by ~0.25 m) within 2 yr of the dam removal. In addition, vegetation became established on the subaerially exposed former impoundment fill the following summer (9 mo), further stabilizing the bank. However, saturated areas of the impoundment fill, connected to upland groundwater seeps, remained sparsely vegetated for 2 yr and experienced rotational slumping (Rumschlag and Peck, 2007). Large woody debris, moved to new channel locations during floods, accelerated lateral erosion in zones 2 and 3. Between the 2008 and 2009 surveys, the left bank at U-2 experienced 2.75 m of lateral erosion when a tree trunk at that location was dislodged and transported downstream (Fig. 4). During the same 2008 winter flood (ninth largest in the 85-yr-long Old Portage record; USGS, 2010), large woody debris was deposited 50 m upstream of transect U-6. This woody debris constricted flow, and by the spring 2009 survey, the adjacent right bank had eroded ~2 m, whereas nearby at U-6, right bank erosion was only 0.5 m.

Downstream Changes

Zone 4 is located downstream of the former Munroe Falls Dam site, and the bed was 74% gravel with low organic and mud content prior to dam removal (Table 1; Fig. 3). Following dam removal, the bed quickly aggraded with moderately sorted, coarse-grained sand eroded from the former impoundment. Four years after removal, the average bed sediment was poorly sorted granules as the finer sediment continued down the river. The greatest amount of aggradation (~0.7 m) occurred midchannel on the D-2 transect within 4 mo postremoval. Continued sedimentation resulted in a total of 1.17 m aggradation by December 2006 (Fig. 4, profile D-2). As deposition constricted the channel and increased stream power per unit area, erosion commenced, and the bar form migrated closer to the left bank (Fig. 4). Since December 2006, the channel at D-2 has incised ~34 cm, with most (24 cm) of that change occurring in 2007. Since 2009, 10 additional downstream transects showed little change to the bed between 0 and −2300 m, whereas downcutting of 0.25–0.5 m occurred between −2300 and −3300 m (Fig. 4, profile D-10). Between −2300 m and −3300 m, the Cuyahoga River meanders and contains floodplain, cut bank, channel, point bar, and chute environments. Preremoval, the chutes were only active during high-discharge events, and little sediment transport occurred in the gravelly zone 4. Postremoval, sand was deposited on point bars and in the floodplain, in addition to aggrading the channel. Postremoval, the chutes were active in all but the lowest flow of the year and developed scour pools, sandy bed forms, and chute bars. As a result of the increased sediment load in zone 4, the chutes served as additional channels in a fashion similar to the braided-channel configurations described by Simon and Rinaldi (2006).

The zone 5–4 boundary also coincides with the upstream limit of the Le Fever dam pool (Fig. 2). In zone 5, the deep-water bed sediment was a poorly sorted, medium-grained sand having high organic mud content (Table 1; Fig. 3). At −4600 m in zone 5, where the river constricted to 30 m width at a series of bridges, flow velocity increases and the bed sediment is gravel (Fig. 3). No profile change has occurred since 2009 in the lower and middle Le Fever dam pool. However, at the upstream end of the Le Fever Dam pool, a large and active midchannel, linguoid bar (i.e., impoundment deltaic feature) has accumulated since the removal of the Munroe Falls Dam (Fig. 6). Small amounts of sand first appeared in this region several months after dam removal. Probing with the rod indicated that by 2009, ~1.35 m of sandy deltaic deposit covered the preremoval substrate. The bar sediment is a moderately sorted, medium-grained sand with migrating 3-D ripple bed forms. Finer sediment from the former Munroe Falls impoundment has been transported farther downstream. Between 5 December 2009, and 21 May 2010, the bar slip face migrated 25 m downstream, and the bar top was eroded (Fig. 6). The change to the bar during winter–spring 2010 likely resulted from the combined effects of enhanced transport during the high-discharge month of March, followed

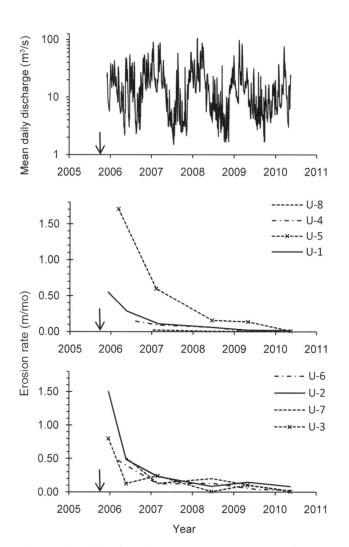

Figure 5. (Top) Cuyahoga River mean daily discharge at Old Portage in cubic meters per second (USGS, 2010). (Middle and Bottom) Lateral erosion rates (m/mo) as determined from cross-section surveys within the former Munroe Falls Dam impoundment. Arrow marks the time of the Munroe Falls Dam removal. Refer to Figure 2 for profile locations.

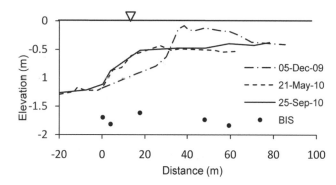

Figure 6. Longitudinal profile (L-1) of the midchannel, transverse bar at the head of the Le Fever Dam pool, plotted with a 28:1 vertical exaggeration. Refer to Figure 2 for profile location. Solid horizontal line represents water level of the Cuyahoga River on December 2009 survey. BIS—base of the impoundment sediment fill.

by erosion during the low-discharge month of April. Based on the 85-yr-long USGS record, the mean daily discharge for March is 25.1 m³/s (USGS, 2010). Between 14 and 17 March 2010, there were four consecutive days having mean discharge greater than 56.6 m³/s and a peak discharge of 76.5 cm (Fig. 5). This large discharge event was followed by 10 d with mean daily discharge below 4.7 m³/s between 15 and 24 April 2010 (USGS, 2010). The fluctuations in both flow velocity and water elevation resulted in a pronounced downstream migration of the deltaic feature. Other studies have documented an important role for floods in transporting sediment downstream following dam removals (Wohl and Cenderelli, 2000; Stanley et al., 2002). During summer 2010, the river was at low flow, daily discharge averaged 9.3 m³/s, there were no floods, and the bar showed little change (Fig. 6).

Le Fever Impoundment Sediment Fill

Throughout the Le Fever impoundment, the fluvial sediment is underlain by the Sharon Formation sandstone, which rises nearly vertically on either side of the channel and, in places, forms ledges confining the river (Fig. 7). In general, the channel width, depth, and thus cross-sectional area increase approaching the Le Fever Dam (Fig. 2; Table 1). Transect D-5,

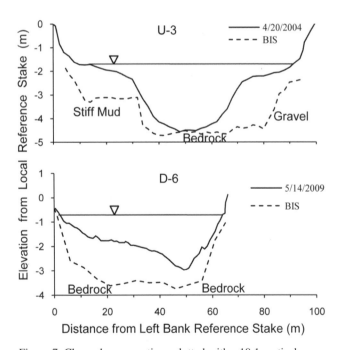

Figure 7. Channel cross sections plotted with a 10:1 vertical exaggeration, showing the extent of impoundment fill behind the former Munroe Falls Dam (top) and Le Fever Dam (bottom). Refer to Figure 2 for profile locations. Solid horizontal lines represent water levels of the Cuyahoga River on the survey date. BIS—base of the impoundment sediment fill. Cross section U-3 is located 166 m upstream of the former Munroe Falls Dam. Cross section D-6 is located 826 m upstream of the Le Fever Dam.

located above the Le Fever dam pool, had a cross-section area of 24.0 m². On a day of similar discharge, transect D-6, located 826 m upstream of the Le Fever Dam within the lower dam pool, had a cross-sectional area of 91.3 m². Between these two transects only minor tributaries, drainage culverts, and groundwater enter the Cuyahoga River; therefore, the large increase in cross-sectional area and accompanying decrease in velocity are the result of the Le Fever Dam. In the vicinity of D-5, high-flow velocity results in active bed-load sand transport as two-dimensional (2-D) small dunes having eroded crests and superimposed 3-D ripples. The slower flow velocity within the impoundment has resulted in ~2 m of mud and sandy mud accumulation in the slack-water shallow margins and <1 m of sand accumulation within the river thalweg (Fig. 7). The thickest fluvial deposit measured by probing was 3.1 m thick. At this location, core CR-09-C6 recovered 2.04 m of sediment before the core barrel could not be hand-pushed into the sediment any farther.

Core CR-09-C6 is representative of the nine shallow-water margin cores collected from the Le Fever dam pool (Fig. 8). Of the total 11.25 m of shallow-water sediment that was cored, 86% was mud, with an average organic content of 19.5%. In core C6, the very poorly sorted, mud layers (d_{50} of 5.4–5.6 Φ) have high organic content (10%–30%; Fig. 8). The basal, poorly sorted, consolidated mud (d_{50} of 3.9 Φ) layer has moderate organic content. The poorly sorted, medium-grained sand layers (d_{50} 1.9 Φ) have low organic content, as does the gravel layer at 90 cm (Fig. 8). The sand layer at the top of the core was likely deposited under high-flow conditions between 2003 and 2008, when five of the top 13 floods, as recorded by the 85-yr-long Old Portage gauge, occurred (USGS, 2010). Within core C-6 and two additional cores that were measured, anthropogenic metal (Cr, Pb, Cu, and Zn) concentrations increased within the mud layers and decreased in the sand and gravel layers (Fig. 8). Elevated metal concentrations in the organic mud are expected due to adsorption by clays, chelation by organics, and bonding with sulfides (Dauvalter, 1997; Lacey et al., 2001; Van Metre and Mahler, 2004; Smol, 2008). The Pb and Zn concentrations exceed the toxic effects range median (ERM) values at which point adverse biologic effects are highly probable (Fig. 8; Long et al., 1995). In addition, one sample from core C-6 has Cr concentrations above the ERM (Fig. 8). Generally, trace-metal concentrations remained uniformly high throughout the core without any noticeable decline toward the base of the core to indicate pre-anthropogenic background conditions. Thus, since sediment began accumulating in the Le Fever dam pool, anthropogenic activities in the watershed have been contributing trace metals to the Cuyahoga River.

Cores collected from the deep-water channel contained predominately poorly sorted, strongly coarse-skewed, medium-grained sand (d_{50} of 1.7 Φ) and lesser amounts of gravel and mud. Deep-water core C-10 had low trace-metal concentrations, and all samples were below the ERM values.

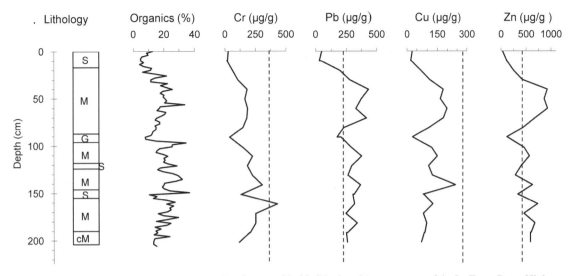

Figure 8. Sedimentological and chemical profiles for core CR-09-C6 taken 54 m upstream of the Le Fever Dam. Highest trace-metal concentrations coincide with organic mud layers (M—mud; S—sand; G—gravel; cM—consolidated mud). U.S. Geological Survey toxic effects range median values for each trace metal are shown as vertical dashed lines (Long et al., 1995).

DISCUSSION

Channel-Evolution Models

Building upon their own and others' research, Doyle et al. (2003) presented a conceptual geomorphic model of channel response to base-level lowering associated with dam removal. They developed this conceptual model both spatially down the longitudinal profile, and temporally at any given location within the postremoval impoundment. In an investigation of the impacts of the IVEX Dam failure in northeast Ohio, Evans (2007) proposed several important modifications to the model by Doyle et al. (2003) to explain that particular site. Indeed, the relative importance of vertical and lateral processes, and differences in channel morphology, depends in part on local differences in the hydrology and geology of each site (Doyle et al., 2003). These local differences include the type of bed and bank sediment, bank cohesion, and the supply of sand from bank erosion (Simon and Rinaldi, 2006). Overall, the effects of the Munroe Falls Dam removal on the middle Cuyahoga River support the Doyle et al. (2003) model. The similarities and several noteworthy, site-specific differences are discussed next.

The first stage (stage A) in the Doyle et al. (2003) model represents the preremoval conditions. At transect U-3, located 166 m upstream of the former Munroe Falls Dam, a deep-water channel occupied the central one-third of the river and was flanked by shallow-water margins (Fig. 7). For a distance of 600 m upstream of the dam, this deep-water channel contained a thin layer (~35 cm) of fluvial sediment on top of bedrock. The low-velocity margins contained thick deposits of fine-grained sediment. This antecedent bathymetry of the Munroe Falls impoundment controlled the location of the stage C postremoval channel and minimized headcutting required to form a channel in the impoundment fill. Similarly, Evans (2007) found that bathymetry variations from preexisting scours in the impoundment sediment possibly influenced later stage C channel incision at the IVEX Dam site.

Stage A is followed by stage B, when dewatering of the impoundment has reduced water depth but not to the point of inducing channel incision in the impoundment fill (Doyle et al., 2003). The sandstone blocks of the Munroe Falls Dam were removed row-by-row over a period of weeks. As base level lowered, flow was confined to the preexisting deep-water channel. During this stage, incision and knickpoint migration occurred along the margins, eroding deltaic deposits at the mouths of minor tributaries and culverts (Rumschlag and Peck, 2007).

Following dam removal, stage C describes incision of a channel having steep banks and migrating upstream by headcut erosion, while impoundment sediment is transported downstream (Doyle et al., 2003). The Munroe Falls impoundment had an existing deep-water channel so the site rapidly progressed to stage D, in which degradation is accompanied by mass wasting and lateral erosion once critical bank height is reached (Doyle et al., 2003). Close to the dam, downcutting through the thin fluvial sediment to the predam substrate occurred within 1 mo of removal. Farther upstream, both downcutting and lateral erosion were significant within the first 2 mo, and downcutting reached the predam substrate within 6 mo. Using flume experiments and modeling, Begin et al. (1981) examined channel response to base-level lowering and found that the rate of degradation attained a peak early after base-level change, and then decreased over time. In addition, the peak rate of degradation was attenuated with increased upstream distance from the outlet. Both of these trends were measured at the Munroe Falls site. Likewise, lateral erosion of the cohesive fine-grained sediment occurred by vertical, block-fall bank failure and was highest during the first 4–9 mo postremoval. As the exposed impoundment sediment dewatered, compacted, and became vegetated, the rate of lateral erosion decreased markedly (Fig. 5). Doyle et al. (2003) reported that bank saturation is more important than bank vegetation in determining bank stability, especially soon after dam removal.

Stage E is attained when the channel bed aggrades with sediment transported from farther upstream, although lateral erosion still continues (Doyle et al., 2003). Stage E was attained for a period of ~2.5 yr, when extensive channel-margin linear bars and channel-floor linguoid bars were present between 800 m and 3700 m upstream of the former dam site. By June 2008, the sand bars in upstream zones 2 and 3 had mostly been transported downstream, in part due to the ninth largest recorded Cuyahoga River flood on 6 February 2008 (NOAA, 2010).

By 2009, the upstream channel had a gravel-armored bed, and sand bed load was transported year round. These conditions represent the quasi-equilibrium stage F for the Munroe Falls site. The limited sediment supply minimized the development of channel meanders and aggradation of point bars and terraces in the former impoundment. The postremoval sinuosity index is ~1.06 above the former dam (Fig. 2), and flow effectively transports sand downstream. Slump blocks at the toe of the vertical bank are eroded, thus providing limited protection to further bank failure (Fig. 4, profile U-2). Large woody debris, moved during high discharge, constricts flow and can accelerate lateral erosion locally by four times compared to that of adjacent banks. The former Munroe Falls impoundment will continue to supply sediment to the next downstream (Le Fever) dam pool for years. In a 10 yr study, it was found that between 32% and 42% of sediment eroded from the IVEX dam impoundment occurred within 2 mo of dam failure and that long-term sediment loading was an important process (Evans, 2007).

Implications for Dam Removals

Models of sedimentary environments are constructed from shared characteristics of many thoroughly studied examples of a specific environment (Walker, 1984). This study of the Munroe Falls Dam provides another example of the long-term sediment response to base-level lowering following dam removal and additional support of Doyle et al.'s (2003) model. Changes observed at Munroe Falls are likely to occur if the Le Fever Dam is removed because the two sites have many characteristics in common. Both dam impoundments have deep-water channels that would contain the postremoval flow and thus influence stage C incision (Fig. 7). The impoundments have similar types and thicknesses of fluvial sediment, underlain by a resistant substrate that would limit degradation. Above the Le Fever Dam, sediment comprises 82.3 m^2 of the D-6 cross section, and upstream of the former Munroe Falls Dam, sediment had covered 87.7 m^2 of the U-3 cross section (Fig. 7). If the Le Fever Dam were to be removed, then a likely scenario of change would be the progression through model stages A to F as described in the previous section. The postremoval water surface slope is 0.53 m km^{-1} upstream of the former Munroe Falls Dam (Rumschlag and Peck, 2007), whereas the postremoval water slope is estimated to be 1.03 m km^{-1} for the Le Fever impoundment (BBC&M Engineering Inc., 2008). The larger postremoval slope at the Le Fever site should accelerate the progression through the model stages because the stream will have greater sediment transporting power at the Le Fever site compared to the Munroe Falls site.

Understanding changes to the impoundment sediment following dam removal is especially important in developed watersheds because the impoundment sediment may be contaminated from various anthropogenic activities. The Le Fever impoundment sediment has Pb and Zn concentrations above the ERM values (Fig. 8). However, not all of the measured metals are readily bioavailable. The total digestion measurement technique employed in this study liberated the metals weakly attached to clays, organics, and sulfides, as well as the metals tightly bound in the mineral structure of the sediment grains. However, the high organic content and reducing conditions present in the impoundment mud suggest that the amount of the metals that are weakly adsorbed, chelated, and sulfide-bound could be significant. Hence, there is a potential that much of the measured metal content could be bioavailable. However, further study, using sequentially extracted metal techniques, for example, would be required to address this issue. The potential of eroded impoundment sediment as a pollution source is not unique to the Le Fever Dam pool and needs to be considered in any dam removal planning process (Rheaume et al., 2001; Van Metre and Mahler, 2004; Peck et al., 2007).

SUMMARY

Within 8.4 river kilometers, four run-of-the-river dams existed on the middle Cuyahoga River in northeastern Ohio. The Munroe Falls Dam was removed in 2005 in order to improve water quality in the low-velocity impoundment, whereas the remaining three dams are under consideration for removal. Geomorphic and sedimentologic studies, conducted both pre- and postremoval, have documented the impacts of the Munroe Falls dam removal on the middle Cuyahoga River. Published dam removal channel-evolution models (Doyle et al., 2003) describe the observed changes to the former Munroe Falls impoundment, with a few minor differences due to local geologic and hydrologic conditions. The deep-water channel, which limited fluvial sediment above bedrock, controlled the location and extent of postremoval channel incision. With 5 yr of postremoval monitoring, it was possible to assign rates at which the Munroe Falls impoundment progressed through the model stages in response to base-level lowering. The many similarities between the Munroe Falls and Le Fever Dam impoundments mean that this study can be used to better understand changes likely to occur should the Le Fever Dam be removed. A better understanding of changes to the Le Fever Dam site is important because the impoundment sediment contains elevated anthropogenic metal concentrations that could impact downstream environments. In addition, this study may be applicable in addressing dam removals from other rivers that are similar to the Cuyahoga River.

ACKNOWLEDGMENTS

We thank past and present students Joe Rumschlag, Nardos Abebe, Christine Devono, Jennifer Court, and Dustin Bates for assistance with field or laboratory work. Tom Quick is gratefully acknowledged for assistance in the laboratory, and Kristopher Mann is thanked for preparing Figure 1. Tony Demasi, City Engineer for the City of Cuyahoga Falls, kindly provided the BBC&M Engineering, Inc., report. Constructive reviews by Todd Grote and Joseph Hannibal improved this paper and are gratefully acknowledged.

REFERENCES CITED

BBC&M Engineering, Inc., 2008, Hydraulic Engineering Report—Cuyahoga River Study, Cuyahoga Falls, Ohio: Cuyahoga Falls, Ohio, Report to City of Cuyahoga Falls, 29 p.
Begin, Z.B., Meyer, D.F., and Schumm, S.A., 1981, Development of longitudinal profiles of alluvial channels in response to base-level lowering: Earth Surface Processes and Landforms, v. 6, p. 49–68, doi:10.1002/esp.3290060106.
Brainwood, M.A., Burgin, S., and Maheshwari, B., 2004, Temporal variations in water quality of farm dams: Impacts of land use and water sources: Agricultural Water Management, v. 70, p. 151–175, doi:10.1016/j.agwat.2004.03.006.
Bronson, C.C., 1866a, Charles Cook Bronson's History of Tallmadge and the Western Reserve 1633–1866, Volume 6: Tallmadge, Ohio, Tallmadge Historical Society (unpublished transcript by Tobi Battista, 1996), 17 p.
Bronson, C.C., 1866b, Charles Cook Bronson's History of Tallmadge and the Western Reserve 1633–1866, Volume 9: Tallmadge, Ohio, Tallmadge Historical Society (unpublished transcript by Tobi Battista, 1996), 103 p.
Dauvalter, V., 1997, Metal concentrations in sediments in acidifying lakes in Finnish Lapland: Boreal: Environmental Research, v. 2, p. 369–379.
Dean, W.E., 1974, Determination of carbonate and organic matter in calcareous sediments and sedimentary rocks by loss-on-ignition: Comparison with other methods: Journal of Sedimentary Petrology, v. 44, p. 242–248.
Doyle, M.W., Stanley, E.H., and Harbor, J.M., 2002, Geomorphic analogies for assessing probable channel response to dam removal: Journal of the American Water Resources Association, v. 38, p. 1567–1579, doi:10.1111/j.1752-1688.2002.tb04365.x.
Doyle, M.W., Stanley, E.H., and Harbor, J.M., 2003, Channel adjustments following two dam removals in Wisconsin: Water Resources Research, v. 39, p. 1011–1027, doi:10.1029/2002WR001714.
Evans, J.E., 2007, Sediment impacts of the 1994 failure of IVEX Dam (Chagrin River, NE Ohio): A test of channel evolution models: Journal of Great Lakes Research, v. 33, p. 90–102, doi:10.3394/0380-1330(2007)33[90:SIOTFO]2.0.CO;2.
Finkbeiner, Pettis, and Strout, Inc., 2002, Munroe Falls Dam Feasibility Study Q407-2: Report to County of Summit, Department of Environmental Services, Ohio, 26 p.
Folk, R.L., 1980, Collection and preparation of samples for analysis, in Petrology of Sedimentary Rocks: Austin, Texas, Hemphill Publishing Co., p. 15–48.
Foos, A.M., 2003, Pennsylvanian Sharon Formation, Past and Present: Sedimentology, Hydrology, and Historical and Environmental Significance: A Field Guide to Gorge Metro Park, Virginia Kendall Ledges in the Cuyahoga Valley National Park, and Other Sites in Northern Ohio: Columbus, Ohio, Division of Geological Survey Guidebook 18, 67 p.
Friedman, G.M., and Sanders, J.E., 1978, Properties of sedimentary particles, in Principles of Sedimentology: New York, John Wiley and Sons, p. 78–80.
Grant, G., 2001, Dam removal: Panacea or Pandora for rivers?: Hydrological Processes, v. 15, p. 1531–1532, doi:10.1002/hyp.473.
Hannibal, J.T., and Foos, A.M., 2003, Historical significance of the Sharon Formation in northeastern Ohio, in Foos, A.M., ed., Pennsylvanian Sharon Formation, Past and Present: Sedimentology, Hydrology, and Historical and Environmental Significance: A Field Guide to Gorge Metro Park, Virginia Kendall Ledges in the Cuyahoga Valley National Park, and Other Sites in Northern Ohio: Columbus, Ohio, Division of Geological Survey Guidebook 18, p. 38–47.
Heinz Center, 2002, Dam Removal: Science and Decision Making: Washington, D.C., Heinz Center for Science, Economics, and the Environment, p. 40–49.
Hunt, C., 1988, Down by the River: The Impact of Federal Water Projects and Policies on Biological Diversity: Washington, D.C., Island Press, p. 17–26.
Kasper, N.R., 2010, An Assessment of the Le Fever Dam Pool, Middle Cuyahoga River, Summit County, Ohio [M.Sc. thesis]: Akron, Ohio, University of Akron, 406 p.
Kondolf, G.M., 1997, Water: Effects of dams and gravel mining on river channels: Environmental Management, v. 21, p. 533–551, doi:10.1007/s002679900048.
Lacey, E.M., King, J.W., Quinn, J.G., Mecray, E.L., Applybe, P.G., and Hunt, A.S., 2001, Sediment quality in Burlington Harbor, Lake Champlain, USA: Water, Air, and Soil Pollution, v. 126, p. 97–120, doi:10.1023/A:1005271101398.
Long, E.R., MacDonald, D.D., Smith, S.L., and Calder, F.D., 1995, Incidence of adverse biological effects within ranges of chemical concentrations in marine and estuarine sediments: Environmental Management, v. 19, p. 81–97, doi:10.1007/BF02472006.
Lown, M.R., 2000, Pennsylvania and Ohio Canal: 160 Years 1840–2000: Munroe Falls, Ohio, Munroe Falls Historical Museum Library, 16 p.
Mecray, E.L., King, J.W., Appleby, P.G., and Hunt, A.S., 2001, Historical trace metal accumulation in the sediments of an urbanized region of the Lake Champlain watershed, Burlington, Vermont: Water, Air, and Soil Pollution, v. 125, p. 201–230, doi:10.1023/A:1005224425075.
Mrakovich, J.V., and Coogan, A.H., 1974, Depositional environment of the Sharon Conglomerate Member of the Pottsville Formation in northeastern Ohio: Journal of Sedimentary Petrology, v. 44, p. 1186–1199.
NOAA (National Oceanic and Atmospheric Administration), 2011, National Weather Service Advanced Hydrologic Prediction Service: http://water.weather.gov/ahps2/hydrograph.php?wfo=cle&gage=olpo1&view=1,1,1,1,1,1,1,1%22 (accessed March 2011).
Ohio Department of Natural Resources Division of Water (ODNR), 1996, Dam Inspection Report: Sonoco Low Head Dam, Summit County: Ohio Department of Natural Resources Division of Water File 1113-035, 7 p.
Ohio Environmental Protection Agency (OEPA), 2000, Total Maximum Daily Loads for the Middle Cuyahoga River Final Report: Columbus, Ohio EPA Division of Surface Water, 32 p.
Peck, J.A., Moore, A., Mullen, A., and Rumschlag, J.H., 2007, The legacy sediment record within the Munroe Falls dam pool, Cuyahoga River, Summit County, Ohio: Journal of Great Lakes Research, v. 33, p. 127–141, doi:10.3394/0380-1330(2007)33[127:TLSRWT]2.0.CO;2.
Pizzuto, J., 2002, Effects of dam removal on river form and process: Bioscience, v. 52, p. 683–691, doi:10.1641/0006-3568(2002)052[0683:EODROR]2.0.CO;2.
Rathburn, S.L., and Wohl, E.E., 2001, One-dimensional sediment transport modeling of pool recovery along a mountain channel after a reservoir sediment release: Regulated Rivers: Research and Management, v. 17, p. 251–273, doi:10.1002/rrr.617.
Raub, E.J., 1984, The Walsh Industries in Cuyahoga Falls: Cuyahoga Falls, Ohio, Cuyahoga Falls Historical Society, 4 p.
Rheaume, S.J., Button, D.T., Myers, D.N., and Hubbell, D.L., 2001, Areal Distribution and Concentrations of Contaminants of Concern in Surficial Streambed and Lakebed Sediments, Lake Erie–Lake Saint Clair Drainages, 1990–97: U.S. Geological Survey Water-Resources Investigations Report 00-4200, 60 p.
Rumschlag, J.H., 2007, The Sediment and Morphologic Response of the Cuyahoga River to the Removal of the Munroe Falls Dam, Summit County, Ohio [M.Sc. thesis]: Akron, Ohio, University of Akron, 296 p.
Rumschlag, J.H., and Peck, J.A., 2007, Short-term sediment and morphologic response of the middle Cuyahoga River to the removal of the Munroe Falls Dam, Summit County, Ohio: Journal of Great Lakes Research, v. 33, p. 142–153, doi:10.3394/0380-1330(2007)33[142:SSAMRO]2.0.CO;2.

Seguin, M., and Seguin, S., 2000, Images of America: Cuyahoga Falls, Ohio: Chicago, Arcadia Publishing, 23 p.

Simon, A., and Rinaldi, M., 2006, Disturbance, stream incision, and channel evolution: The roles of excess transport capacity and boundary materials in controlling channel response: Geomorphology, v. 79, p. 361–383, doi:10.1016/j.geomorph.2006.06.037.

Smol, J.P., 2008, Pollution of Lakes and Rivers: A Paleoenvironmental Perspective: Malden, Massachusetts, Blackwell Publishing, p. 172–174.

Stanley, E.H., and Doyle, M.W., 2003, Trading off: The ecological effects of dam removal: Frontiers in Ecology and the Environment, v. 1, p. 15–22, doi:10.1890/1540-9295(2003)001[0015:TOTEEO]2.0.CO;2.

Stanley, E.H., Luebke, M.A., Doyle, M.W., and Marshall, D.W., 2002, Short-term changes in channel form and macroinvertebrate communities following low-head dam removal: Journal of the North American Benthological Society, v. 21, p. 172–187, doi:10.2307/1468307.

Syvitski, J.P.M., Vörösmarty, C.J., Kettner, A.J., and Green, P., 2005, Impacts of humans on the flux of terrestrial sediment to the global ocean: Science, v. 308, p. 376–380, doi:10.1126/science.1109454.

Tuckerman, S., and Zawiski, B., 2007, Case studies of dam removal and TMDLs: Process and results: Journal of Great Lakes Research, v. 33, p. 103–116, doi:10.3394/0380-1330(2007)33[103:CSODRA]2.0.CO;2.

USGS (United States Geological Survey), 2010, Stream Flow Data for the Cuyahoga River at USGS 04206000 (Old Portage): http://waterdata.usgs.gov/oh/nwis/uv?site_no_=04206000 (accessed October 2010).

Van Metre, P.C., and Mahler, B.J., 2004, Contaminant trends in reservoir sediment cores as records of influent stream quality: Environmental Science & Technology, v. 38, p. 2978–2986, doi:10.1021/es049859x.

Walker, R.G., 1984, General introduction: Facies, facies sequences and facies models, in Walker, R.G., ed., Facies Models (2nd ed.): Toronto, Canada, Geological Association of Canada, Geoscience Canada Reprint Series 1, p. 1–9.

Walter, R.C., and Merritts, D.J., 2008, Natural streams and the legacy of water-powered mills: Science, v. 319, p. 299–304, doi:10.1126/science.1151716.

Wohl, E.E., and Cenderelli, D.A., 2000, Sediment deposition and transport patterns following a reservoir sediment release: Water Resources Research, v. 36, p. 319–333, doi:10.1029/1999WR900272.

MANUSCRIPT ACCEPTED BY THE SOCIETY 29 JUNE 2012

The Geological Society of America
Reviews in Engineering Geology XXI
2013

Sediment impacts from the Savage Rapids Dam removal, Rogue River, Oregon

Jennifer A. Bountry
Yong G. Lai
Timothy J. Randle
Bureau of Reclamation, Technical Service Center, Denver, Colorado 80225, USA

ABSTRACT

Before a dam removal project is implemented, engineers are often asked to estimate the potential for impacts from the release of reservoir sediment. Field measurements, numerical models, and physical models are typically used to develop sediment impact estimates. This information helps decision makers to make informed decisions about when and how to remove the dam, whether to allow the river to erode the reservoir sediment, or to remove or stabilize the reservoir sediment prior to dam removal, or whether mitigation of the effects is needed. Although numerous dams have been removed, mostly small in size, few case studies on sediment impacts have been documented. Because there are limited case studies, dam removal regulators and stakeholders often err on the side of caution when selecting the level of preremoval analysis or determining whether the reservoir sediment needs to be removed prior to dam removal.

The purpose of this paper is to increase our knowledge base for application to future dam removals. The chapter discusses sediment impacts associated with the removal of the 11.9-m-high Savage Rapids Dam on the Rogue River near Grants Pass, Oregon. A unique factor to the Savage Rapids project was the construction and operation of a new diversion facility and water intake located immediately downstream of the dam, which introduced additional consequences associated with the release of reservoir sediment.

BACKGROUND

Three main-stem dams have been removed on the Rogue River in southwest Oregon to improve salmon and steelhead fish passage: Savage Rapids Dam at river kilometer (RK) 172 in 2009, Gold Hill Dam at RK 195 in 2008, and Gold Ray Dam at RK 203 in 2010 (Fig. 1). Savage Rapids Dam was 11.9 m high and 152 m long, Gold Hill Dam was 2.4 m high and 274 m long, and Gold Ray Dam was 11.6 m high and 110 m long. The only remaining main-stem dam, Lost Creek Dam, located at RK 246, serves as flood control for the upstream-most 30% of the basin. Savage Rapids Dam, the focus of this paper, was built in 1921 by the Grants Pass Irrigation District (GPID) to divert river flows for irrigation to canals located on both sides of the Rogue

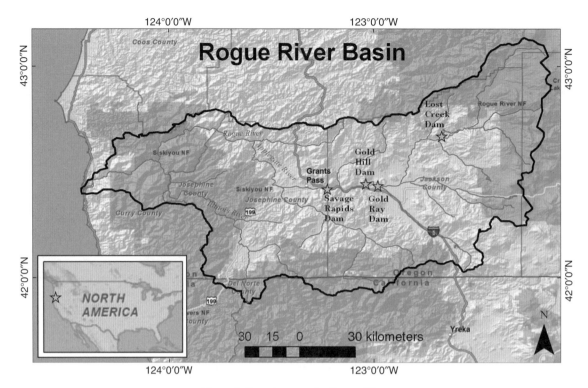

Figure 1. Location map for Rogue River.

River (Fig. 2). The dam created a permanent backwater pool that extended 0.8 km upstream. During the irrigation season (mid-April through October), wood stop logs were used to raise the reservoir 3.35 m, which created a reservoir pool 4.0 km in length. The reservoir was also utilized for recreational boating and fishing, mostly during the irrigation season. The dam did not provide any flood control or power generation.

An environmental impact statement (EIS) was accomplished in 1995, and it concluded that dam removal was the preferred alternative to reduce impacts to fisheries at Savage Rapids Dam. An amended 2001 Federal Court Consent Decree required GPID to cease using Savage Rapids Dam for water diversions by the end of the 2008 irrigation season. Congress authorized the U.S. Bureau of Reclamation to construct a new pumping plant to allow continued water supply to the district, which had to be completed prior to dam removal (Fig. 2). Following construction of the new pumping plant and intake during 2006–2009, a cofferdam was constructed, and the majority of the dam was removed between April and October 2009. A portion of the dam was left in place on the north side to save costs and help protect the new intake from high-velocity flood waters. The total project cost, including dam removal and construction of the new pumping plant and intake, was approximately $39.3 million.

Rather than excavating the sediment, the river was allowed to erode and transport the reservoir sediment downstream of the dam. This chapter presents the approach used to develop reach-scale estimates of the timing and magnitude of sediment impacts associated with dam removal, along with additional analysis done to help minimize future sediment impacts at the new GPID intake structure. Postremoval monitoring data are used to compare the preremoval estimates with observed sediment impacts.

RESERVOIR SEDIMENT CHARACTERIZATION

The first step in estimating potential sediment impacts was determination of the reservoir sediment volume and size gradation, and whether there were any contaminants. Initial estimates of reservoir sediment volume done in the 1990s ranged between 400,000 m^3 and 750,000 m^3. These estimates were based on the simple assumption that sediment had deposited in a wedge shape along the entire 4-km-long reservoir (irrigation pool). Because these estimates indicated potentially large sediment impacts, a more accurate estimate was needed. In 1999, drill cores and dive inspections along with a bathymetric survey were conducted to refine the thickness and spatial distribution of the sediment deposit. The new data indicated that the reservoir had actually only trapped sediment in the 0.8 km reach upstream from the dam. This occurred because the majority of sediment is transported through the Rogue River in the nonirrigation season, so no measurable deposition accumulated in the extended portion of the reservoir during irrigation season operations. In addition, drill-hole data indicated that rather than a uniform sloped channel segment, the area of reservoir sediment deposition contained a long rapid with a deep pool upstream. The previous

Figure 2. (A) A year prior to and (B) immediately after removal of Savage Rapids Dam (river runs from top to bottom).

volume estimate assumed a uniform river slope for computing a wedge-shaped sediment deposit. Based on these new data, the volume of reservoir sediment was estimated to be much smaller at 150,000 m³.

Savage Rapids Reservoir likely filled with sediment in the first few floods following its construction in the 1920s. In the early 1900s, gold mining occurred in the upstream Rogue River Basin, which indicated there was a potential for finding contaminants in the reservoir sediment. GPID requested the chemical composition of reservoir sediment be tested to ensure that if it were released as a result of dam removal, it would not pose any risk to water quality, fish and wildlife, or human uses. During sediment collection by the drill rig in 1999, 25 samples were collected, which after analysis, showed that the reservoir sediment deposit consisted of 2% fines (silt- and clay-sized particles), 71% sand, and 27% gravel overall. Chemical testing confirmed that the sediment was equal to or less than background levels for arsenic, cadmium, mercury, copper, lead, mercury, iron, and zinc (U.S. Bureau of Reclamation, 2001).

PREREMOVAL ESTIMATES OF RESERVOIR SEDIMENT EROSION AND DOWNSTREAM TRANSPORT

GPID was concerned about how long it would take for coarse reservoir sediments to erode and potentially impact the new water intake. Fish regulating agencies were also interested in the length of time it would take the reservoir to be restored to ensure there were no fish passage barriers. Savage Rapids Reservoir was relatively narrow due to bedrock controls, i.e., only two to three times wider than the river. As a result of the narrow reservoir and insignificant amount of cohesive material, it was expected that following dam removal, nearly all of the reservoir sediment would be easily eroded by the river during floods.

To quantify the rate and extent of reservoir sediment erosion, a one-dimensional (1-D) hydraulic and sediment transport model (HEC6t) was utilized (USACE-HEC, 1993). Because the hydrology following dam removal was uncertain, both a wet and dry hydrology were used for modeling based on historical data at the Grants Pass gauging station located 8 km downstream (no major tributaries enter in this reach). With the dry hydrology, two floods were modeled in the first winter, with a mean daily flow rate of ~283 m³/s. Drilling data were used to estimate the predam riverbed, below which the model was not allowed to erode. The 1-D model with dry hydrology estimated that 66% of 150,000 m³ of reservoir sediment would be eroded from the first flood, and 75% of the reservoir sediment would erode within 1 yr (Fig. 3; U.S. Bureau of Reclamation, 2001). A wet hydrology was estimated to erode the reservoir sediment at an even faster rate. Both hydrologic scenarios indicated a small portion of reservoir sediment would remain in place even after a decade of simulation.

Fish regulating agencies and downstream landowners were concerned about where the eroded reservoir sediment would be deposited in the downstream river channel and how long it might be prevalent. The reservoir sediment estimate of 150,000 m³ was equivalent to only 1–2 yr of average annual sediment load. The average annual sediment load was computed by determining the sediment transport capacity at a typical alluvial section, and reducing it by 30% to account for the proportion of the basin above Lost Creek Dam (assuming all coarse sediment is trapped in this reservoir). There are 18 pools of varying size between Savage Rapids Dam and the first major tributary and sediment source located 19.3 km downstream (the Applegate River shown in Fig. 1). The pools have an estimated cumulative storage capacity of 400,000 m³, which is slightly more than double the estimated reservoir sediment volume. The first two major pools in the 2.6 km below the new intake and former dam are some of the larger pools and have an estimated capacity of ~115,000 m³. Below the Applegate River, the river enters a canyon reach, which, based on a total stream power assessment, has a relatively high sediment transport capacity. Based on these simple calculations, it was expected that deposition from dam removal would only be detectable in the 19.3 km reach between the dam and the Applegate River and would largely occur in pools until flushed by high flow events.

Josephine County requested a flood impact analysis to verify that sediment aggradation would not result in increased flood stage. To accommodate the county's request for a flood impact study, the 1-D sediment transport model used for reservoir erosion was also utilized to estimate downstream river aggradation (U.S. Bureau of Reclamation, 2001). Initial 1-D model results indicated eroded reservoir sediment would be permanently deposited in downstream pools and not be scoured out even after a simulation of a 10 yr hydrograph including substantial floods (Fig. 4). Permanent filling of downstream pools from eroded reservoir sediment did not seem reasonable for long-term estimates because the Rogue River has existing deep pools. Further, field measurements from the U.S. Geological Survey (USGS) gauging station at Grants Pass showed a pool filling and scouring up to 2 m during a large flood in the 1990s (U.S. Bureau of Reclamation, 2001). Depth-averaged sediment transport models have a tendency to fill pools with sediment because of utilization of a single depth-averaged velocity at each cross section, which does not represent scour processes that can occur during floods. To account for this model limitation, the 1-D model was "primed"

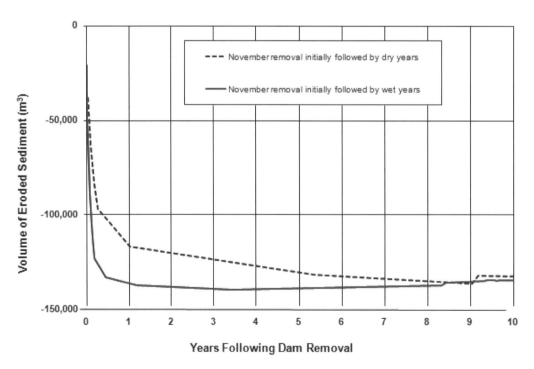

Figure 3. Modeled duration of reservoir sediment erosion following dam removal starting in November followed by a wet and dry hydrology.

to allow the pools to fill with background sediment load, and then the reservoir sediment erosion and downstream transport were simulated. The approach of limiting pool storage resulted in temporary deposition of only a few feet in downstream pools, which was eroded in the model simulation during high flows. The riverbed for ~1.6 km downstream of the dam was modeled as a continuous riffle because survey data were not available at the time of modeling, so limited deposition was estimated to occur. With the modified approach, reservoir sediment was estimated to be transported past the Applegate River (19.3 km downstream) within a 10 yr period. Dam removal followed by dry hydrology (infrequent, smaller storms) was estimated to take up to 10 yr to transport all of the sediment, whereas a wet hydrology took only a few years.

SEDIMENT BREACHING

The river was restored to its original position on 9 October 2009 by breaching the reservoir sediment deposit. October was selected because it was just before the start of the Endangered Species Act–listed Southern Oregon–Northern California Coastal Coho run into the Rogue River. Removing the dam in October would also provide an entire winter flood season to erode and transport reservoir sediments past the new GPID intake before irrigation needed to start in the spring. The contractor built a pilot channel with heavy equipment, which quickly restored the river position and limited potential fish stranding. Had there been no risks of blocking fish passage, a pilot channel would not have been required to initiate river erosion processes. In the month following dam removal, both natural and man-made debris was observed to erode from or become exposed in the former reservoir, including tires, logs, steel parts, and various other materials. A timber crib structure became exposed ~91 m upstream of the dam that was subsequently removed by the contractor after noticing it was causing a hydraulic drop within the pilot channel.

An unexpected public reaction during the sediment breaching was the local excitement to be the first boaters to travel through the restored river section. Although there were posted signs and interviews in the paper recommending avoiding the river during construction, several jet boats and dories went through the site on the day the sediment breaching occurred. Because of the rapidly changing channel, exposed debris, and extremely turbid conditions, it was difficult to see river bottom conditions, making the conditions unsafe. Tragically, one fatality occurred as a jet boat traveled down the channel and hit something on the river bottom, causing the passengers to be thrown from the boat. The following day, a large group of around 100 people floated through the site. The river was much less turbid, and the group was able to carefully observe downstream conditions at the site, and there were no additional accidents, although one boat was not able to maintain the recommended path and was swamped. Large numbers of people were also observed to access the newly exposed bedrock to attempt to mine for gold that may have been in the sediment deposits.

RESERVOIR EROSION AND RIVER DEPOSITION MONITORING RESULTS

Monitoring data were collected to assess how reservoir sediment erosion and downstream deposition impacts compared with preremoval estimates. Time-lapse cameras and simple field observations proved very useful for documenting the speed and spatial extent of the initial reservoir sediment erosion. The river flow during the initial sediment breaching was ~37 m^3/s, and remained less than 57 m^3/s without any high flows through the end of December. A large portion of pilot channel erosion occurred within the first few days (Fig. 5). By the end of October, pilot channel erosion had slowed such that lateral widening was not detectable in the repeat photographs. Large portions of reservoir sediment remained on either side of the enlarged pilot channel.

On 1 January 2010, the first post–dam-removal flood peak of 300 m^3/s occurred (Fig. 6). Based on a March 2010 survey, ~50% (75,000 m^3) of reservoir sediment was eroded by the river from the initial headcut processes and the January flood. The eroded volume of 50% is slightly less than the 1-D model estimate of 66% for the first winter using a dry hydrology. The smaller volume of erosion may in part be due to the mean daily flow for the 1 January 2010 flood, which was ~200 m^3/s smaller than the first modeled mean daily flood value of 283 m^3/s. Other contributors are uncertainty in predam topography and the limitations of using a 1-D model to accurately compute the headcut and widening processes following dam removal, particularly in pool environments. Another small flood of similar magnitude occurred in June 2010, but only minimal reservoir sediment erosion occurred based on visual observations and a survey done in August 2010.

Following the January flood, "Savage Rapids" was fully restored above the dam, but to a higher elevation than estimated from drilling data (Fig. 7). The exposed rapid contains large sporadic boulders that were likely not captured in the drill-hole data. This highlights the difficulty in determining robust postdam removal topography in a pool-riffle system with irregular bedrock outcrops and boulders. Drill-hole or probing data also penetrate to "refusal" or rock, which can sometimes go below the predam riverbed into the alluvial sediment layer. The reservoir sediment extended above Savage Rapids and had completely filled a historical pool. This pool was still full of reservoir sediment as of August 2010, even after the January and June floods (Figs. 7 and 8). The first winter floods were small—about half the magnitude of a 2 yr flood. During the second winter after dam removal (2010–2011), high flows reached a 2 yr flood frequency. It will be informative to see if these floods were of sufficient magnitude and duration to partially or completely scour the pool within the former reservoir, or if the pool sediment will require an even larger flood to erode.

Just downstream of the former dam, an ~5-m-deep scour hole had been created from water flowing over the dam (Fig. 7). Monitoring data show that the scour hole simply filled with sediment and did not cause an upstream bed lowering, largely

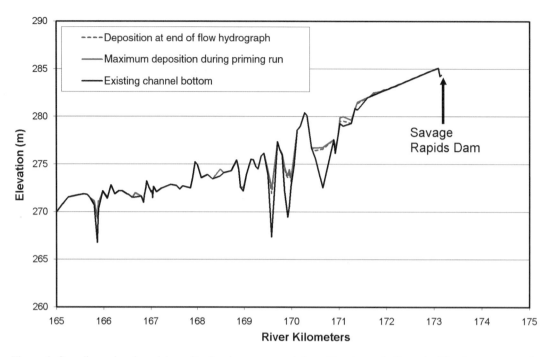

Figure 4. One-dimensional model results showing permanent deposition in pools if no modification to approach is included.

Figure 5. Looking upstream on 9 October 2009 at pilot channel erosion.

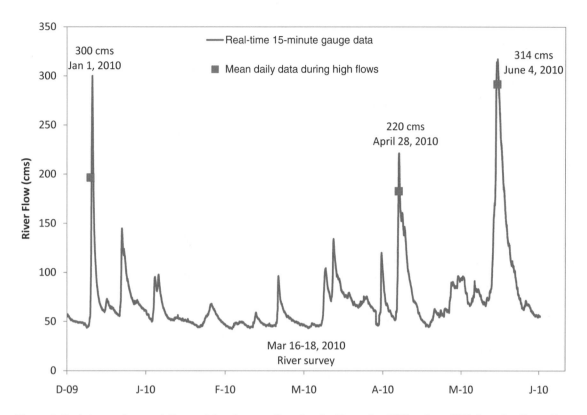

Figure 6. Real-time and mean daily provisional stream flow data for December 2009 to June 2010 from the Grants Pass U.S. Geological Survey gauge (14361500) ~8 km downstream of Savage Rapids Dam. cms—cubic meters per second.

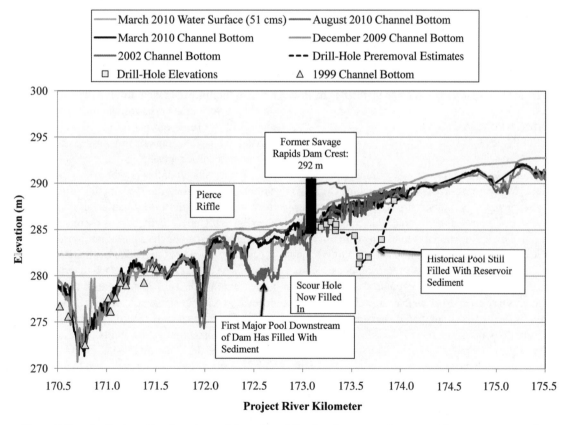

Figure 7. Longitudinal profile of erosion and deposition following dam removal. cms—cubic meters per second.

Figure 8. Sediment conditions at Savage Rapids Dam as of July 2010.

because of the bedrock controls in Savage Rapids. Further downstream of the dam, the first large river pool (also 5 m deep), which had partially filled with sediment from the low-flow reservoir sediment erosion (October through December), had nearly completely filled with reservoir sediment after the January 2010 flood (as measured in March 2010; Fig. 7). After the June 2010 flood, the first pool completely filled in, but filling of pools further downstream was not detected. This first major pool is estimated to have a storage capacity of 40,000 m^3, accounting for about half of the measured reservoir sediment erosion. Downstream riffle crests remained free of deposition as estimated by the 1-D model.

LOCAL SEDIMENT IMPACTS AT THE PUMPING PLANT

GPID considered locations for the new pumping plant and intake within their existing property ownership, which extended ~0.5 km downstream and ~1 km upstream. A location upstream of the majority of reservoir sediment would have reduced the potential for burial and turbidity impacts following dam removal. However, because the pumping plant and intake had to be constructed and be operational prior to dam removal, constructing the plant upstream of the dam would have required building large cofferdams in the reservoir pool that were cost prohibitive. Further downstream of the dam, private land ownership was prevalent, and there were no options for construction sites. As a result, GPID selected a single pumping plant location on the south (left) side of the river immediately downstream of the dam, with a pipe bridge to carry water for delivery to a canal on the opposite side.

The greatest challenge in designing the intake ahead of dam removal was the uncertainty in postremoval channel topography. The selected pump intake location was in the immediate path of the eroding reservoir sediment. Bedrock outcrops that had the potential to cause eddies and deposition in front of the intake were buried under the reservoir sediment, making their configuration unknown. The U.S. Bureau of Reclamation design engineers requested a two-dimensional (2-D) hydraulic model, SRH-2D, be utilized to refine the alignment and position of the pumping plant in the river to avoid potential eddies from the bedrock outcrops (U.S. Bureau of Reclamation, 2006; Lai, 2008, 2010). Additional topographic data were gathered to fill data gaps from the prior 1-D sediment transport modeling in the 1.6 km downstream of the dam. Drilling data were used to develop a postremoval topographic surface in the reservoir, assuming for modeling purposes that all of the reservoir sediment had already been eroded and transported past the intake. Only limited drilling data were available in the bedrock areas.

Despite uncertainties in the exact bedrock configuration within the reservoir, the 2-D model estimates confirmed that there was a potential for hydraulic eddies at the selected intake site (Fig. 9). Conceptually, it was assumed that a uniform flow line was needed along the south shoreline to limit potential future deposition in front of the intake. Evaluation of a number of alternatives with the 2-D model indicated that shifting the intake ~9 m farther out into the river would reduce the potential for sediment deposition without substantially blocking navigation or fish passage in the river channel. Shifting the intake into the main river flow would also improve the ability to meet fish sweeping velocity criteria of 0.61 m/s established by fish regulating agencies. The sweeping velocity is the flow velocity component parallel to the screen face with the pump turned off, which is designed to prevent fish from being swept into the intake. Excavation of bedrock outcrops as part of dam removal was also considered as a modeling alternative, but this was not adopted because there

Figure 9. Two-dimensional model velocity results showing depth-averaged velocity (ft/s) at 238 cms (cubic meters per second) and estimated eddy upstream of intake (left bank) prior to shifting the intake farther out into the river. Ufts is the model-computed directional velocity vector in ft/s.

was a lot of uncertainty in the extent of bedrock that could not be resolved until the dam was removed.

Because of the uncertainty in dam removal, modeling reports noted that some excavation of reservoir sediment may be needed in front of the intake if the first winter flood season did not scour the sediment. However, the need for potential excavation was not highlighted in summary discussions to stakeholders, and no adaptive management plan was developed. Once the sediment was breached in October 2009, the intake was closely monitored to ensure that operation could again start in the spring of 2010. In the 3 mo low-flow period following dam removal, no sediment impacts occurred at the intake, and all deposition was limited to the opposite (north) side of the river channel. The January 2010 flood just reached the threshold of sediment transport for this reach of the Rogue River. As a result, eroded reservoir sediment was not transported very far and was deposited immediately in front of the GPID intake. To ensure intake operation started up in April, deposited sediment was excavated to clear out a channel in front of the intake (Fig. 10). Because no prior adaptive management plan had been developed, negotiations had to occur expediently with regulating agencies while ensuring there was limited or no impact to fish.

It is estimated that 7500–11,500 m³ of sediment were excavated and relocated in February 2010, ~5%–8% of the total 150,000 m³ of reservoir sediment volume (Fig. 10). To eliminate potential fish stranding pools, a portion of the excavated sediment was used to fill in irregular bedrock floodplain topography immediately upstream and downstream of the remaining dam. The remaining excavated sediment was pushed into the riffle downstream of the intake. As the reservoir sediment was excavated, it was observed that the native bedrock extended farther out into the river channel than expected, creating eddies along the south shoreline where the intake was located (Fig. 10). Discharge and velocity measurements accomplished by the USGS Medford, Oregon, office indicated that large boulders (presumably native) within Savage Rapids and along the south shoreline resulted in more river flow being directed to the north, away from the intake. To encourage more flow to be directed toward the intake, a small portion of the boulders in Savage Rapids was relocated and used to armor the fill area upstream of the remaining north

Figure 10. Looking upstream near the end of the February 2010 excavation work, noting bedrock that extended farther out than estimated.

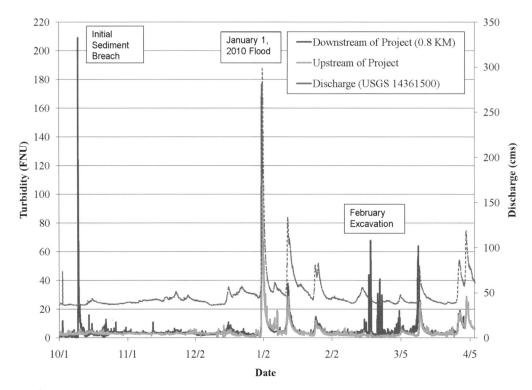

Figure 11. Measured turbidity following dam removal. cms—cubic meters per second. FNU is the turbidity measurement unit, and stands for Formazin Nephelometric Unit, measured scattered light at 90° from the incident light beam with an infrared light source.

dam section. Additionally, a small section of the exposed bedrock was chipped off to encourage a more uniform flow path along the face of the intake.

In April 2010, an additional flood of 220 m³/s occurred that was even smaller in magnitude than the January flood (Fig. 6). This flood reworked additional reservoir sediment, and dive teams and remote cameras revealed that a small amount of sediment had been deposited in the pool adjacent to the intake. The amount of deposition from the April 2010 flood was visibly not as widespread as had occurred from the January 2010 event. About 1500 m³ of sediment (1% of total reservoir volume) were again excavated and placed in the downstream riffle to ensure the intake could properly function at the end of April. The combined cost of the February and April excavation was less than 0.5% of the total project cost.

At the end of the 2010 irrigation season, the intake was closed shut for the winter. The larger floods that occurred during the 2010–2011 winter season and removal of the upstream Gold Ray Dam in 2010 have the potential to contribute additional coarse sediment load past the GPID intake. GPID has recently purchased an excavator to help with potential future excavation needs at the intake. It is expected that future deposition impacts will diminish relative to the initial impacts immediately following dam removal.

WATER QUALITY

Because of the limited amount of fine sediment and expected short duration, preremoval estimates of water-quality impacts were limited to a conceptual model. Immediately after dam removal, suspended sediment concentrations were expected to be high relative to background levels, particularly because the sediment breaching occurred during a low-flow period. However, concentrations were expected to quickly subside and subsequently increase in diminishing levels above background and only with a larger flood than had occurred previously. Best practice measures were included during construction activities to minimize impacts. Silt fences were placed downstream of areas being disturbed that could result in increased suspended sediment concentrations.

Monitoring included turbidity measurements to provide an indication of the duration of water-quality impacts upstream and downstream of the dam removal (Fig. 11). The initial turbidity caused by the pilot channel erosion reached 200 times background levels only a couple hours after breaching the sediment. By 5 p.m., turbidity had lowered to less than 50 times background levels, and by the next morning, less than 20 times background levels. Within 3 days following sediment breaching, turbidity had returned to near background levels, all of which occurred during low-flow conditions.

During the January 2010 flood, the turbidity was ~1.6 times background levels. Turbidity was generated at the project site during the February 2010 excavation work that occurred in low flow, but it was quickly diminished after excavation stopped. GPID has not had any operational impacts as a result of the elevated turbidity, largely because the sediment breaching and reservoir sediment erosion occurred during nonirrigation periods. The City of Grants Pass operates a water treatment plant ~8 km downstream of Savage Rapids year-round but was able to effectively treat elevated suspended sediment concentrations caused by this project.

The turbidity recorder downstream of the dam was located in the first major pool, which became completely filled with eroded reservoir sediment. In future projects, it is recommended to have at least two downstream turbidity recorders, one near the project site and one farther downstream out of the potential sediment deposition area.

SUMMARY AND CONCLUSIONS

The removal of Savage Rapids Dam was a unique learning opportunity to compare preremoval estimates of sediment impacts with post–dam-removal monitoring data. Had a water intake not been constructed immediately downstream of the dam, predam removal estimates of sediment impacts could have been adequately addressed with a conceptual model and simple mass balance computations. Mitigation for short-term, localized sediment impacts following dam removal would also not have been needed. Summary discussion is presented here on developing the reservoir sediment volume estimates, estimating reservoir erosion and downstream deposition following dam removal, suspended sediment impacts, and incorporation of an adaptive management plan and safety issues immediately following dam removal and throughout the first flood season.

Computing an accurate reservoir sediment volume is crucial in estimating potential reservoir sediment impacts. Using simple wedge-based approximations of reservoir sediment volume without consideration of seasonal operations of the dam and predam river morphology resulted in conservatively high estimates. Additional field data and incorporating knowledge of seasonal reservoir operations reduced uncertainty of the sediment volume estimate. However, even with state-of-the-art drilling equipment, uncertainty was still present due to irregular bedrock outcrops and difficulty in distinguishing the top of predam riverbed alluvium from reservoir sediment deposits.

The 1-D sediment transport model utilized was able to quantify timing of reservoir sediment erosion, but the uncertainty in hydrology following dam removal made it difficult to estimate short-term, localized impacts. The 1-D model accurately predicted that deposition in the downstream river would not occur on hydraulic controls (riffles), and, therefore, would not increase future flood stage as a result of released reservoir sediment. However, the 1-D model uses a depth-averaged approach that could not accurately predict the magnitude or duration of deposition in downstream pools. For this project, a conceptual model with simple mass balance calculations of reservoir erosion and downstream deposition would have sufficed for analyzing general trends and potential impacts. Because of the pumping plant intake that was immediately downstream of the dam, this particular project required additional data collection and modeling for design that would not have been required based solely on concerns regarding reach-scale sediment impacts. In future projects with features that have a high risk to sediment impacts in close proximity to the dam, it is recommended that state-of-the-art tools be considered. Incorporation of a physical model, field test, or more advanced numerical modeling tool such as 2-D sediment models may improve the estimating capability (Lai and Greimann, 2010).

Savage Rapids would have benefited from the use of an adaptive management plan because of the inherent uncertainties in short-term, localized estimates. The potential for sediment mitigation work in the first winter following dam removal was alluded to in technical reports, but it was not clearly communicated, and, therefore, it was unexpected by stakeholders and regulators. Although the excavation work required was a fraction of the total project cost (less than 0.5%) and relatively easy to implement, it temporarily reduced the credibility of the project and required special provisions to implement from a regulatory perspective. Had an adaptive management plan been collaboratively developed prior to dam removal, it may have lessened the difficulties in implementing relatively minor and short-term mitigation resulting from a dry winter.

Because of the small percentage of fines in the reservoir sediment, suspended sediment impacts were of short duration. Measured turbidity associated with dam removal lasted only a few hours to days and dissipated quickly in the downstream direction. The timing of the reservoir sediment breach was planned to occur when limited salmon and steelhead were present in the downstream river in order to minimize any potential effects to fisheries. Short-term turbidity impacts occurred during post-removal sediment excavation activities and during high flows, but turbidity impacts are diminishing with each subsequent flood. For similar dam removal projects with limited fine sediment and a reservoir sediment volume that is similar to the annual coarse sediment load of the river, extensive impact analysis prior to dam removal or more elaborate monitoring of suspended sediment may not be needed.

In future dam removal projects, more aggressive education and collaboration on safety concerns during and immediately following a dam removal project would benefit local boaters, particularly when the site is easily accessible and in public view. Education is needed for all boater levels of the dynamic and unknown conditions specific to a dam removal project, including uncertainties that may arise due to unexpected debris and turbidity conditions that prevent observation of flow conditions. Posting information at entry points to the river could be incorporated in addition to direct communication at the site itself.

ACKNOWLEDGMENTS

We would like to acknowledge the funding support from U.S. Bureau of Reclamation's Science and Technology program and field observations and monitoring support from Grants Pass Irrigation District (GPID). We would also like to thank the U.S. Geological Survey office staff in Medford, Oregon, for participation and input. Several Bureau of Reclamation staff also provided valuable technical data, review, and assistance with time-lapse photography and topographic surveys that contributed to this paper. Oregon State University contributed valuable survey data that were critical to developing model input data and post–dam-removal information. We would also like to thank Bob Hunter for his support of the project and input on boater safety lessons learned.

REFERENCES CITED

Lai, Y.G., 2008, SRH-2D Version 2: Theory and User's Manual: Denver, Colorado, Technical Service Center, Bureau of Reclamation, 101 p.

Lai, Y.G., 2010, Two-dimensional depth-averaged flow modeling with an unstructured hybrid mesh: Journal of Hydraulic Engineering, v. 136, no. 1, p. 12–23, doi:10.1061/(ASCE)HY.1943-7900.0000134.

Lai, Y.G., and Greimann, B.P., 2010, Predicting contraction scour with a two-dimensional depth-averaged model: Journal of Hydraulic Research, v. 48, no. 3, p. 383–387, doi:10.1080/00221686.2010.481846.

U.S. Army Corps of Engineers–Hydrologic Engineering Center (USACE HEC), August 1993, HEC-6 Scour and Deposition in Rivers and Reservoirs, User's Manual, Version 4.1: Davis, California, U.S. Army Corps of Engineers, 161 p.

U.S. Bureau of Reclamation, 2001, Savage Rapids Dam Sediment Evaluation Study: prepared by the Pacific Northwest Regional Office and the Denver Technical Service Center: Denver, Colorado, U.S. Bureau of Reclamation; 24 p.

U.S. Bureau of Reclamation, 2006, Numerical Modeling of Flow Hydraulics in Support of the Savage Rapids Dam Removal: prepared by the Denver Technical Service Center: Denver, Colorado, U.S. Bureau of Reclamation, 105 p.

Manuscript Accepted by the Society 29 June 2012

The Geological Society of America
Reviews in Engineering Geology XXI
2013

Changes in biotic and habitat indices in response to dam removals in Ohio

Kenneth A. Krieger
National Center for Water Quality Research, Heidelberg University, Tiffin, Ohio 44883, USA

Bill Zawiski
Ohio Environmental Protection Agency, Division of Surface Water, Northeast District Office, Twinsburg, Ohio 44087, USA

ABSTRACT

Dams on rivers modify habitat and water chemistry, resulting in degradation of fish and macroinvertebrate community integrity within and, in some cases, downstream of the dam pools. Thus, removal of a dam is usually accompanied by the expectation of improved habitat quality and biotic integrity. The Ohio Environmental Protection Agency applies a Qualitative Habitat Evaluation Index, an Index of Biotic Integrity (IBI, fishes), and an Invertebrate Community Index (ICI) to assess stream habitat quality and the habitat-dependent structural and functional integrity of the fish and invertebrate communities. Our objective was to demonstrate that these three indices reliably detect differences in the quality of habitat and fish and macroinvertebrate communities between dam pools and free-flowing reaches and that they are sensitive to changes in habitat and biotic condition following dam removal. Data from 21 stream reaches in Ohio containing dams showed that habitat and biota in dam pools possess lower quality than nearby upstream and downstream reaches. Case studies of dam removals on the Cuyahoga, Olentangy, and Sandusky Rivers confirmed that the indices are sensitive to the rapid changes in habitat and biotic communities that accompany return of dam pools to free-flowing conditions. IBI and ICI scores indicated that the former dam pools had met or exceeded the designated aquatic life use criteria within 1 yr following dam removal. We conclude that the IBI and ICI are valuable tools for measuring the rapidity and extent of changes in the fish and macroinvertebrate communities, respectively, following dam removal.

INTRODUCTION

Ecosystem impacts associated with dams are well documented and include changes in habitat and biotic communities within the dam impoundments as well as downstream of the dams (Bednarek, 2001; Hart et al., 2002; Heinz Center, 2002).

The removal of most dams is neither preceded nor followed by scientific study (Doyle et al., 2005). In Ohio, for example, removal of a dam is often selected as an implementation measure as part of a total maximum daily load (TMDL) program required by the federal Clean Water Act (Section 303[d]) when a water body is not meeting appropriate federal or state water-quality goals. Ohio applies biological water quality standards in addition to the more common chemical criteria (Ohio Administrative Code Chapter 3745–1-7, Table 7–15). Environmental conditions in an impounded area often result in the impairment of aquatic life as defined for river ecosystems and thereby trigger the TMDL study requirement (Ohio Administrative Code Chapter 3745–2-12, Clean Water Act Section 303[d]). Dam removal can be a cost-effective means to restore biological water quality (Doyle et al., 2005).

Dams in Ohio serve a variety of functions. Tributaries to the Ohio River such as the Tuscarawas and Mahoning Rivers have numerous flood-control structures along their main stems and tributaries. Dams in both the Ohio River and Lake Erie basins provide public water supplies. Many dams on rivers draining to Lake Erie are remnant milldams from the bygone era of water-borne power generation. Of the hundreds of dams in Ohio, many are obsolete and considered hazardous, and thus they should be considered for removal (Roberts et al., 2007).

Numerous studies have demonstrated that the integrity of aquatic biological communities of streams is dependent on the habitat quality of those streams. Most if not all dam pools on rivers reduce the integrity of fish and macroinvertebrate communities by altering the stream hydrology and certain physicochemical factors, such as lowering dissolved oxygen concentrations and permitting sedimentation of riffles within the dam pool (Rumschlag and Peck, 2007). Therefore, the removal of a dam is usually predicted to result in improved biotic integrity.

The removal of dams on rivers results in hydrologic, sedimentologic, and morphologic changes that are specific to each site (Rumschlag and Peck, 2007), and thus removal of a dam does not necessarily imply a return of river hydrology to a predisturbance state, the nature of which is often unknown (Gottgens and Evans, 2007). Therefore, the response by stream biota to dam removal may not be in the direction of improved ecological and community integrity or to its original ecological state. In some cases, dam removal results in increased movement of streambed sediments in response to increased channel velocities and the initiation of equilibration to a new base level (Rumschlag and Peck, 2007), and, thus, riffles downstream of the former impoundment may become buried, at least temporarily.

Dam removal reconnects stream reaches, thereby permitting native fishes to resume historic upstream migrations to former spawning habitat (Jones et al., 2003) and perhaps permitting renewed migration by some invertebrates such as crayfish, snails, and unionid mussels, the latter of which are dependent on fish hosts for large-scale dispersal. Renewal of channel continuity also can potentially reestablish gene flow in an upstream direction (Haponski et al., 2007). However, removal of dams also could permit invasive fishes, such as the round goby (*Neogobius melanostomus*) in Lake Erie (Gottgens, 2009) and the parasitic sea lamprey (*Petromyzon marinus*) in the Laurentian Great Lakes (Lavis et al., 2003), to migrate to previously unavailable habitat upstream. Furthermore, dam removal on some streams potentially could result in deleterious effects on upstream members of the aquatic-terrestrial food web from toxic contaminants bioaccumulated in anadromous fish (Freeman and Bowerman, 2002).

Stream community structure and function within impounded streams respond rapidly following dam removal by shifting from a more depositional, hypoxia-tolerant community to a lotic community resembling those upstream and downstream of the former impoundment. Many examples have been documented. In Ohio, degraded fish and macroinvertebrate communities showed marked improvements within less than 1 yr following removal of dam pools on the Cuyahoga River where the upstream watershed area is 349 km^2 (Tuckerman and Zawiski, 2007), and on the Sandusky River where it drains a 1999 km^2 watershed (Krieger and Stearns, 2007). In Wisconsin, macroinvertebrate communities in a former impoundment near the lower end of an ~1700 km^2 watershed became similar to the communities at upstream and downstream unimpounded sites within 1 yr (Stanley et al., 2002).

Many ecological and biological end points, or metrics, have been proposed to assess environmental impacts, including those of dams, on the integrity of fish and invertebrate communities of streams (Barbour et al., 1995). The Index of Biotic Integrity (IBI) for fishes (Simon and Lyons, 1995) and the Invertebrate Community Index (ICI) (DeShon, 1995) as developed and applied by the Ohio Environmental Protection Agency (Ohio EPA) over the past three decades (Ohio EPA, 1987b, 1989a, 1989b) both incorporate

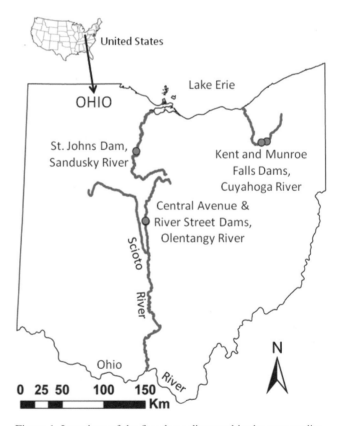

Figure 1. Locations of the five dams discussed in the case studies. The Olentangy River is a tributary of the Scioto River, which flows into the Ohio River.

a variety of metrics that reflect community structure, function, and condition and thereby serve as tools for detecting changes in biotic integrity. Similarly, the Qualitative Habitat Evaluation Index (QHEI) is composed of metrics that measure habitat characteristics associated with biotic community integrity (Rankin, 1989, 1995). Ohio EPA applies these three indices in a rigorous biological assessment program to determine compliance with biological water-quality standards (Yoder, 1995).

Although the biological and habitat database accumulated by Ohio EPA since the late 1970s has been used by researchers because of its detail and size (e.g., Walton et al., 2007; Manolakos et al., 2007), assessment of dams and impoundments was not a direct objective of those studies. More recently, however, Ohio EPA has begun to incorporate monitoring sites that include upstream, impounded, and downstream reaches around dams during their seasonal survey work to document the widely held view that dam removal is a feasible method to restore water quality (Ohio EPA, 2008, 2010). The purpose of the present study was to examine the utility of the IBI, ICI, and QHEI for detecting and quantifying the impact of dam pools on biotic integrity and habitat quality and the changes in biotic integrity and habitat quality subsequent to dam removal.

STUDY AREAS

We present two case studies for examination of the response of IBI scores following dam removal: (1) The Kent and Munroe Falls Dams on the Cuyahoga River in northeast Ohio, and (2) the River Street and Central Avenue Dams on the Olentangy River in central Ohio. In a third case study, we examine the response of the ICI to removal of a dam on the Sandusky River in north-central Ohio (Fig. 1). Rumschlag and Peck (2007) provided a detailed description of the Munroe Falls Dam on the Cuyahoga River and photographs of the dam site prior to, during, and following removal of the dam, as did Tuckerman and Zawiski (2007) for the Kent Dam on the Cuyahoga River. Characteristics of the Sandusky and Olentangy Rivers and their watersheds are described in Ohio EPA reports (Ohio EPA, 2003, 2007). Characteristics of each of the five dams, their impoundments, and the upstream watersheds are summarized in Table 1.

METHODS

Fish and invertebrate surveys are usually conducted from 15 June through 30 September (ICI) or from mid-June to early

TABLE 1. CHARACTERISTICS OF THE DAMS AND THE WATERSHEDS PRESENTED AS CASE STUDIES (RUMSCHLAG AND PECK, 2007; TUCKERMAN AND ZAWISKI, 2007; KRIEGER AND STEARNS, 2007)

	Cuyahoga River	Cuyahoga River	Olentangy River	Olentangy River	Sandusky River
Dam name	Kent	Munroe Falls	Central Avenue	River Street	St. Johns
Dam latitude (°N)	41.1534	41.1416	40.3007	40.2940	41.0300
Dam longitude (°W)	81.3597	81.4368	83.0632	83.0604	83.2161
Distance upstream of river mouth (km)	88.2	80.3	41.7	41.0	80.8
Ecoregion	Erie-Ontario Lake Plain	Erie-Ontario Lake Plain	Eastern Corn Belt Plains	Eastern Corn Belt Plains	Eastern Corn Belt Plains
Dam length (m)	38	44	61	56	46
Dam height (m)	4.5	3.9	1.2	1.1	2.2
Length of dam pool (km)	1.6	5.9	1.9	0.8	11.5
Dam type	Stone arch, concrete retaining wall	Masonry	Masonry	Concrete arch	Concrete arch
Year built	1836	1903	1955	1950s	1901
Year removed	2004	2005	2008	2005	2003
Watershed area above dam (km^2)	754	840	1090	1116	1999

October (IBI), when in most years stream flows are low, pollution stresses are greatest, and seasonal changes in the fish community are minimal (Ohio EPA, 1989a). To assess fish community ecological integrity, Ohio EPA personnel collect fish with pulsed DC (direct current) electrofishing methods and apply resulting data to the IBI and a Modified Index of Well-being (Iwb) (Ohio EPA, 1987b). A complete discussion of the IBI and Iwb appears in Ohio EPA's biocriteria documents (Ohio EPA, 1987a, 1987b; Rankin, 1989), and only the IBI will be discussed further herein. The IBI scores 12 assessment categories, or metrics (Table 2), to render a possible maximum score of 60. Most metric scores are scaled on the basis of watershed area, allowing comparison of streams in watersheds of various sizes.

Because the IBI methodology integrates multiple metrics into a single, comparable score, it is ideally suited to compare fish communities in impounded river segments to those found upstream and downstream of the impoundment. The IBI is calibrated to lotic systems. However, a dam creates an impoundment (dam pool) that deviates from a lotic ecosystem while not quite being a lentic ecosystem; that is, the riffle-run-pool sequence of the river channel is interrupted, stream velocity is reduced, and dissolved oxygen depletion may accompany base-flow conditions, among other effects of impoundment. This deviation should be detected by the IBI, and an IBI similar to the one used by Ohio EPA has been used to track changes in the Milwaukee River, Wisconsin, following removal of a dam (Kanehl et al., 1997).

TABLE 2. METRICS INCLUDED IN THE INDEX OF BIOTIC INTEGRITY (IBI) AT NONHEADWATER SITES (DRAINAGE AREA ≥ 51.80 KM^2), THE INVERTEBRATE COMMUNITY INDEX (ICI), AND THE QUALITATIVE HABITAT EVALUATION INDEX (OHIO EPA, 1987b; RANKIN, 1989)

Index of Biotic Integrity (IBI). Each metric is scored 1, 3, or 5.

1	Number of native species
2	Number of darter species (wading sites) or percent round-bodied suckers (boat sites)
3	Number of sunfish species
4	Number of sucker species
5	Number of intolerant species
6	Percent abundance of tolerant species
7	Percent abundance of omnivores
8	Percent abundance of insectivores
9	Percent abundance of top carnivores
10	Number of individuals in sample (excluding tolerant species)
11	Percent abundance of simple lithophilic spawners
12	Percent of individuals with DELT (deformities, eroded fins, lesions, tumors) anomalies

Maximum total IBI score = 60

Invertebrate Community Index (ICI). Each metric is scored 0, 2, 4, or 6.

1	Total number of taxa
2	Total number of mayfly taxa
3	Total number of caddisfly taxa
4	Total number of Dipteran taxa
5	Percent mayfly composition
6	Percent caddisfly composition
7	Percent Tribe Tanytarsini midge composition
8	Percent other Diptera and noninsect composition
9	Percent tolerant organisms
10	Total number of qualitative EPT (Ephemeroptera, Plecoptera, Trichoptera) taxa

Maximum total ICI score = 60

Qualitative Habitat Evaluation Index (QHEI). Maximum metric scores are in parentheses.

1	Substrate—type, quality (20)
2	In-stream cover—type, amount (20)
3	Channel morphology—sinuosity, development, channelization, stability (20)
4	Riparian zone—width, quality, bank erosion (10)
5a	Pool quality—maximum depth, current, morphology (12)
5b	Riffle quality—depth, substrate stability, substrate embeddedness (8)
6	Gradient (10)

Maximum total QHEI score = 100

A description of the ICI and development of the associated water-quality criteria is described in detail elsewhere (Ohio EPA, 1987a, 1987b; DeShon, 1995). The ICI utilizes 10 metrics (Table 2) to render a possible maximum score of 60. The scores for nine metrics are derived from a quantitative Hester-Dendy artificial substrate sample. The tenth metric is based on the number of taxa (kinds) of mayflies, stoneflies, and caddisflies (the "EPT" taxa: Ephemeroptera, Plecoptera, Trichoptera) in a qualitative sample obtained by using dip nets and forceps to sample all habitat types within the stream reach where the artificial-substrate sampler was deployed. As with the IBI, the scores of most ICI metrics are affected by watershed area because of effects of differing stream habitat characteristics on the structural and functional characteristics of aquatic biological communities as drainage area changes (OEPA, 1987b).

The minimum ICI and IBI scores required for designation of attainment of a specific aquatic life use designation, such as "warm-water habitat," differ slightly among the five primary ecoregions in Ohio (Yoder and Rankin, 1995). Warm-water fish and invertebrate communities are typical of most streams throughout Ohio (Ohio EPA, 1987b).

The QHEI utilizes six primary metrics that render a maximum possible score of 100. Five metrics each consist of multiple submetrics (Table 2). The QHEI metrics are observed and scored either during the IBI survey or independently (Rankin, 1989). In general, a QHEI score of 60 or greater indicates that a sufficiently diverse physical habitat is available to support a typical warm-water fish community. We selected the case studies in the Cuyahoga, Olentangy, and Sandusky watersheds on the basis of our judgment that the dams and their impoundments could be considered the lone or dominant alteration in each stream, thereby reducing the potential for confounding factors such as point source discharges.

In addition, we analyzed 21 sites (Table 3) from the Ohio EPA biological assessment database to investigate whether the IBI, ICI, and QHEI are sensitive on a broad scale to differences in stream biological and habitat quality above, within, and downstream of dam impoundments. The data set generally was limited to assessments after the year 2000, as during that time many streams in Ohio demonstrated that biological and chemical integrity had improved following upgrades to wastewater treatment facilities during the 1980s and 1990s.

RESULTS

Database Comparisons

Box plots of the IBI, ICI, and QHEI scores of the 21 sites (Table 3) show that scores for dam pools are generally lower than scores for upstream and downstream sites (Fig. 2). As discussed already, the literature generally indicates negative ecological impacts from dams and their impoundments, and the data presented here agree with this.

Case Study 1—Fishes in the Cuyahoga River

IBI scores for the Cuyahoga River (Lake Erie Basin) improved following alterations in 2005 to the Kent Dam in Portage County (river kilometer [RK] 88.2) and removal in 2006 of the Munroe Falls Dam located in Summit County (RK 80.3) (Fig. 3). A score from 2010 indicates that fish communities at the site of the former Munroe Falls Dam have recovered to the point of meeting Ohio's warm-water habitat criterion; that is, the IBI score of 38 was within the "area of insignificant departure" (4 points) below an IBI score of 40 required to meet the warm-water habitat criterion for wading sites (Ohio EPA, 1987b).

Recovery of the fish community in the former Kent Dam pool occurred fairly quickly and to a greater extent than in the former Munroe Falls Dam pool (Fig. 3), and this can be explained at least in part from differences in habitat. The Kent Dam impounded a stream reach with a bedrock-dominated substrate and narrow channel. Coupled with the narrower channel and slightly greater gradient, habitat recovery and readjustment of bed load occurred quicker in the former Kent Dam pool. By contrast, a lower gradient and greater sediment accumulation in the former Munroe Falls Dam pool required more time to reestablish suitable fish habitat, and recovery of the fish community is still continuing. The amount of cover (QHEI metric 2; Table 2) in the former Munroe Falls Dam pool was limited and was recorded as sparse to moderate.

Case Study 2—Fishes in the Olentangy River

Two dams were removed from the Olentangy River (Ohio River Basin) in Delaware, Ohio, between RK 40.9 and RK 41.8: the River Street Dam in late 2005 and the Central Avenue Dam in 2008. Biological water-quality evaluations were conducted in 2005 prior to removal and in 2009 following removal. Although initial Olentangy River IBI scores were higher when compared to those in the Cuyahoga River, postremoval monitoring showed improvements in the IBI scores (Fig. 3). The lower IBI score of 30 at RK 45.4 was in a dam pool that was still intact at the time of the 2009 sampling. IBI scores at the dam pool sites improved following removal from partial attainment in the impounded sections to full attainment in the now free-flowing sections.

Case Study 3—Macroinvertebrates in the Sandusky River

Removal of a dam on the Sandusky River in a mostly agricultural watershed in north-central Ohio (Lake Erie Basin) provided a case study of the rapidity and extent of responses of the macroinvertebrate community and the sensitivity of the ICI to habitat improvement. The St. Johns Dam on the Sandusky River at RK 80.8 in rural Seneca County, Ohio, was constructed in 1901 at the same location as two previous dams dating to around 1837. The dam spanned ~46 m and was 2.2 m high, maintaining a dam pool ~11.5 km long (Table 1). Following the decision to remove the dam, a 4.6-m-long breach in the upper 0.7–0.9 m of

TABLE 3. OHIO ENVIRONMENTAL PROTECTION AGENCY ASSESSMENT RESULTS IN SELECTED RIVERS UPSTREAM, WITHIN, AND DOWNSTREAM OF DAM POOLS

Stream	Year	Upstream				Dam Pool				Downstream			
		RK	IBI	ICI	QHEI	RK	IBI	ICI	QHEI	RK	IBI	ICI	QHEI
Blanchard River	2005	115.7	40	47	51	93.0	36	12	46	80.1	38	46	61.5
Cuyahoga River*	1996	87.5	28	44	70	89.6	28	N.D.†	51	80.0	34	42	83
Cuyahoga River	2008	38.8	34	50	83.5	33.5	26	24	56	27.8	48	48	83
Jonathan Creek	2008	5.3	51	38	87.5	1.8	44	32	65	1.4	46	47	76.5
Mahoning River	2006	150.0	38	34	59	137.6	30	12	55	N.D.	N.D.	N.D.	N.D.
Mahoning River	2006	N.D.	N.D.	N.D.	N.D.	101.1	33	14	41.5	91.0	45	26	60.5
Mahoning River	2006	N.D.	N.D.	N.D.	N.D.	73.6	40	20	48.5	N.D.	N.D.	N.D.	N.D.
Muskingum River	2006	156.7	44	54	86	140.0	36	42	60.5	136.2	46	48	79
Muskingum River	2006	N.D.	N.D.	N.D.	N.D.	129.1	40	42	60.5	122.0	43	50	82
Muskingum River	2006	108.3	46	48	85	102.5	42	44	57	90.8	44	38	75.5
Muskingum River	2006	N.D.	N.D.	N.D.	N.D.	83.8	44	40	61	78.5	50	44	81
Muskingum River	2006	N.D.	N.D.	N.D.	N.D.	69.5	38	44	63.5	63.6	46	48	84.5
Muskingum River	2006	N.D.	N.D.	N.D.	N.D.	58.3	44	40	63.5	53.9	47	44	84
Muskingum River	2006	N.D.	N.D.	N.D.	N.D.	33.6	40	50	62	22.4	42	42	86
Muskingum River	2006	N.D.	N.D.	N.D.	N.D.	15.1	44	36	60	9.0	38	50	86
Olentangy River	2004	6.3	50	44	71	3.4	38	10	32.5	2.9	45	40	76
Olentangy River	2005	44.3	42	42	81	41.8	38	26	45.5	40.9	46	50	84
Olentangy River	2005	N.D.	N.D.	N.D.	N.D.	45.4	30	20	55.5	N.D.	N.D.	N.D.	N.D.
Olentangy River	2005	N.D.	N.D.	N.D.	N.D.	41.5	34	32	49	N.D.	N.D.	N.D.	N.D.
Vermilion River	2002	47.0	41	54	80	38.5	34	23	48.5	36.2	45	50	79
West Branch Mahoning	2006	33.7	49	52	82	5.1	29	10	34.5	0.6	46	42	78.5

Note: Shown for each biotic survey are year of the survey, location in river kilometers (RK) upstream of the river mouth, and scores of the Index of Biotic Integrity (IBI), Invertebrate Community Index (ICI), and Qualitative Habitat Evaluation Index (QHEI).
*Upstream and downstream sites represent Munroe Falls Dam. Dam pool site was the Kent Dam impoundment (Table 1).
†N.D.—no data.

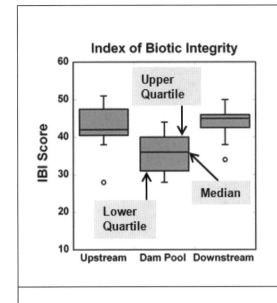

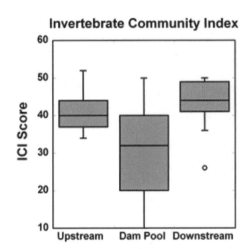

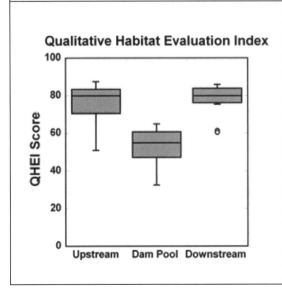

Figure 2. Box and whisker plots of the Index of Biotic Integrity (IBI), Invertebrate Community Index (ICI), and Qualitative Habitat Evaluation Index (QHEI) scores of the 21 river reaches listed in Table 3. The upper and lower limits of each box mark the upper (75th) quartile (UQ) and lower (25th) quartile (LQ) of the data, respectively, and the horizontal line within each box represents the median. The "whiskers" (vertical lines) indicate the maximum and minimum values. The interquartile distance (IQD) is the difference between the upper and lower quartiles (UQ – LQ). Points for which values are either greater than UQ + 1.5 * IQD or less than LQ – 1.5 * IQD are shown as outliers (circles).

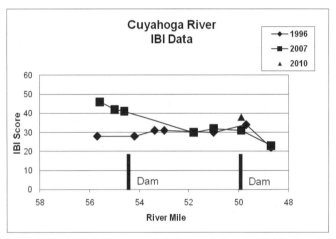

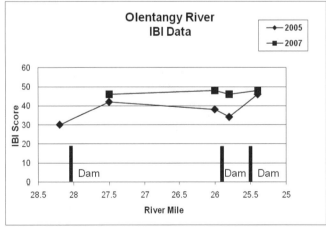

Figure 3. Index of Biotic Integrity (IBI) scores for the Cuyahoga and Olentangy Rivers before and after dam removal and dam modification projects. On the Cuyahoga River, the Kent Dam was located at river mile (RM) 54.8 (river kilometer, RK, 88.2), and the Munroe Falls Dam was located at RM 49.9 (RK 80.3). On the Olentangy River, the Central Avenue Dam was located at RM 25.9 (RK 41.7), and the River Street Dam was located at RM 25.5 (RK 41.0). Ohio Environmental Protection Agency reports river miles (1.000 mile = 1.609 km).

TABLE 4. COMPARISON OF MACROINVERTEBRATE DATA FROM SEVEN STUDY SITES ON THE SANDUSKY RIVER AT INCREASING RIVER KILOMETERS (RK) UPSTREAM IN 1990 AND 1999 (OHIO EPA, 2000), 2001 (OHIO EPA, 2003), IMMEDIATELY BEFORE DAM REMOVAL IN NOVEMBER 2003, AND FOLLOWING DAM REMOVAL IN SUMMER 2004 AND SUMMER 2005 (REPORTED HEREIN)

Site	RK	Year	Density number/ 0.0929 m^2	Total taxa	Qualitative EPT	ICI score	Warm-water habitat criterion	Narrative evaluation
Below St. Johns Dam	76.8/ 76.9	1990	848	55	16	48	EWH met	Exceptional
		1999	1734	55	11	38	Met	Good
		2001	–	–	–	54	EWH met	Exceptional
		2004	3103	37	12	36	Met	Good
		2005	5306	60	14	52	EWH met	Exceptional
At (behind) St. Johns Dam, west	80.8	2003 (Nov.)	752	33	11*	24	Not met	Fair/dam pool
		2004	2950	36	11	44	Met	Good
		2005	5163	45	15	50	EWH met	Exceptional
At (behind) St. Johns Dam, east	80.8	2003 (Nov.)	1083	25	9*	20	Not met	Fair/dam pool
		2004	3740	35	10	48	EWH met	Exceptional
		2005	4462	55	12	50	EWH met	Exceptional
Midpool	84.0	2001	–	–	–	12	Not met	Poor/dam pool
		2003 (Nov.)	988	36	8*	24	Not met	Fair/dam pool
		2004	716	58	5	34	Not met	Fair
		2005	573	55	11	38	Met	Good
Upstream end of pool	92.2/ 92.4	1990	546	64	15	48	EWH met	Exceptional
		1999	411	44	1	18†§	Not met	Fair/dam pool
		2001	–	–	–	48	EWH met	Exceptional
		2004	3984	54	11	44	Met	Good
		2005	3004	52	13	48	EWH met	Exceptional
Upstream of pool	98.3/ 98.5	1999	632	68	15	40	Met	Good
		2004	3914	42	3	36	Met	Good
		2005	2019	54	11	52	EWH met	Exceptional
Upstream of pool	104.6/ 104.8	1990	641	53	14	48	EWH met	Exceptional
		2001	–	–	–	E#		Exceptional
		2004	3230	45	5	44	Met	Good
		2005	1670	60	17	46	EWH met	Exceptional

Note: Numerical scores of the Invertebrate Community Index (ICI) and narrative descriptions of the macroinvertebrate communities are shown. ICI scores of 36 and 46 were required, respectively, to meet the warm-water habitat (WWH) and exceptional warm-water habitat (EWH) criteria (Ohio EPA, 1987b). Metric scores are also shown for total taxa (total taxa identified in both quantitative and qualitative samples) and qualitative EPT (Ephemeroptera [mayfly], Plecoptera [stonefly], and Trichoptera [caddisfly]) taxa collected only in the qualitative sample.

*The number of EPT taxa in the quantitative sample was substituted because no qualitative sample was collected.
†Dam pool conditions with no detectable current.
§Significant departure from warm-water habit biocriterion.
#Ohio EPA provided a narrative evaluation in lieu of an ICI score (E = Exceptional).

the dam near the west bank was created in March 2003, and the dam was completely removed in November 2003.

In its report on biological sampling of the Sandusky River in 2001, Ohio EPA stated that even though the St. Johns Dam permitted small boat recreation, it severely impaired the attainment of the designated warm-water habitat use. The report predicted "significant improvement in aquatic life use attainment" following dam removal but also noted that the insecticide dieldrin had been detected in sediments in the dam pool at concentrations exceeding toxicity guidelines (Ohio EPA, 2003).

Macroinvertebrates were sampled according to Ohio EPA methods (Ohio EPA, 1987b, 1989a, 1989b) at three sites in the dam pool at the time of dam removal in November 2003 and at those sites and five additional sites in the summers of 2004 and 2005 above and within the former dam pool and below the dam (Krieger and Stearns, 2007). The sample data for each site were converted to scores for each of the ICI metrics, and the metric scores were summed to provide a total ICI score for the site (Table 4).

ICI scores for November 2003 at the three dam-pool sites ranged from 20 to 24, yielding a narrative evaluation at all three sites of "fair/dam pool" (Table 4). None of the sites met the warm-water habitat ICI scoring criterion of 36 for the "Eastern Corn Belt Plains" ecoregion (Ohio EPA,1989b). Several mitigating factors may have resulted in scores that were lower than they otherwise would have been: (1) The Hester-Dendy samplers were retrieved after the water level receded quickly during dam removal, leaving them out of the water for a few hours, so it is likely that some invertebrates abandoned the assemblies before they were retrieved. (2) No qualitative samples were collected in conjunction with the quantitative samples in November. Metric 10, total number of qualitative EPT taxa, could not therefore be determined directly, and the number of EPT taxa present in the quantitative samples was used instead, in accordance with Ohio EPA's procedure for such occasions. (3) November is after the designated season to collect samples for determination of ICI values. Seasonal differences in life cycles may have increased or decreased the numbers of taxa and numbers of individuals within particular taxa that were found in November 2003, and therefore the ICI metric scores. Even with those complicating factors, however, the ICI score attained at the midpool (RK 84.0) site in November 2003 was higher than the score calculated by Ohio EPA for that site for the summer of 2001 (Table 4).

In August 2004, 9 months after dam removal and during the prescribed sampling season, the same three sites appeared to have improved greatly (Table 4), with scores of 34 at the midpool site (just below the 36 needed to meet the warm-water habitat criterion), and 44 on the west side of the river formerly less than 100 m upstream of the dam (RK 80.8). On the opposite side at that location, an ICI score of 48 met the exceptional warm-water habitat criterion (requiring an ICI ≥46). A site within and near the upper end of the former dam pool also met the warm-water habitat criterion with an ICI value of 44. In 2005, every site had a higher ICI score than in 2004 (Table 4), and all met the exceptional warm-water habitat criterion except the midpool site (RK 84.0), which met the warm-water habitat criterion.

Interpretation of the biotic index score is strengthened by examining the scores of the individual metrics that comprise it. In this case study using the ICI, metric 1, total number of taxa, scored slightly lower at one site in 2004 (after dam removal) but was unchanged at the other two sites (Table 4). The number of taxa within the lower reaches of the impoundment (RK 80.8, RK 84.0) was depressed in November 2003 (Fig. 4). Following removal of the dam, the number of taxa near the former dam (RK 80.8) increased slightly in 2004, while at RK 84.0, which had been in the dam pool, the number of taxa was higher than at any other site. Overall, the macroinvertebrate communities at RK 80.8 and RK 84.0, formerly within the dam pool, experienced a progressive increase in the number of taxa from 2003 through 2005, indicating improved environmental quality.

DISCUSSION

Five of the sites sampled in the Sandusky River case study were also sampled by Ohio EPA in 1990, 1999 (Ohio EPA, 2000), and 2001 (Ohio EPA, 2003). The ICI scores from those earlier surveys were compared with the 2003, 2004, and 2005 scores (Table 4). The ICI scores prior to dam removal varied considerably at most sites over the range of years each was sampled, and their narrative evaluations varied slightly between "exceptional" and "good" at four sites that were either downstream of the dam, near the upstream end of the dam pool, or upstream of the dam pool (Table 4). For example, at RK 76.8/76.9 below St. Johns Dam, the ICI score varied from 38 ("good") to 54 ("exceptional")

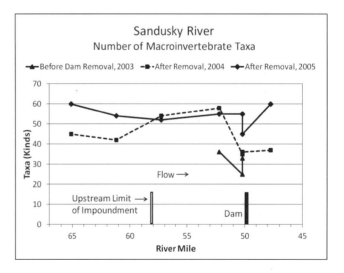

Figure 4. Number of taxa (kinds) of macroinvertebrates found in combined quantitative and qualitative samples in November 2003 and the summers of 2004 and 2005 at each of the St. Johns Dam study sites on the Sandusky River. Two sites were on opposite sides of the river immediately upstream of the dam at river mile 50.2 (river kilometer 80.8). Ohio Environmental Protection Agency reports river miles (1.000 mile = 1.609 km).

in only a 2 yr period between 1999 and 2001. Within the dam pool (prior to dam removal), the midpool site (RK 84.0) received an exceptionally low ICI score of 12 ("poor/dam pool") in 2001 but received a score of 24 ("fair/dam pool") in 2003, and near the upstream end of the dam pool (RK 92.2/92.4), the ICI score was 18 ("fair/dam pool") in 1999 but was 48 ("exceptional") in both 1990 and 2001. Some degree of variation in IBI and ICI scores is expected at a site from year to year, even when the physical habitat remains relatively unchanged, and scores that indicate attainment of the designated biological use (such as warm-water habitat) one year but not another need to be interpreted within a long-term context and in light of stream flow, preceding weather conditions, and other factors influencing each survey (Yoder and Rankin, 1995). Such factors are likely to be magnified within dam pools as opposed to free-flowing stream reaches.

Notably, macroinvertebrate surveys on opposite sides of the channel at RK 80.8 yielded ICI scores within four points of each other both before and after dam removal and were identical in 2005. These results indicate that the ICI is not strongly influenced by variations in microhabitat within a sampling reach (length ~100 m in this case study).

As reflected by the macroinvertebrate community, the biological quality of the formerly impounded reach of the Sandusky River responded rapidly to the changes in habitat quality that resulted from removal of the St. Johns Dam, and this result is consistent with previous studies (Hart et al., 2002; Maloney et al., 2008; Stanley et al., 2002). On the basis of ICI scores, two of three sites (both at RK 80.8) in the dam pool sampled before dam removal improved from ratings of "fair" in November 2003 to "good" and "exceptional" in August 2004. The third, midpool site (RK 84.0) received a rating of "fair" both before and after removal, though the score improved from 24 to 34 (Table 4). The lack of attainment of a "good" rating at that site within a period of a few months may be attributable to its location in an extensive natural pool of the river with no riffles upstream or downstream for several kilometers. Because the ICI methodology is based on invertebrate communities of riffles, the location of that site probably resulted in a lower rating at least in part because of the limited number of taxa and individuals typical of riffles that were available to colonize the site.

Additional sampling at all sites in the summer of 2005 revealed continued improvement in the quality of the Sandusky River within the reach formerly occupied by the St. Johns Dam pool. Furthermore, the entire study area exhibited a higher stream quality in 2005 than in 2004 in that six sites met the "exceptional warm-water habitat" criterion (ICI ≥46), and the seventh site (RK 84.0) improved from "fair" to "good." The recovery of the macroinvertebrate community in the Sandusky River after removal of the St. Johns Dam followed the pattern observed on the Cuyahoga River where the ICI was applied following dam modification or removal (Tuckerman and Zawiski, 2007).

Doyle et al. (2005) presented two conceptual frameworks, or scenarios, for ecosystem recovery following dam removal. One scenario assumes that all components of the stream ecosystem will recover to predam conditions, largely because the channel morphology (habitat) upon which the ecosystem depends returns to the predam state, and the stream ecosystem as a whole is unchanged from the time of dam construction. The second scenario assumes that some components of the ecosystem will recover completely but that other components will only partly recover or will reach alternate states compared to the predam condition. Individual components may not return to the predam state because of irreversible modification of channel morphology or changes in components of stream ecosystem structure and function in other parts of the watershed after dam construction.

The frameworks presented by Doyle et al. (2005) also provide a useful tool upon which to base realistic restoration expectations, such as the time required for a dam removal project to result in improved fish and macroinvertebrate communities. For example, a slight decrease in the IBI was documented in 2007 at RK 80.3 on the Cuyahoga River at the former Munroe Falls Dam site, and Doyle et al. (2005) predicted a temporary decline in fish community integrity accompanying the downstream migration of accumulated sediment. Similarly, Maloney et al. (2008) reported a decrease in the downstream IBI and a lag in fish community recovery following a dam breach on the Fox River, Illinois. In our study, the number of macroinvertebrate taxa downstream of the former St. Johns Dam (RK 76.8) on the Sandusky River was slightly lower than at most of the other sites, even those within the former impoundment, in the summer following dam removal; however, the next summer (2005), that site and one of the two sites immediately upstream of the former dam had as many or more taxa than most of the sites further upstream. Downstream migration of sediment may have caused the temporary declines in the quality of the fish and macroinvertebrate communities in the Cuyahoga and Sandusky studies but was not documented.

Doyle et al. (2005) noted that a dam removal project may be declared to have successfully restored a stream ecosystem on the basis of the recovery of a few species or taxonomic groups (e.g., fish or macroinvertebrates) even though other groups (e.g., vegetation) may not have recovered fully. In our study, we have presented the recovery only of fish and macroinvertebrate communities as expressed by the IBI and ICI, and these two indices should not be presumed to demonstrate recovery of other components of the stream ecosystem. Nevertheless, in the Ohio case studies presented, the IBI and ICI did reveal a rapid response of the fish and macroinvertebrate communities to the reestablishment of free-flowing conditions. The presence and proportional contributions of individual taxa of the fish and macroinvertebrate communities may differ somewhat after dam removal as opposed to predam conditions, but by integrating multiple structural and functional community characteristics, the IBI and ICI effectively assess the extent of change in those communities.

SUMMARY AND CONCLUSIONS

Our results confirm that, as measured by the IBI and ICI used by Ohio EPA, the integrity of fish and macroinvertebrate communities generally is compromised in dam pools in river channels. The case studies presented here combined with a growing literature indicate that the biological integrity of streams measurably improves following dam removal. Furthermore, the case studies show that the changes in IBI and ICI scores reflected rapid improvements in the biological communities that occurred in response to dam removal and subsequent restoration of the habitat to free-flowing conditions. We conclude that both the IBI and the ICI provide sensitive tools that are as useful for measuring biotic responses to dam removal as they are for measuring attainment of biological criteria, such as meeting warm-water habitat criteria. As Doyle et al. (2005, p. 242) stated, ". . . dam removal represents a very powerful tool for restoring streams to more natural conditions." The need continues to monitor changes in stream habitat conditions for specific dam removal projects, especially with regard to potential impacts of sedimentation downstream of dam removal sites.

ACKNOWLEDGMENTS

The Sandusky River case study was supported by grants to Heidelberg University from the Ohio Department of Natural Resources, Division of Natural Areas and Preserves. Fieldwork and sample processing and analysis for the Sandusky study were assisted by Robert Vargo, Scenic River coordinator for northwest Ohio, and numerous Heidelberg University students. Staff of the Ohio Environmental Protection Agency Division of Surface Water, Ecological Assessment Unit assisted in identifications of chironomids in the Sandusky River samples. Johan Gottgens and an anonymous reviewer provided suggestions that greatly improved this manuscript.

REFERENCES CITED

Barbour, M.T., Stribling, J.B., and Karr, J.R., 1995, Multimetric approach for establishing biocriteria and measuring biological condition, *in* Davis, W.S., and Simon, T.P., eds., Biological Assessment and Criteria: Tools for Water Resource Planning and Decision Making: Boca Raton, Florida, Lewis Publishers, p. 63–77.

Bednarek, A.T., 2001, Undamming rivers: A review of the ecological impacts of dam removal: Environmental Management, v. 27, p. 803–814, doi:10.1007/s002670010189.

DeShon, J.E., 1995, Development and application of the Invertebrate Community Index, *in* Davis, W.S., and Simon, T.P., eds., Biological Assessment and Criteria: Tools for Water Resource Planning and Decision Making: Boca Raton, Florida, Lewis Publishers, p. 217–243.

Doyle, M.W., Stanley, E.H., Orr, C.H., Selle, A.R., Sethi, S.A., and Harbor, J.M., 2005, Stream ecosystem response to small dam removal: Lessons from the Heartland: Geomorphology, v. 71, p. 227–244, doi:10.1016/j.geomorph.2004.04.011.

Freeman, R., and Bowerman, W., 2002, Opening rivers to Trojan fish: The ecological dilemma of dam removal in the Great Lakes: Conservation in Practice, v. 3, no. 4, p. 35–39, doi:10.1111/j.1526-4629.2002.tb00045.x.

Gottgens, J.F., 2009, Impact of the Removal of the Secor Road Dam on the Fish Community Structure and Composition in the Ottawa River, Ohio: Final Report to Toledo Metropolitan Area Council of Governments Project 06(h) EPA 10, 16 p.

Gottgens, J.F., and Evans, J.E., 2007, Dam removals and river channel changes in northern Ohio: Implications for Lake Erie sediment budgets and water quality: Journal of Great Lakes Research, v. 33, Special Issue 2, p. 87–89.

Haponski, A.E., Marth, T.A., and Stepien, C.A., 2007, Genetic divergence across a low-head dam: A preliminary analysis using logperch and greenside darters: Journal of Great Lakes Research, v. 33, Special Issue 2, p. 117–126.

Hart, D.D., Johnson, T.E., Bushaw-Newton, K.L., Horwitz, R.J., Bednarek, A.T., Charles, D.F., Kreeger, D.A., and Velinsky, D.J., 2002, Dam removal: Challenges and opportunities for ecological research and river restoration: Bioscience, v. 52, p. 669–681, doi:10.1641/0006-3568(2002)052[0669:DRCAOF]2.0.CO;2.

Heinz Center, 2002, Dam Removal: Science and Decision Making: Washington, D.C., The H. John Heinz III Center for Science, Economics, and the Environment, 221 p.

Jones, M.L., Netto, J.K., Stockwell, J.D., and Mion, J.B., 2003, Does the value of newly accessible spawning habitat for walleye (*Stizostedion vitreum*) depend on its location relative to nursery habitats?: Canadian Journal of Fisheries and Aquatic Sciences, v. 60, p. 1527–1538, doi:10.1139/f03-130.

Kanehl, P.D., Lyons, J., and Nelson, J.E., 1997, Changes in the habitat and fish community of the Milwaukee River, Wisconsin, following removal of the Woolen Mills Dam: North American Journal of Fisheries Management, v. 17, p. 387–400, doi:10.1577/1548-8675(1997)017<0387:CITHAF>2.3.CO;2.

Krieger, K.A., and Stearns, A.M., 2007, Response of the Macroinvertebrate Community of the Sandusky River in Seneca and Wyandot Counties, Ohio, to Removal of the St. Johns Dam: Tiffin, Ohio: Final Report to Ohio Department of Natural Resources Division of Natural Areas and Preserves, 43 p.

Lavis, D.L., Hallett, A., Koon, E.M., and McAuley, T.C., 2003, History of and advances in barriers as an alternative method to suppress sea lampreys in the Great Lakes: Journal of Great Lakes Research, v. 29, supplement 1, p. 362–372, doi:10.1016/S0380-1330(03)70500-0.

Maloney, K.O., Dodd, H.R., Butler, S.E., and Wahl, D.H., 2008, Changes in macroinvertebrate and fish assemblages in a medium-sized river following a breach of a low-head dam: Freshwater Biology, v. 53, p. 1055–1068, doi:10.1111/j.1365-2427.2008.01956.x.

Manolakos, E., Virani, H., and Novatny, V., 2007, Extracting knowledge on the links between the water body stressors and biotic integrity: Water Research, v. 41, p. 4041–4050, doi:10.1016/j.watres.2007.05.002.

Ohio Environmental Protection Agency (Ohio EPA), 1987a (updated 1988), Biological Criteria for the Protection of Aquatic Life: Volume I: The Role of Biological Data in Water Quality Assessment: Columbus, Ohio, Ecological Assessment Section, Division of Water Quality Planning and Assessment, Surface Water Section, 44 p.

Ohio Environmental Protection Agency, 1987b (updated 1988), Biological Criteria for the Protection of Aquatic Life: Volume II: Users Manual for Biological Field Assessment of Ohio Surface Waters: Columbus, Ohio, Ecological Assessment Section, Division of Water Quality Planning and Assessment, Surface Water Section, variously paginated.

Ohio Environmental Protection Agency, 1989a, Biological Criteria for the Protection of Aquatic Life: Volume III: Standardized Biological Field Sampling and Laboratory Methods for Assessing Fish and Macroinvertebrate Communities: Columbus, Ohio, Ecological Assessment Section, Division of Water Quality Planning and Assessment, Surface Water Section, variously paginated.

Ohio Environmental Protection Agency, 1989b, Addendum to Biological Criteria for the Protection of Aquatic Life: Volume II: Users Manual for Biological Field Assessment of Ohio Surface Waters: Columbus, Ohio, Surface Water Section, Division of Water Quality Planning and Assessment, 21 p.

Ohio Environmental Protection Agency, 2000, Biological and Water Quality Study of Sycamore Creek and the Sandusky River 1999, Wyandot and Seneca Counties, Ohio: Division of Surface Water, Ohio Environmental Protection Agency Technical Report EAS/2000-6-3, 42 p. plus appendices.

Ohio Environmental Protection Agency, 2003, Biological and Water Quality Study of the Sandusky River and Selected Tributaries 2001, Seneca, Wyandot, and Crawford Counties, Ohio: Division of Surface Water, Ohio Environmental Protection Agency Technical Report EAS/2003-4-6, 132 p. plus appendices.

Ohio Environmental Protection Agency, 2007, Total Maximum Daily Loads for the Olentangy River Watershed, Final Report: Columbus, Ohio Environmental Protection Agency, Division of Surface Water, 106 p.

Ohio Environmental Protection Agency, 2008, Cuyahoga River Aquatic Life Use Attainment Following the Kent and Munroe Falls Dam Modifications: Ohio Environmental Protection Agency Biological and Water Quality Report NEDO/2008–08-01, 49 p. plus appendices.

Ohio Environmental Protection Agency, 2010, Biological and Habitat Study of the Olentangy River, 2005 and 2009 City of Delaware Dam Removals: Ohio Environmental Protection Agency Technical Report EAS/2010–5-8, 12 p. plus appendices.

Rankin, E.T., 1989, The Qualitative Habitat Evaluation Index (QHEI): Rationale, Methods, and Application: Columbus, Ohio, Ohio Environmental Protection Agency, Division of Surface Water, Ecological Assessment Section, 54 p. plus appendices.

Rankin, E.T., 1995, Habitat indices in water resource quality assessments, *in* Davis, W.S., and Simon, T.P., eds., Biological Assessment and Criteria: Tools for Water Resource Planning and Decision Making: Boca Raton, Florida, Lewis Publishers, p. 181–208.

Roberts, S.J., Gottgens, J.F., Spongberg, A.L., Evans, J.E., and Levine, N.S., 2007, Assessing potential removal of low-head dams in urban settings: An example from the Ottawa River, NW Ohio: Environmental Management, v. 39, p. 113–124, doi:10.1007/s00267-005-0091-8.

Rumschlag, J.H., and Peck, J.A., 2007, Short-term sediment and morphologic response of the middle Cuyahoga River to the removal of the Munroe Falls Dam, Summit County, Ohio: Journal of Great Lakes Research, v. 33, Special Issue 2, p. 142–153.

Simon, T.P., and Lyons, J., 1995, Application of the Index of Biotic Integrity to evaluate water resource integrity in freshwater ecosystems, *in* Davis, W.S., and Simon, T.P., eds., Biological Assessment and Criteria: Tools for Water Resource Planning and Decision Making: Boca Raton, Florida, Lewis Publishers, p. 245–262.

Stanley, E.H., Luebke, M.A., Doyle, M.W., and Marshall, D.W., 2002, Short-term changes in channel form and macroinvertebrate communities following low-head dam removal: Journal of the North American Benthological Society, v. 21, p. 172–187, doi:10.2307/1468307.

Tuckerman, S., and Zawiski, B., 2007, Case studies of dam removal and TMDLs: Process and results: Journal of Great Lakes Research, v. 33, Special Issue 2, p. 103–116.

Walton, B.M., Salling, M., Wyles, J., and Wolin, J., 2007, Biological integrity in urban streams: Toward resolving multiple dimensions of urbanization: Landscape and Urban Planning, v. 79, p. 110–123, doi:10.1016/j.landurbplan.2005.10.004.

Yoder, C.O., 1995, Policy issues and management applications of biological criteria, *in* Davis, W.S., and Simon, T.P., eds., Biological Assessment and Criteria: Tools for Water Resource Planning and Decision Making: Boca Raton, Florida, Lewis Publishers, p. 327–343.

Yoder, C.O., and Rankin, E.T., 1995, Biological criteria program development and implementation in Ohio, *in* Davis, W.S., and Simon, T.P., eds., Biological Assessment and Criteria: Tools for Water Resource Planning and Decision Making: Boca Raton, Florida, Lewis Publishers, p. 109–144.

MANUSCRIPT ACCEPTED BY THE SOCIETY 29 JUNE 2012

Using airborne remote-sensing imagery to assess flow releases from a dam in order to maximize renaturalization of a regulated gravel-bed river

M.S. Lorang
F.R. Hauer
D.C. Whited
P.L. Matson
Flathead Lake Biological Station, The University of Montana, 32125 Bio Station Lane, Polson, Montana 59860-6815, USA

ABSTRACT

Gravel-bed river floodplains are dynamic landscapes that support a high level of ecosystem biocomplexity and biodiversity in large part because of the continual physical turnover of habitat. We evaluated the potential of a gravel-bed river to do geomorphic work on a series of floodplains below a dam by linking airborne hyperspectral imagery with corresponding ground-truth measures of flow velocity, water depth, floodplain surface topography, and vegetative cover. These data were analyzed in a geographic information system to map the spatial distribution of potential stream power over a range of discharge regimes. Nodes of flow separation at specific discharges that co-occurred with zones of high stream power were used as a metric to determine potential geomorphic threshold levels and location of channel avulsions. In order to address discharge duration as a factor affecting geomorphic change on the study floodplains, we established the relationship between discharge and total cumulative power applied to a single key floodplain and then used that relationship to examine historical discharge records and changes in flow release in terms of total cumulative power. We used the assumption that similar levels of total cumulative stream power, above a minimum geomorphic threshold, would produce similar levels of geomorphic work as a higher-magnitude, short-duration flood event. These results form the basis of an objective approach to evaluating flow releases needed from dams to maintain the dynamic structure and ecological function of gravel-bed river floodplains. Moreover, the methodologies presented herein lend themselves to quantitative investigation of potential geomorphic changes related to complete dam removal and return of normalized flow of water and materials through the river system.

INTRODUCTION

The Shifting Habitat Mosaic of River Floodplains: A Theoretical Framework for Dam Removal

Rivers and their accompanying floodplains form landscapes composed of patches or mosaics of biophysical space that are used by a wide variety of different species. The mosaic of floodplain habitat changes over time, primarily in response to flooding, which drives channel avulsion and cut-and-fill alluviation processes (i.e., channel abandonment and the erosion and deposition of fine and coarse sediments, respectively). These processes result in changes in the pattern and rate of surface flow, hyporheic groundwater exchange, and associated biogeochemical processes. The habitat mosaic of a river floodplain is further modified by the recruitment and succession of riparian forests. Hence, many organisms (e.g., trees, mammals, fish, birds, aquatic and terrestrial insects) have evolved to become facultative or obligate users of floodplain habitats to fulfill all or part of their life histories. This complex biophysical system produces a shifting habitat mosaic that is a fundamental feature of river ecosystems and forms a theoretical framework to underpin restoration activities from dam removal to flow reregulation (Stanford et al., 2005).

Maintaining the biogeomorphic functioning of river systems requires hydraulic processes that form new channels and bars while enabling recruitment of riparian trees (Rood et al., 1999; Nilsson and Bergeron, 2000; Rood et al., 2005). This is crucial to developing successful approaches to river management, ranging from flow regulation to removal of bank stabilization and ultimately to complete decommissioning of dams (Shafroth et al., 2002). Moreover, change in the pattern and composition of downstream floodplain habitat (e.g., vegetation cover type, channels, bars) becomes a fundamental metric that can be measured as part of a monitoring protocol. The key element is the ability to predict geomorphic change across discharges, thus forming a specific basis for dams to produce normative flows (i.e., flows that attempt to mimic predam conditions). Ultimately, anything impacting primary drivers of hydrology and geomorphology (e.g., dams, bank stabilization) will have the greatest impact on the whole river system. Hence, approaches to river restoration and management activities first need to focus on these primary drivers as a priority for understanding floodplain ecosystems and for management of human activities that impact rivers and their floodplains (Petts, 1984; Stanford et al., 1996; Hauer et al., 2003; Burke et al., 2009).

In the research presented herein, we develop an assessment protocol that employs remote-sensing techniques using airborne imagery, acoustic Doppler depth and velocity profiling, and flood inundation modeling. Results are evaluated within a framework of the shifting habitat mosaic concept of river floodplains (Stanford et al., 2005). This research was done on floodplain reaches of the Snake River below Palisades Dam, Idaho. The dam was constructed in the 1950s by the Bureau of Reclamation (BoR), a U.S. federal agency, to provide irrigation water and hydropower.

Our research objective was to determine the hydrologic regimes necessary to reestablish and sustain improved ecological integrity of the river and its floodplains, yet allow for flow regimes within the contractual irrigation obligations of the BoR. We address the following research question: What discharge regimes are necessary to produce sufficient stream power to realize cut-and-fill alluviation and channel avulsion that would emulate that of predam conditions?

The methodological approach developed is based on assessment of cumulative floodplain power on five different downstream floodplains with predictions based on the historical record of river discharge, a 2002 data set of airborne imagery and ground-truthing, and an assessment of channel change conducted in 2005 following implementation of recommended reregulated flow releases.

STUDY SITE BACKGROUND

Physical Geography and Water Demand

This study was conducted on the Snake River between Palisades Dam and the confluence of the Snake River with the Henry's Fork in southeastern Idaho (Fig. 1). The Upper Snake River drains the primarily montane landscape of western Wyoming and eastern Idaho. The Upper Snake River originates in Yellowstone National Park and flows south through Grand Teton National Park and Jackson Lake. A combined concrete and earth-fill dam regulates the outlet of Jackson Lake. Jackson Lake Dam impounds ~770 million m^3 and draws from a drainage area at the dam site of 4724 km^2. The Snake River then flows ~110 km to Palisades Reservoir (Fig. 1).

The river corridor below Palisades Dam is typical of the Interior West, where the river sequentially alternates between confined canyon and unconfined floodplain reaches (Stanford and Ward, 1993). The Upper Snake River (sixth order) has a drainage basin of 15,048 km^2 above the Palisades Dam (Fig. 1). The authorized purposes of Palisades Dam are flood control, irrigation, and hydropower generation. Construction of the dam was completed in 1957. The dam provides a supplemental water supply to ~271,149 ha of irrigated land. The dam creates a reservoir of 1.7×10^9 m^3, and the capacity of the outlet works for Palisades Dam is 935 m^3/s (cubic meters/second).

Natural Discharge Regime versus Flow Regulation

Historically, the hydrologic regime of the Upper Snake River basin was characterized by low flow during summer, fall, and winter with a rising limb during spring snowmelt, which typically began in late March and early April. Peak discharge occurred in late May or early June. Discharge patterns in the study area of the Snake River since construction of Palisades Dam have changed dramatically, particularly among high-discharge years. In general, dam operators increased discharge from the dam prior to onset of spring snowmelt, as early as late February and early

March, representing initiation of a rising spring hydrograph 30–45 days earlier compared to predam regimes. This change in hydrographic pattern after dam construction is the result of operators' anticipation of high discharge coming from the upper basin. Hence, the intent of spilling water in late winter is to maximize water storage capacity in the Palisades Reservoir and minimize risk of flooding downstream of the dam during peak runoff. During the 46 yr of record prior to Palisades Dam operation, the Snake River achieved or exceeded 935 m^3/s 12 times. In contrast, during the 46 yr of record since dam operation, 935 m^3/s has been exceeded only once, in 1997 when snowmelt floodwaters exceeded the holding capacity of the reservoir.

Study Area Floodplains

Five major floodplain reaches occur in the Snake River between Palisades Dam and the confluence with Henry's Fork (Fig. 1). These floodplains are referenced throughout this paper, beginning with the floodplain closest to the dam, as: Swan, Conant, Fisher, Heise, Upper Twin, and Lower Twin (collectively Twin; Fig. 1). Each floodplain contains both complex parafluvial (i.e., regularly inundated and frequently scoured gravel- and cobble-bed sediments) and orthofluvial (i.e., periodically inundated, but generally not scoured floodplain sediments) zones (Stanford et al., 2005; Lorang and Hauer, 2006). The orthofluvial region of each floodplain is characterized by riparian vegetation, particularly a mature or maturing gallery forest. Both the parafluvial and the orthofluvial zones of the floodplains contain a variety of aquatic habitats (e.g., spring brooks, ponds, inundation channels) that are a function of the legacy of past geomorphic processes.

Ecologically Based Systems Management

The methodology and analysis presented herein were completed in 2003 with discharge recommendations from Palisades Dam as part of the Ecologically Based Systems Management (EBSM) plan presented to BoR (Hauer et al., 2004). The objective of EBSM was to provide guidelines for discharge magnitude and regimes needed to accomplish geomorphic work (e.g., channel avulsion, cut-and-fill alluviation) on the five floodplains in the study area. The criteria for changes in flow releases were the following, as related to climate condition for the basin:

1. In ultrawet years, 4 yr of the last 45 yr, where total annual discharge is >8.6 × 10^9 m^3—approach 935 m^3/s for as long as possible with flows >700 m^3/s for 12–15 d.
2. In wet years, 11 yr of the last 45 yr, where total annual discharge is 7.2–8.6 × 10^9 m^3—exceed 700 m^3/s for 8 to 12 d.

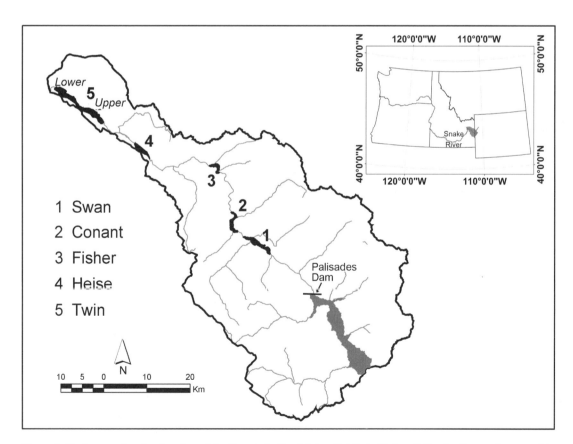

Figure 1. Study area showing location of the five study floodplains distributed below Palisades Dam and geographic location of the study area along the Snake River.

3. In moderate years, 17 yr of the last 45 yr, where total annual discharge is 4.2–7.2 × 10^9 m^3—exceed 538–700 m^3/s and sustain as long as volume allows.
4. In dry years—stay with existing protocols.

These protocols were followed for the years 2003, 2004, and 2005, at which time new aerial images were collected after the 2005 spring runoff to assess what levels of geomorphic work were completed relative to the model predictions of the EBSM flows.

METHODS

Aquatic Field Methods

A Sontek RS3000 Acoustic Doppler Velocity Profiler (ADP) was used to acquire detailed water depth and vertical profile measurements of flow velocity along channel reaches within the study floodplains. We deployed the ADP from the front of a small jet boat and the aluminum frame of a cat raft with both velocity profile data and depth data spatially linked with a global positioning system (GPS) receiver collocated with the ADP. During data acquisition, the ADP was maneuvered across the channel to obtain data from as full an array of depths and velocities as possible. Both the ADP and GPS data were recorded simultaneously on a field laptop computer. The ADP data were then processed to create an integrated velocity value (average velocity for an individual ADP profile), as well as a depth value for each GPS location.

Four ADP surveys were collected in summer and fall of 2002 for each floodplain reach (20–22 June, 17–20 August, 24–26 September, and 25–26 November). The ADP data were obtained at 42, 142, 226, and 325 m^3/s, respectively, thereby covering a broad discharge range, including levels approaching bankfull conditions where all of the exposed gravel bars of the parafluvial zone were inundated. Hence, the highest flow data were collected during a "parafluvial-full" condition, where scour and widespread bed-load motion begins. Over 25,000 discrete measures of depth and flow velocity were recorded during the ADP surveys.

Remote-Sensing Methods

Airborne remotely sensed data were collected with an AISA hyperspectral imagery system from Spectral Imaging, Oulu, Finland. The AISA hyperspectral sensor is operated from aircraft at the height (1000 m) and speed (87 knots) and sensor recycling frequency (30 ms) required to generate a 1 m × 1 m pixel resolution. Waveband configuration for digital data acquisition is from 256 individual spectral wavebands (400–950 nm) arrayed into 20 aggregate bands. The remote-sensing data were collected along predetermined flight lines oriented with the long axis of each study floodplain and having flight line overlaps of 40%–50%. All data were collected within a time period of 1.5 h either side of solar noon. We selected the clearest days possible during a sampling interval spanning several weeks to capture flow and vegetation attributes that were targeted for the particular season and to maximize the quality of the imagery data.

Individual flight lines were cross-referenced with existing digital orthophoto quadrangles (DOQs) to examine the spatial positioning of each flight line. If an individual flight line needed further georectification, then additional ground-control points were added to improve the rectification in a given flight line. All georectified flight lines had a mean root mean square error of less than 4 m. Once all flight lines were georectified for a given reach, they were then stitched together to create a final mosaic. All georectification and minor color balancing between flight lines were completed in ERDAS Imagine 8.5.

In addition to rectification errors, rapid turbulence experienced during data acquisition occasionally caused the aircraft to roll at a rate faster than the GPS/IMU data stream. Turbulence induced error (TIE) during image acquisition resulted in image distortion for some areas. These distortions were highly localized and appeared as waves in the imagery. Rectification errors as well as errors caused by aircraft turbulence affect accuracy assessments, causing portions of the image to be spatially offset from the true location. Rectification, color balancing, and reflectance anomalies are inherent in virtually all remotely sensed data and have been widely reported and discussed in terms of applications to river channel classification for over a decade (Winterbottom and Gilvear, 1997; Bryant and Gilvear, 1999; Aspinall et al., 2002; Marcus, 2002; Marcus et al., 2003; Gilvear et al., 2004; Legleiter et al., 2004; Leckie et al., 2005; Marcus and Fonstad, 2008). Less than 5% of water surface pixels contained recognizable errors associated with rectification, reflectance anomalies, seams between data lines, or shadows cast on water surfaces by trees.

We used depth and mean velocity data from the ADP to classify the hyperspectral imagery data corresponding to water bodies for all study reaches. An unsupervised classification approach (ISODATA, Iterative Self-Ordering Data Analysis; Tou and Gonzalez, 1977) was used to identify areas of similar spectral reflectance. Once an unsupervised classification of spectral reflectance was generated, the ADP data were distributed in a geographic information system (GIS) environment to aggregate classes and assign unique depth and velocity categories following protocol developed in past work (Whited et al., 2003; Lorang et al., 2005).

Modeling Methods

The modeling approach described herein begins with the creation of a digital elevation model (DEM) by relating vegetation cover types created from remote-sensing classification to their relative elevation above base-flow conditions as measured with total station topographic surveys. The September imagery was used for land-cover classification because of the high contrast between vegetation types during autumnal senescence. First, an unsupervised classification was used to discriminate between vegetative cover and nonvegetative cover (i.e., vegetation vs.

cobble and water). This was followed by a supervised classification approach for the vegetative cover. To help discriminate among different vegetation types, homogeneous stands of the varying cover types (e.g., cottonwood, willow, reed canary grass, dry grass) were identified and associated with specific hyperspectral signatures. These specific imagery signatures were used as "training areas" to classify the image into different land-cover types. Mean spectral signatures were calculated for each cover type and subsequently used in a supervised classification to produce a land-cover map for each reach.

Using the spectral signatures, the Mixed Tune Matched Filtering (MTMF) algorithm in ENVI (RSI, 2000) was then applied to the vegetative component of the imagery to discriminate the varying vegetation types. For each reach, a final land-cover map was produced consisting of eight dominant cover types (i.e., water, cobble, deciduous–predominately cottonwood, willow, mixed grasses, dry grasses, reed canary grass, and shadows). In the Twin reach, willows were not easily differentiated from cottonwood; therefore, cottonwood and willow were aggregated into a single coverage identified as a "deciduous" category. A pasture category was also added in the Conant reach.

We then produced a detailed floodplain DEM from the classified hyperspectral imagery and ground-based topographic surveys in a manner similar to that used by Lorang et al. (2005). Topographic surveys were conducted along transects that extended across the floodplain. These transects were chosen to include a broad range of topography (e.g., slope, elevation) across as many classified cover type features as possible. Other features captured by these surveys included relative elevations and slopes between gravel bars, water surface, and bank top elevations throughout the floodplain reach. The survey data were then overlaid on the various classified cover types and assigned a relative elevation to the main channel, as well as a typical slope value, to characterize the transition from one cover type to the next. For example, water-surface elevation in the main channel was set to zero in all cross sections, and all other cover types were assigned relative elevations (i.e., ± change in elevation from the main channel). Hence, relative elevations and slopes, both across and between cover types, were assigned to the identified major land-cover features classified from the hyperspectral imagery. With this combination of data (i.e., survey data and hyperspectral imagery), we were able to produce a DEM of each floodplain of adequate resolution to satisfy inundation modeling at the floodplain scale.

We used the constructed DEM to model inundation patterns for each floodplain at 10 m³ step intervals. We then compared modeled floodplain inundation with actual spatial inundation extent for each reach using georectified aerial imagery collected on 20 June 2002 (~312 m³/s) and 12 April 2003 (~42 m³/s). Similarly, we used airborne video taken on 17 June 1997 (~1048 m³/s) to generate flood inundation extent maps for each reach. These three inundation maps were then used to calibrate stage-discharge relationships for each reach.

We estimated flow velocity at varying discharges by establishing a relationship between velocity and river stage for all reaches. This relationship was developed by conducting multiple ADP floats, thereby producing measures of flow and depth over repeated locations and at various discharge levels. In the inundation model, velocity was then increased according to the regression equations developed from depth-velocity relationships measured in the ADP surveys and data collected from a handheld ADV (Acoustic Doppler Velocimeter). The handheld ADV was used exclusively in shallow waters (<0.8 m) where the boat-mounted ADP loses signal. Flow velocity increase with modeled step increments was controlled by a depth versus Froude number relationship developed from the combined ADV and ADP data.

Once spatially explicit estimates of the energy slope (S), water depth (h), and flow velocity (V) were determined for each water pixel, and where ρ and g are the density of water and the gravitational constant, respectively (Bagnold 1966), stream power (P, W m^{-2}) was then calculated with the following equation:

$$P = (\rho g h S)V. \qquad (1)$$

We used a 30-m-resolution U.S. Geological Survey (USGS) DEM to estimate the slope of the water surface. This is only a rough estimate of S that captures the general slope inflections along the river and is certainly not true for higher discharges. Some of the steeper slopes associated with rapids at the ends of bars and riffle-pool sequences are not completely resolved at the resolution of a 30 m DEM. To resolve some of the inherent variation and loss of detail associated with the coarse spatial resolution of the USGS 30 m DEM, we focused additional effort on the Fisher floodplain, where we conducted detailed surveys of water slope with a Leica survey total station. Surveying the water surface slope for all of the floodplains was not feasible. The Fisher floodplain was then used to establish the cumulative stream power–discharge relationship discussed later herein.

We developed two criteria to determine the corresponding discharge from the flood inundation modeling to a minimum geomorphic threshold for potential parafluvial avulsion and potential orthofluvial avulsion. The criteria are similar for both types:

(1) formation of nodes of flow separation due to connection of parafluvial secondary channels or orthofluvial flood channels with the main channel, and
(2) correlation of zones of high stream power associated with those nodes of flow separation.

The potential parafluvial geomorphic threshold is the discharge when secondary channels, within the parafluvial channel, begin to form and connect with the main channel and are also collocated with zones of relative high stream power. Likewise, a potential orthofluvial geomorphic threshold occurs when orthofluvial flood channels connect to the main channel and are collocated with nodes of flow separation and high stream power.

We then produced graphic representations of stream power at 10 m³ flood-stage intervals from 42 m³/s to 1048 m³/s. We recorded the discharge levels for each signature pattern of power that met our criteria for each floodplain at both parafluvial and orthofluvial levels of discharge. Discharge levels, where further increase in stage did not produce significant changes in the first two criteria, established an upper threshold of avulsion potential. The established cumulative stream power–discharge relationship from the Fisher floodplain was used to investigate how changes in historical flow releases would increase the cumulative power expressed on a floodplain while at the same time not allowing discharge to exceed the 935 m^3/s limit.

RESULTS

Aquatic Data Analysis

We generated velocity versus depth regression Equations 2 and 3 from depth-velocity relationships measured in the ADP surveys (Fig. 2) and the data collected from a handheld ADV (Acoustic Doppler Velocimeter; Fig. 3). Equation 2 was used to simulate velocity for water depths >0.8 m, and Equation 3 was used for water depths <0.8 m, including newly inundated areas as stage increased, where x is the water depth at a given stage (Figs. 2 and 3).

$$y = 0.4493 \ln(x) + 1.3986, \qquad (2)$$

$$y = 1.789(x) - 0.2042. \qquad (3)$$

Flow velocity was not allowed to increase unchecked, with incremental stage increases following Equations 2 and 3, but rather was controlled by a depth versus Froude number ($Fr = V/\sqrt{gh}$) relationship developed from the ADP data, where V is the mean velocity, and h is the average depth determined from each ADP measured velocity profile (Fig. 4, dotted gray line). This Fr number upper limit (hereafter referred to as the Fr cap) was also dependent on the starting depth and velocity, where, for example, shallow rapids could reach a maximum value not to exceed $Fr = 1$ and then would change according to the decreasing depth–Fr cap relationship (gray dashed line, Fig. 4). This is true for all water types, such that after initial velocity was estimated (from classified imagery described later herein) for a given depth, an upper velocity limit for an individual pixel would be obtained according to the Fr cap. For example, a pixel with a starting water depth of 4 m and velocity of, say, 2.5 m/s would reach a maximum flow velocity (based on Eq. 2 following each 10 cm incremental stage increase) governed by a maximum Fr value of 0.4, and then

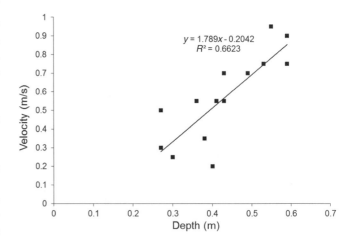

Figure 3. Correlation between measured water depth and flow velocity for water depths <0.8 m across in a single shallow riffle area.

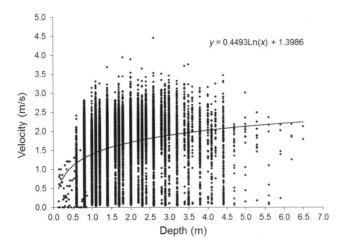

Figure 2. A plot of measured water depth and flow velocity for Acoustic Doppler Velocity Profiler (ADP) profiles (25,308 depth-velocity measures) spread spatially along the length of the river from all floodplains in the study over five discharge levels. The log regression curve of these data was used to determine variation in flow velocity with change in stage.

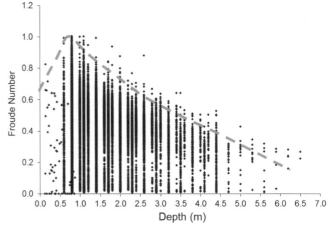

Figure 4. A plot of Froude number versus water depth for all Acoustic Doppler Velocity Profiler (ADP) measures. The dashed gray line represents the Froude maximum used in the geographic information system modeling of flow velocities set slightly lower than the complete data envelope to ensure a conservative estimate in velocity based on depth in the stream power modeling.

the Froude cap would decrease slightly with an increase in water depth (gray line, Fig. 4).

Remote-Sensing Results

The unsupervised classification approach identified areas of similar spectral reflectance for each of the five floodplains (Fig. 5A). All reaches were classified into five depth categories (<0.5, 0.5–1, 1–1.5, 1.5–2.0, and >2.0 m) and five velocity categories (<0.5, 0.5–1.0, 1.0–1.5, 1.5–2, and >2.0 m/s; Figs. 5B and 5C). These initial classifications of water depth and flow velocity provided the basis for modeling depths and velocities at higher stage levels as explained previously. The ranges for each category were a function of the range of depths and flow velocities obtained with the ADP and the resolution that can be achieved from the hyperspectral imagery.

Two methods were used to assess the accuracy of these depth and velocity estimates. Traditionally, the accuracy of a classification is assessed by comparing the reference data (e.g., ADP survey data) with values on the classification map. This method is referred to as the "pure" accuracy assessment because it makes a pixel to pixel comparison of modeling versus ground-truth data. However, in spatial representations of continuous data (e.g., depth and velocity data) where sharp boundaries between classes rarely occur, it is preferable to apply a "fuzzy" assessment of classification accuracy (Gopal and Woodcock, 1994; Muller et al., 1998). The fuzzy assessment allows determination of variance within the reference data and its departure from that classified in adjacent classes (i.e., one class above or one class below the depth or velocity classification being tested). Error matrices were generated for each floodplain and include both the pure and fuzzy assessments (Table 1).

Some of the error between measured and classified depths and velocities was related to rectification error and the distortions discussed previously caused by TIE, as well as error associated with the relative accuracy of the GPS. The accuracy of real-time GPS data varies as a function of the number of satellites available and their position. In addition, both velocity and depth are recorded as the average velocity over a 5 s interval. Thus, depending on flow and geomorphic conditions, an individual profile could be an average of multiple flow and depth conditions for a given GPS location. Hence, the true ADP position can be as much as 3–4 m away from the GPS recorded position, resulting in variance between the measured ADP profile and the hyperspectral imagery and, thus, lower pure class accuracy.

Rectification errors, aircraft turbulence distortions, and GPS errors all contribute to potential misclassifications in the accuracy assessments spread across all five floodplains (Table 1). These errors account for much of the error in the pure accuracy assessment. However, the use of the fuzzy assessment homogenizes these effects by evaluating classification within the context of neighboring classes. While the fuzzy assessment may overestimate the classification accuracy, the pure assessment underestimates it. Despite the various sources of potential error, hydrologic and geomorphic structure (i.e., depth and velocity), and the associated aquatic habitats (i.e., pools, riffles, rapids, and shallows) all appear in appropriate juxtapositions and orientations in river channels and distributed across the floodplain in logical places,

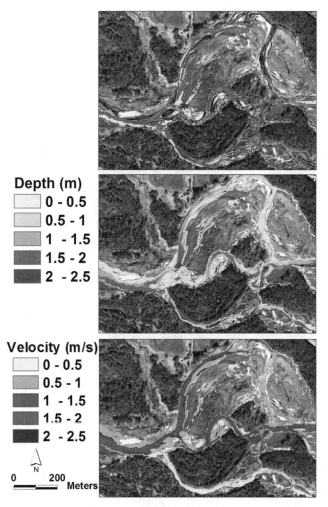

Figure 5. (A) Unsupervised classification of hyperspectral data extracting spectral reflectance characteristics of water. These data illustrate the variation in spectral reflectance used to classify hydraulic characteristics. Acoustic Doppler Velocity Profiler (ADP) data were distributed in the geographic information system environment to aggregate classes and assign unique depth (B) and velocity (C) categories.

TABLE 1. ACCURACY ASSESSMENT FOR ALL REACHES AT 142 m^3/s; SUMMARIZED AS PURE AND FUZZY PERCENTAGES

Floodplain	Depth		Velocity	
	Pure (%)	Fuzzy (%)	Pure (%)	Fuzzy (%)
Swan	60	97	33	74
Conant	53	89	43	85
Fisher	61	91	62	88
Heise	72	94	52	90
Twin	47	86	50	87

which we confirmed through direct observation in the field (Figs. 5B and 5C).

Modeling Results

Estimates of flow velocity for the flooding scenarios were based on the initial depth and velocity classification generated from the September low-flow imagery (142 m^3/s) and then increased according to the regression Equations 2 and 3. Therefore, in our modeling approach, each pixel starts with a given depth and velocity based on the classification of the starting imagery discharge of 142 m^3/s, and then using 10 cm stage increments, velocity is estimated to a discharge of 1048 m^3/s (Fig. 6).

We analyzed the accuracy of the modeled velocities and depths by comparing the measured depths and velocities obtained from the ADP surveys at 325 m^3/s (Tables 2 and 3). Using the stage-discharge relationships, the 325 m^3/s discharge corresponded to a stage increase on average for all floodplains of 0.5 m. Using the modeled depths and velocities for a 0.5 m stage increase, we developed error matrices from the ADP survey

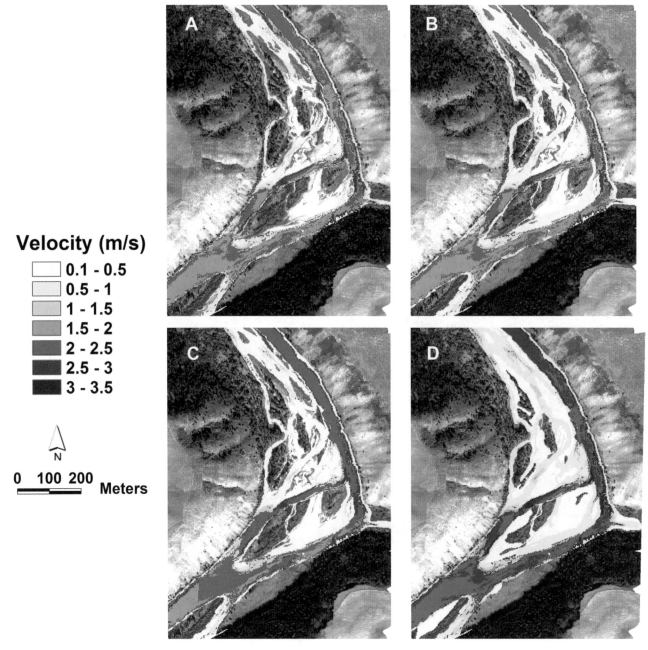

Figure 6. Plot of the spatial distribution of modeled flow velocity for discharges of (A) 226 m^3/s, (B) 325 m^3/s, (C) 566 m^3/s, and (D) 1048 m^3/s in the lower part of the Conant floodplain.

TABLE 2. AN EXAMPLE OF ERROR ASSESSMENT TABLES FOR EACH CLASS OF MODELED DEPTH VERSUS MEASURED ACOUSTIC DOPPLER VELOCITY PROFILER DATA FOR THE CONANT VALLEY FLOODPLAIN

Conant depth 325 m³/s Classified (m)	Reference (m)				Classified total	User's accuracy pure (% correct)	User's accuracy fuzzy (% correct)
	0–1	1–1.5	1.5–2	>2			
0–1	19	7	11	3	40	47.50	65.00
1–1.5	17	10	3		30	33.33	100.00
1.5–2	25	63	77	0	174	44.25	85.63
>2	1	15	131	196	343	57.14	95.34
Reference total	62	95	222	208	587		
					0.51448		
Producer's accuracy pure (% correct)	30.65	10.53	34.68	94.23	532		
					0.9063		
Producer's accuracy fuzzy (% correct)	58.06	84.21	95.05	98.56			

Overall classification pure = 51.45%.
Overall classification fuzzy = 90.63%.

TABLE 3. AN EXAMPLE OF ERROR ASSESSMENT TABLES FOR EACH CLASS OF MODELED VELOCITY VERSUS MEASURED ACOUSTIC DOPPLER VELOCITY PROFILER DATA FOR THE CONANT VALLEY FLOODPLAIN

Conant velocity 325 m³/s Classified (m/s)	Reference (m/s)					Classified total	User's accuracy pure (% correct)	User's accuracy fuzzy (% correct)
	0–0.5	0.5–1	1–1.5	1.5–2	>2			
0–0.5	2		4			13	69.23	69.23
0.5–1		1	3	3	1	8	12.50	50.00
1–1.5	2	6	5	6	3	21	23.81	80.95
1.5–2		2	33	95	167	297	31.99	99.33
>2		6	18	59	164	247	66.40	90.28
Reference total	11	15	63	163	334	586		
						0.4676		
Producer's accuracy pure (% correct)	81.82	6.67	7.97	38.28	49.10	548		
						0.93515		
Producer's accuracy fuzzy (% correct)	81.82	46.67	65.08	98.16	99.10			

Overall classification pure = 46.76%.
Overall classification fuzzy = 93.52%.

obtained at 325 m³/s to validate the modeled results (Tables 2 and 3). Our modeled estimates of flow velocity are in the same range of accuracy we found for the original classification of the hyperspectral imagery at 42 m³/s. Unfortunately, we were not able to access the accuracy of the modeled flows above 325 m³/s as this was the highest flow measured with the ADP. However, we feel that the error is kept to an acceptable minimum scaled against the broad range in depth/flow categories and the Fr cap constraints and errors associated with collection and classification of the hyperspectral imagery.

This approach allows us to make the best estimates possible of velocity based on a step-wise increment in stage to arrive at constrained yet reasonable values for stream power for each within-channel pixel of the hyperspectral image based on the classification of depth, velocity, and DEM-derived slope. Combining our two avulsion node criteria with the step-wise stream power inundation modeling resulted in plots of parafluvial and orthofluvial avulsion discharge levels for each floodplain based on their established stage-discharge relationships (Fig. 7).

These types of plots were made for 10 cm incremental increases in stage for each floodplain. Because we calibrated each individual floodplain stage and discharge relationship, we were able determine the spatial distribution of potential geomorphic work for each corresponding discharge. Integral to our thesis of stream power and geomorphic work is that when nodes of flow separation correlate with zones of high stream power, avulsion potential exists. Therefore, we can quantify important geomorphic thresholds by determining what discharge level is necessary to initiate a channel avulsion event, be it creation of a secondary

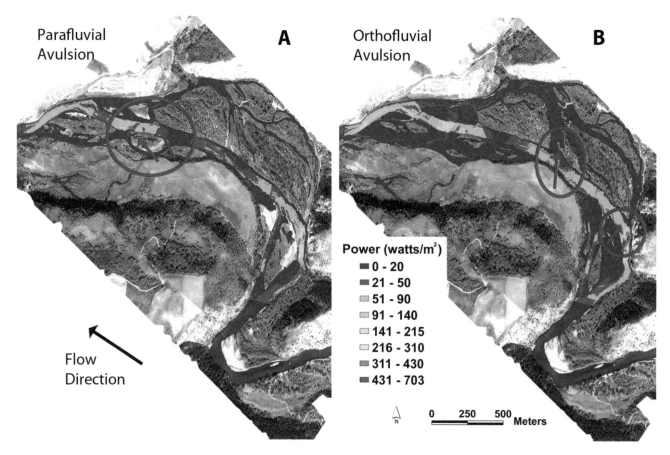

Figure 7. Plot of the spatial distribution of stream power (watts/m^2) for the Fisher floodplain at discharge levels of 566 m^3/s (A) and 962 m^3/s (B) required to create and maintain secondary channels throughout the parafluvial (A) and orthofluvial (B) zones. The red circle indicates areas where secondary channels are actively flowing and the main channel has relatively high stream power associated with those nodes of flow separation (red arrows).

channel in the parafluvial zone or a major avulsion involving the orthofluvial zone (Table 4).

In order to address discharge duration as a factor affecting geomorphic change on the study floodplains, we examined the relationship between discharge and total cumulative power applied to the floodplain. We determined the relationship of discharge to cumulative power (the summation of stream power among all pixels comprising the channels of the floodplain) for ~25 m^3/s increments from 42 m^3/s to 1048 m^3/s for the Fisher floodplain, where we had the total station level of detail for slope measures (Fig. 8). We then used a best-fit, third-degree polynomial regression equation to correlate stream power across discharges:

$$Y = (3 \times 10^{-8})x^3 - (5 \times 10^{-5})x^2 + 0.0461x + 0.1296 \ r^2 = 0.99. \quad (4)$$

Based on our earlier analysis of the discharges necessary for parafluvial avulsion, we estimated daily power for the Fisher floodplain by comparing the annual distribution of power for various historical discharge events as discharge exceeded 425 m^3/s. We selected 425 m^3/s as geomorphic threshold discharge because this was the average level determined to initiate parafluvial avulsions for all floodplains (Table 4).

In order to evaluate the cumulative power approach, we applied the EBSM criteria to postdam discharge regimes and compared those with predam discharge regimes. The EBSM criteria call for changing the flow release from the dam, which historically was characterized by early releases of water prior to the natural flood season, and adjusted that release of water to

TABLE 4. COMPARISON OF THE RANGE OF DISCHARGES NECESSARY TO PROMOTE PARAFLUVIAL AND ORTHOFLUVIAL GEOMORPHIC WORK IN THE FORM OF CHANNEL AVULSION

Floodplain	Parafluvial (m^3/s)	Orthofluvial (m^3/s)
Swan	142 to 538 (340)	566 to 1048 (807)
Conant	311 to 538 (425)	566 to 736 (651)
Fisher	255 to 680 (467)	736 to 963 (850)
Heise	142 to 595 (468)	595 to 1048 (821)
Twin	468 to 509 (439)	623 to 963 (793)
Total means	255 to 566 (425)	623 to 963 (793)

Note: The discharge numbers in parentheses are the mean values.

occur during the natural flood season (i.e., late May early June) in an attempt to mimic a more "normative" seasonal flow (Fig. 9). Timing of flow release patterns was held to the variance of water flowing into the reservoir, with the only limit being that maximum discharge was held below the 935 m³/s level to satisfy contractual obligations and outlet design.

For predam versus postdam comparison, we applied the power analysis (summation of results from Eq. 4) to the daily discharge for all years to generate an annual cumulative power value. We selected the 20 water years with the highest total annual discharge prior to dam construction (Fig. 10, black squares) and then the 10 water years with the highest total annual discharge after dam construction (Fig. 10, gray triangles). Although total volumes of water in the before dam and after dam years were similar, there was a significant reduction in power available to do geomorphic work in the after-dam years (Fig. 10). Clearly, this was mainly due to the successful flood-control strategies that have been integral to the management practices and general operation of Palisades Dam.

We next examined how duration of the annual flood volume of water can be used to maximize the available stream power to do the geomorphic work by first applying the EBSM criteria (i.e., Fig. 9 example) to each of the 10 postdam discharge regimes. These results could then be compared to the total annual cumulative power, when 425 m³/s is achieved or exceeded among each of the 20 highest discharge volume water years prior to dam operation (Fig. 10, black squares) and with the 10 largest flood years for the dam operation after dam construction (Fig. 10, open circles).

Finally, to assess how the river behaved with reference to the EBSM protocols applied in 2004 and 2005, we collected airborne imagery on 28 September 2005 and georectified that data to directly compare channel form and spatial change for each of the five floodplains (Fig. 11). Although both 2004 and 2005 were relatively dry water years, the minimum threshold, which we predicted would achieve geomorphic change, was achieved. We observed four distinct avulsion locations on the Twin floodplain (two are highlighted in Fig. 11). Avulsion processes occurred at sites that were identified in Hauer et al. (2004) as likely sites of avulsion if discharge would approach a minimum geomorphic threshold discharge of 425 m³/s to initiate gravel-bed movement and begin creating parafluvial avulsions (Fig. 11). Several avulsion sites were observed in the Twin floodplain (Fig. 11, lower panel) and illustrate corresponding geomorphic work possible for stream power levels reaching ~566 m³/s (upper panels) compared with the realized change in channel configuration that occurred between September 2002 and September 2005.

DISCUSSION

Natural unregulated floodplains are among the most dynamic landforms on the planet. Not only are natural floodplains among

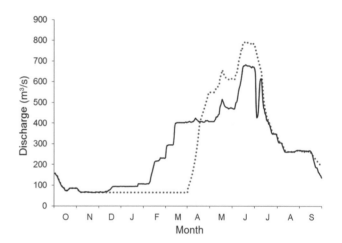

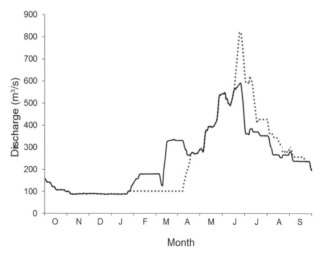

Figure 9. Two plots showing examples of postdam hydrographs (solid lines) compared with Ecologically Based Systems Management (EBSM) altered flow regimes (dotted lines) illustrating a shift to more normative flow regimes by releasing water later in April rather than in January and February.

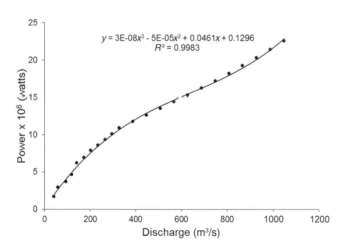

Figure 8. Relationship between discharge and total available stream power on the Fisher floodplain.

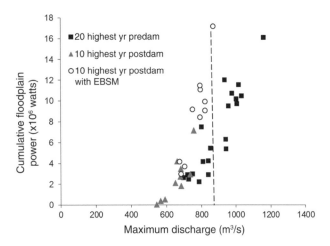

Figure 10. Cumulative annual power (watts/floodplain) on the Fisher floodplain for the 20 yr of the highest water volume discharge prior to dam construction (black square data points), for 10 yr of the highest water volume discharge after dam construction (gray triangle data points), and for 10 yr of the highest water volume discharge after dam construction with altered Ecologically Based Systems Management (EBSM) discharge scenarios applied to the annual hydrographic regime (open circle data points). Dotted line represents 850 m³/s.

the most biologically complex and diverse landscapes on Earth, but their total worldwide economic value has been calculated at nearly U.S.$4 trillion annually (Tockner and Stanford, 2002). With increasing human population along river corridors, floodplains are rapidly being degraded or lost altogether. They are now among the most threatened landscapes worldwide, in part because of the construction and operation of dams. Dams have been built to harness rivers by storing water and controlling floods. Hence, the ability of the river to do geomorphic work on a floodplain, which has an essential role in maintaining the shifting habitat mosaic fundamental to the ecological integrity and sustainability of the river, has been greatly diminished. The ability to reregulate the flow of water from dams to best accommodate normative flow (Stanford et al., 1996; Poff et al., 1997) while staying within the contractual obligations of operation provides an opportunity to allow rivers to do the work of restoration within the existing societal constraints.

A fundamental factor to that objective is the ability to determine the role of flood duration through the concept of cumulative stream power as presented herein. Using the airborne remote-sensing data coupled with ADP measures of water depth and flow, we have developed an approach to modeling of stream power to determine more precisely the minimum threshold levels of discharge necessary to do the geomorphic work that drives a shifting habitat mosaic and the level of work that can be accomplished through extending the duration of those geomorphic threshold-crossing flows. We assume that the greater the stream power, the greater the amount of geomorphic work, resulting in the regeneration of new gravel surfaces and subsequently new riparian forests, as well as the production of new aquatic habitat (e.g., channels, spring brooks, ponds).

Hence, we developed the modeling approach to identify two geomorphic threshold levels defined by potential channel avulsion activity, first in the parafluvial zone and second in the orthofluvial zone of the floodplain (Fig. 7). Parafluvial avulsion events result in the formation and maintenance of secondary channels within the annually scoured zone of the river. Parafluvial cut-and-fill alluviation and avulsion are the first and minimum geomorphic threshold events needed as part of the long-term sustaining processes of a viable shifting habitat mosaic (Stanford et al., 2005). Orthofluvial avulsion events result from complete abandonment of the main channel, or a splitting of the channel into a new main or secondary channel that flows through orthofluvial region. These events leave the abandoned parafluvial zone to become new off-channel and spring brook habitats of the future orthofluvial zone of the floodplain and set new surfaces to be colonized by riparian vegetation.

In our modeling approach, large, coarse-scale patterns of stream power are partially driven, and hence spatially set, by the 30 m USGS DEM. However, within zones of similar slope, we have great variability in stream power predictions because of the more accurate and spatially explicit modeled estimates of water depth and flow velocity (Fig. 6). We also increased the resolution of our estimates of lateral flooding by comparing modeled flood inundation with actual extent of a flood-wetted floodplain at three different discharge levels from 226 m³/s to 1048 m³/s. From our analysis, we concluded that for all floodplains, on average, a minimum level of 425 m³/s is required for parafluvial avulsion potential to initiate, and that at 566 m³/s, the highest avulsion potential was reached (Table 4). Our orthofluvial threshold analysis concluded that on average among all floodplains, a minimum level of 793 m³/s is required for avulsion events to potentially initiate. At 962 m³/s, the highest potential for orthofluvial avulsion was reached (Table 4). Increasing discharge above both maximum potentials would simply increase the rate or intensity at which geomorphic work could be preformed and potentially produce gravel bars with slightly higher elevations. A discharge level of 962 m³/s is above the design of the outlet works for Palisades Dam (935 m³/s) and contractual obligation of BoR to protect against flooding. However, this level of discharge was only a 5 yr return interval event prior to dam construction. Clearly, there is the ability to maintain the shifting habitat mosaic of the Snake River floodplains from a water availability and policy perspective.

Our results show that there is variation in threshold discharge between floodplains (Table 4) and that each floodplain has a slightly different balance among the available stream power to transport sediment (i.e., capacity), the supply of sediment, and limitations of lateral floodplain expansion. The more confined a floodplain, the smaller is the lateral extent and the higher is the stream power for a given discharge. A confined reach results in limited lateral expansion and an increase in the transport capacity of the river above the supply of sediment. As soon as supply becomes greater than transport capacity, gravel bars form, having

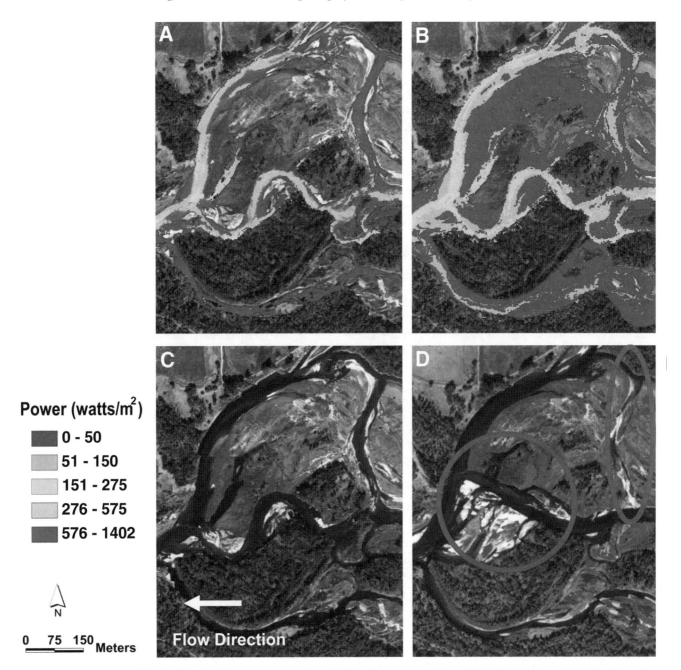

Figure 11. Spatially explicit classification of stream power (watts/m^2) on the upper portion of Twin floodplain at 142 m^3/s (A) and at 566 m^3/s (B). Lower images are from hyperspectral image captured in 2002 (C) and 2005 (D). Red circle indicates sites of specific cut-and-fill channel reconfiguration, and the red ellipse indicates channel avulsion site.

the potential to precipitate further channel avulsions. The Swan and Twin floodplains have broad floodplains due to the unconfined nature of the valley in that river segment, whereas Conant, Fisher, and Heise floodplains have greater degrees of valley confinement and reflect that in the wider range over which secondary channels connect with the main channel.

Increasing the amount of power available to do work on the floodplain, by allowing higher discharges and increasing the duration of discharge above geomorphic threshold levels, results in the creation of more diversity of floodplain habitat. However, this only applies for floodplains that are not limited in lateral expansion by levees and riprap. The Upper Twin, Lower Twin, and the Heise floodplains have the greatest amount of geomorphic modification due to levees. The nodal-stream power analysis showed that these floodplains have limitations on potential orthofluvial avulsion events due to the construction of rock levees that prevent the river from avulsing into old orthofluvial flood channels, even at high discharge. It is generally true for most of the river and channel bottom that stream power increases as discharge increases. However, this may not be true everywhere on the floodplain, nor across all discharges. In some areas, as discharge increases, flow resistance (e.g., large wood jam, newly formed gravel bar) can alter the local energy gradient, reduce velocity, and locally reduce the available stream power. For alluvial channels, this would result in sediment deposition and the formation of gravel bars (Fig. 11).

In the previous sections, we discussed the minimum discharge levels necessary to initiate mobilization of the gravel/cobble-bed sediments of the floodplains in the study area. Geomorphic work not only depends on the peak flows above threshold discharge levels, but also depends on the duration of these discharges, and thus the length of time over which work is done. Costa and O'Connor (1995) demonstrated that for some stream systems in the Cascade Mountains, high discharges of short duration did not do significant amounts of geomorphic work compared to lower-level discharges that were maintained over much longer durations. They concluded that this was mainly a characteristic of well-vegetated floodplains exposed to flashy discharge regimes (e.g., event-driven spates of high peak discharge but very short duration).

Peak discharge among postdam years (except for 1997) has remained under 850 m^3/s with annual cumulative power between 2 and 4×10^6 watts (Fig. 9). Also, among postdam water years, peak discharge achieved the maximum discharges only in the very wet years, approaching the 792 m^3/s threshold required for orthofluvial avulsion, and that level is a mean threshold for the potential to initiate orthofluvial avulsions across all the study floodplains (Table 4). A 962 m^3/s discharge level is required to achieve the *highest* potential for orthofluvial avulsion, which has only been achieved once since dam construction (1997). In contrast, there were 7 years between 1911 and 1956 that exceeded or achieved 962 m^3/s. On the Fisher floodplain, these high and sustained discharges generated annual cumulative power between 9 and 12×10^6 watts (Fig. 10). This likely represents an approximate level of threshold cumulative power, adjusted for floodplain area, necessary to achieve high potential for orthofluvial avulsions among all the floodplains in the study area.

To more fully address the issue of peak flow and duration of flow, we further evaluated the flow regimes of the 10 highest discharge volume years after 1957. To accomplish this, we adjusted the spring discharges occurring before 1 April to be redistributed during the natural spring freshet (Fig. 9). These results demonstrate that power generated from these redistributed flows would have significantly increased geomorphic work and that the cumulative power necessary to sustain parafluvial avulsions would have been easily achieved. Furthermore, potential for orthofluvial avulsion could have been realized without having to release or sustain discharges above 850 m^3/s (Fig. 10, black squares). To reiterate; in this illustration of the data (Fig. 10), the postdam high-water-volume years and the power they derived are expressed as gray triangle data points. The same water years with redistributed discharge regimes, using the same volume of water, generate power to do geomorphic work illustrated by the open circle data points (Fig. 10). By making these adjustments to the discharge regime in Palisades operations, the Snake River achieves discharges in wet and moderately wet years that group between 9 and 12×10^6 watts, similar to the power generated in wet years prior to dam operations (Fig. 10; black square data points). We believe that this is a central issue in reregulated flow regimes and critical to the restoration and sustainability of the Snake River shifting habitat mosaic and the organisms that are dependent on its dynamics.

Hydrologic Regimes in 2003–2005 and Measured Channel Change

Discharge regimes in 2004 and 2005 were based on fairly dry water years in the lower 15% among all years of total annual water volume since construction of Palisades Dam. In 2004, although water volume was very low, a concerted effort was made to follow EBSM flow protocols. In 2004, maximum discharge was 540 m^3/s Similar dry years in the recent past prior to 2004 had seen maximum discharges between 368 and 422 m^3/s. However, note that wet year 2003 preceded EBSM analysis and recommendations in Hauer et al. (2004). The following years sustained flood durations that were conducted with the highest discharge held near 425 m^3/s, resulting in significant channel zone migration in the form of bank erosion, bar formation, and channel avulsions (Fig. 11).

In this geospatial analysis, channel changes were observed in areas associated with the greatest levels of modeled stream power and adjacent to nodes of flow separation and overflow channel connections. These monitoring photos clearly showed that over the 3 yr with the application of EBSM and redistributed flows (Hauer et al., 2004), parafluvial avulsion occurred (May 2004) as our results predicted. This serves as a strong validation to our geomorphic modeling approach and predictions at least for this section of the Snake River. Higher flows for several days are integral to reestablishing the shifting habitat mosaic in the South Fork of the Snake River below Palisades Dam and central to an EBSM approach to river resource management.

CONCLUSIONS

We identified two geomorphic threshold levels defined by potential channel avulsion in the South Fork of the Snake River, first in the parafluvial region and second in the orthofluvial region of the floodplain. From our analysis, we concluded that, on average, for all floodplains, a minimum level of 425 m^3/s is required for parafluvial avulsion potential to initiate, and that at 566 m^3/s, the highest potential was reached. Our orthofluvial threshold analysis concluded that on average among all floodplains, a minimum level of 623 m^3/s is required for avulsion events to potentially initiate, and at 963 m^3/s, the highest potential for orthofluvial avulsion was reached. This level of discharge has only been achieved once since dam construction (1997). In contrast, there were 7 predam years (1911–1956) that achieved or nearly achieved the 963 m^3/s level where the highest potential for orthofluvial geomorphic work is obtained. On the Fisher floodplain, these high and sustained discharges generate annual cumulative power between 9 and 12×10^6 watts. This likely represents an approximate level of threshold cumulative power, adjusted for floodplain area, necessary to achieve the orthofluvial avulsions for all the floodplains in the study area.

Our results demonstrate that parafluvial avulsions could be easily achieved and orthofluvial avulsion could be realized without having to release or sustain discharges above 850 m^3/s, the flood stage above which the BoR is required to restrain. By making adjustments to the discharge regime that fall within the current legal constraints of dam operations and demands on water for irrigation and acceptable levels of flood protection, the Snake River can achieve discharges, 793 m^3/s on average, in wet and moderately wet years that group between 9 and 12×10^6 watts, similar to the power generated in wet years prior to dam operations. We believe that this is a central issue in reregulated flow regimes and critical to the restoration and sustainability of the Snake River shifting habitat mosaic and the organisms that are dependent on its dynamics. These results form the basis of an objective methodology to determine minimum flow releases from dams aimed at reconnecting fluvial processes that maintain the dynamic structure and ecological function of gravel-bed river floodplains while at the same time staying within the current legal constraints of dam operations and demands on water for irrigation and acceptable levels of flood protection. Moreover, the approach could be used to assess downstream fluvial changes following dam removal.

ACKNOWLEDGMENTS

This manuscript benefited greatly from the comments and editorial advice of Desiree Tullos, James E. Evans, and Christopher T. Robinson. We thank Chris Jansen Lute, Program Manager (retired), Water Resources Management, Pacific Northwest Region, U.S. Bureau of Reclamation, for her foresight and diligent management of the research program within the BoR from which this project was funded with grants to F.R. Hauer and M.S. Lorang.

REFERENCES CITED

Aspinall, R.J., Marcus, W.A., and Boardman, J.W., 2002, Considerations in collecting, processing, and analyzing high spatial resolution, hyperspectral data for environmental investigations: Journal of Geographical Systems, v. 4, p. 15–29, doi:10.1007/s101090100071.

Bagnold, R.A., 1966, An Approach to the Sediment Transport Problem from General Physics: U.S. Geological Survey Professional Paper 422-I, 42 p.

Bryant, R.G., and Gilvear, D.J., 1999, Quantifying geomorphic and riparian land cover changes either side of a large flood event using airborne remote sensing: River Tay, Scotland: Geomorphology, v. 29, p. 307–321, doi:10.1016/S0169-555X(99)00023-9.

Burke, M., Jorde, K., and Buffington, J.M., 2009, Application of a hierarchical framework for assessing environmental impacts of dam operation: Changes in streamflow, bed mobility and recruitment of riparian trees in a western North American river: Journal of Environmental Management, v. 90, supplement 3, p. S224–S236, doi:10.1016/j.jenvman.2008.07.022.

Costa, J.E., and O'Connor, J.E., 1995, Geomorphically effective floods, in Costa, J.E., Miller, A.J., Potter, K.W., and Wilcock, P.R., eds., Natural and Anthropogenic Influences in Fluvial Geomorphology: American Geophysical Union Geophysical Monograph 89, p. 45–56.

Gilvear, D.J., Davids, C., and Tyler, A.N., 2004, The use of remotely sensed data to detect channel hydromorphology; River Tummel, Scotland: River Research and Applications, v. 20, p. 795–811, doi:10.1002/rra.792.

Gopal, S., and Woodcock, C., 1994, Theory and methods for accuracy assessment of thematic maps using fuzzy sets: Photogrammetric Engineering and Remote Sensing, v. 60, no. 2, p. 181–188.

Hauer, F.R., Dahm, C.N., Lamberti, G.A., and Stanford, J.A., 2003, Landscapes and ecological variability of rivers in North America: Factors affecting restoration strategies, in Wissmar, R.C., and Bisson, P.A., eds., Strategies for Restoring River Ecosystems: Sources of Variability and Uncertainty in Natural and Managed Systems: Bethesda, Maryland, American Fisheries Society, p. 81–105.

Hauer, F.R., Lorang, M.S., Whited, D., and Matson, P., 2004, Ecologically based systems management: The Snake River—Palisades Dam to Henrys Fork, Final Report to U.S. Bureau of Reclamation Boise, Idaho: Polson, Montana, Flathead Lake Biological Station, Division of Biological Sciences, The University of Montana, 133 p.

Leckie, D.G., Cloney, E., Jay, C., and Paradine, D., 2005, Automated mapping of stream features with high-resolution multispectral imagery: An example of the capabilities: Photogrammetric Engineering and Remote Sensing, v. 71, p. 145–155.

Legleiter, C.J., Roberts, D.A., Marcus, W.A., and Fonstad, M.A., 2004, Passive optical remote sensing of river channel morphology and in-stream habitat: Physical basis and feasibility: Remote Sensing of Environment, v. 93, p. 493–510, doi:10.1016/j.rse.2004.07.019.

Lorang, M.S., and Hauer, F.R., 2006, Fluvial geomorphic processes, in Hauer, F.R., and Lamberti, G.A., eds., Methods in Stream Ecology: San Diego, California, Elsevier, p. 145–168.

Lorang, M.S., Whited, D.C., Hauer, F.R., Kimball, J.S., and Stanford, J.A., 2005, Using airborne multispectral imagery to evaluate geomorphic work across flood plains of gravel-bed rivers: Ecological Applications, v. 15, p. 1209–1222, doi:10.1890/03-5290.

Marcus, W.A., 2002, Mapping of stream microhabitats with high spatial resolution hyperspectral imagery: Journal of Geographical Systems, v. 4, p. 113–126, doi:10.1007/s101090100079.

Marcus, W.A., and Fonstad, M.A., 2008, Optical remote mapping of rivers at sub-meter resolutions and watershed extents: Earth Surface Processes and Landforms, v. 33, p. 4–24, doi:10.1002/esp.1637.

Marcus, W.A., Legleiter, C.J., Aspinall, R.J., Boardman, J.W., and Crabtree, R.L., 2003, High spatial resolution hyperspectral mapping of in-stream habitats, depths, and woody debris in mountain streams: Geomorphology, v. 55, no. 1–4, p. 363–380, doi:10.1016/S0169-555X(03)00150-8.

Muller, S.V., Walker, D.A., Nelson, F.E., Auerbach, N.A., Bockheim, J.G., Guyer, S., and Sherpa, D., 1998, Accuracy assessment of a land-cover map of the Kuparuk River basin, Alaska: Considerations for remote regions: Photogrammetric Engineering and Remote Sensing, v. 64, no. 6, p. 619–628.

Nilsson, C., and Bergeron, K., 2000, Alteration of riparian ecosystems caused by river regulation: Bioscience, v. 50, p. 783–792, doi:10.1641/0006-3568 (2000)050[0783:AORECB]2.0.CO;2.

Petts, G., 1984, Impounded Rivers: Perspective for Ecological Management: Chichester, UK, John Wiley & Sons, 326 p.

Poff, N.L., Allan, J.D., Bain, M.B., Karr, J.R., Prestergaard, K.L., Richter, B., Sparks, R., and Stromberg, J., 1997, The natural flow regime: A paradigm for river conservation and restoration: Bioscience, v. 47, p. 769–784, doi:10.2307/1313099.

Rood, S.B., Taboulchanas, K., Bradley, C.E., and Kalischuk, A.R., 1999, Influence of flow regulation on channel dynamics and riparian cottonwoods along the Bow River, Alberta: Rivers, v. 7, no. 1, p. 33–48.

Rood, S.B., Samuelson, G.M., Braatne, J.H., Gourley, C.R., Hughes, F.M.R., and Mahoney, J.M., 2005, Managing river flows to restore floodplain forests: Frontiers in Ecology and the Environment, v. 3, no. 4, p. 193–201, doi:10.1890/1540-9295(2005)003[0193:MRFTRF]2.0.CO;2.

RSI, 2000, ENVI 3.5 User's Guide, Research Systems: Boulder, Colorado, ENVI, 930 p.

Shafroth, P.B., Friedman, J.M., Auble, G.T., Scott, M.L., and Braatne, J.H., 2002, Potential responses of riparian vegetation to dam removal: Bioscience, v. 52, no. 8, p. 703–712, doi:10.1641/0006-3568(2002)052[0703:PRORVT]2.0.CO;2.

Stanford, J.A., and Ward, J.V., 1993, An ecosystem perspective of alluvial rivers: Connectivity and the hyporheic corridor: Journal of the North American Benthological Society, v. 12, no. 1, p. 48–60, doi:10.2307/1467685.

Stanford, J.A., Ward, J.V., Liss, W.J., Frissell, C.A., Williams, R.N., Lichatowich, J.A., and Coutant, C.C., 1996, A general protocol for restoration of regulated rivers: Regulated Rivers: Research and Management, v. 12, p. 391–413, doi:10.1002/(SICI)1099-1646(199607)12:4/5<391::AID-RRR436>3.0.CO;2-4.

Stanford, J.A., Lorang, M.S., and Hauer, F.R., 2005, The shifting habitat mosaic of river ecosystems: Verhandlungen—Internationale Vereinigung für Theoretische und Angewandte Limnologie, v. 29, no. 1, p. 123–136.

Tockner, K., and Stanford, J.A., 2002, Riverine flood plains: Present state and future trends: Environmental Conservation, v. 29, no. 3, p. 308–330, doi:10.1017/S037689290200022X.

Tou, J.T., and Gonzalez, R.C., 1977, Pattern Recognition Principles: Reading, Massachusetts, Addison-Wesley, 377 p.

Whited, D.C., Stanford, J.A., and Kimball, J.S., 2003, Application of airborne multispectral digital imagery to characterize the riverine habitat: Verhandlungen—Internationale Vereinigung für Theoretische und Angewandte Limnologie, v. 28, p. 1373–1380.

Winterbottom, S.J., and Gilvear, D.J., 1997, Quantification of channel bed morphology in gravel-bed rivers using airborne multispectral imagery and aerial photography: Regulated Rivers: Research and Management, v. 13, p. 489–499, doi:10.1002/(SICI)1099-1646(199711/12)13:6<489::AID-RRR471>3.0.CO;2-X.

Manuscript Accepted by the Society 29 June 2012

Assessing stream restoration potential of recreational enhancements on an urban stream, Springfield, Ohio

John Ritter
Kelly Shaw
Aaron Evelsizor
Katherine Minter
Chad Rigsby
Kristen Shearer

Departments of Geology and Biology, Wittenberg University, P.O. Box 720, Springfield, Ohio 45501, USA

ABSTRACT

The stream restoration potential of recreational modifications made to lowhead dams on an urban reach of Buck Creek, in Springfield, Ohio, is dependent on constraints imposed by the urban infrastructure on stream grade. A privately led initiative to improve the recreational potential of a 9 km reach of Buck Creek and its tributary Beaver Creek includes the modification of four lowhead dams. The hydraulic heights of these dams will be replaced with a series of v-shaped drop structures engineered to create hydraulics conducive to kayak play. The drop structure is a constructed channel constriction composed of a hard step in the long stream profile immediately upstream of a scour pool, forming a morphologic sequence of constriction, step, and pool. In this study, we assess the potential benefits of these changes for urban stream restoration. Two of the dams have been modified to date. Stream quality, as measured by the qualitative habitat evaluation index (QHEI), dissolved oxygen of surface and substrate water, and the pollution tolerance index (PTI), increased at the Snyder Park site but decreased at the Art Museum site. Stream quality increased at the Snyder Park site, where stream grade could be lowered upstream of the lowhead dam, but decreased at the Art Museum site, where grade upstream of the lowhead dam had to be maintained because of water and wastewater utilities buried in the channel bed. Where stream grade is lowered in the former impoundment, sand and gravel deposits upstream of the constriction are not embedded with finer particles and organic matter. Increased QHEI values, particularly the substrate metric, and greater abundance and diversity of pollution-intolerant macroinvertebrates, supported by higher dissolved oxygen in the substrate water, characterize the Snyder Park site. At the

Art Museum site, the v-shaped constriction increased the upstream impounded area. The substrate has become embedded with fine sands, silts, and organics, lowering QHEI values, dissolved oxygen is critically low in the substrate, and macroinvertebrate populations are more pollution-tolerant. The results highlight the significance of stream grade if stream restoration is to be incorporated into the engineering design of in-stream recreational features.

INTRODUCTION

Many urban stream reaches in the United States suffer from contaminated sediments, degraded water quality, and lost habitat, conditions exacerbated by lowhead dams. These conditions limit recreational and other economic uses of urban streams. Removal or modification of dams along urban streams has the potential to not only provide recreational opportunities but also to restore stream function and the environmental quality of the stream corridor. Dam removal has emerged as a major environmental issue during the past decade. With a typical life expectancy of 50 yr (American Rivers and Trout Unlimited, 2002), dams built from 1950 to 1970, a period of accelerated dam building, are increasingly being removed rather than rehabilitated (Doyle et al., 2003a). This is especially true for small, lowhead dams (i.e., generally having hydraulic heights <7.6 m) because the cost of removal is typically three times less than the cost of repair (Born et al., 1998; ICF Consulting, 2005). The most commonly cited reason for lowhead dam removal is ecology, from restoring fish and wildlife habitat and fish passage to improving water quality and remediating environmental problems (ICF Consulting, 2005). For small, lowhead dams in urban settings, removal may be complex, even prohibitive, because of the urban infrastructure. For example, roads and built structures may necessitate a static stream planform, and utilities buried in the stream channel (e.g., water and wastewater) may require a stable grade. Stream restoration based on natural channel design (e.g., Rosgen, 1996) is not feasible in this environment with these constraints. However, as cities across the country invest in innovative places such as river corridors to create green space (Harnik, 2010), the potential for recreational modifications of lowhead dams to serve the dual purpose of stream recreation and restoration needs to be examined. In-stream recreational features can be designed to benefit stream process in the absence of natural stream form.

A privately led initiative to improve the recreational potential of a 9 km reach of Buck Creek and its tributary Beaver Creek, in Springfield, Ohio, includes modification of four lowhead dams with hydraulic heights up to 2 m (Fig. 1). The four lowhead dams are referred to by their location as Snyder Park, Art Museum, and Municipal Stadium on Buck Creek, and Beaver Creek on Beaver Creek (Fig. 1). The hydraulic heights of these dams will be replaced with a series of v-shaped drop structures engineered to create hydraulics conducive to kayak play. The drop structure is a constructed channel constriction composed of a hard step in the long stream profile immediately upstream of a scour pool, forming a morphologic sequence of constriction, step, and pool. The modifications range from complete removal and replacement by drop structures in order to maintain local stream grade to reshaping existing structures to constrict flow to a central point on the dam and locally modify grade to allow for safe passage through the structure. Modification has recently been completed at two of the lowhead dam sites, which form the basis of this study. The objectives of our work are (1) to measure changes in riparian and in-stream habitat, water quality, and macroinvertebrate populations resulting from the lowhead dam modifications; and (2) to assess the stream restoration potential of recreational enhancements on an urban stream.

STUDY AREA

Buck Creek has been an integral part of Springfield's early history, powering most of the 65 mills early in the twentieth century along its reach through Springfield (Prince, 1922), and, until 1958, providing freshwater for the city's public water supply. The lowhead dams associated with the mills have been removed and the remaining dams were built to protect water and wastewater utilities crossing Buck Creek and Beaver Creek, buried in the channel bed upstream of the dams or encased in the concrete comprising the dam. Three of the four dams, Art Museum, Municipal Stadium, and Beaver Creek (Fig. 1A), currently protect utilities. At these sites, a stable stream grade upstream of the dam is critical. The Snyder Park dam (Fig. 1A) protected utilities in the past, but they are no longer located at the dam site; however, a stable grade is still required to protect a wastewater utility ~335 m upstream of the dam. Engineering designs for the dam modifications focused on safety, navigability, and recreation, and stream environment and function were a secondary consideration. The Snyder Park and Art Museum dams were modified in 2010 and are the focus of this paper. Modifications to Municipal Stadium were completed in 2012 and Beaver Creek will be modified in 2013. The study reach is at the downstream end of the Buck Creek watershed (Figs. 1A and 1B).

The contributing area of the Buck Creek watershed upstream of the Snyder Park lowhead dam is 362.9 km^2 (Fig. 1B; Table 1). It is composed of three subwatersheds: the area draining into and from the C.J. Brown Reservoir on Buck Creek upstream of its confluence with Beaver Creek, referred to as the upper Buck Creek subwatershed, the Beaver Creek subwatershed, and the Springfield subwatershed (Fig. 1B; Table 1). The upper Buck

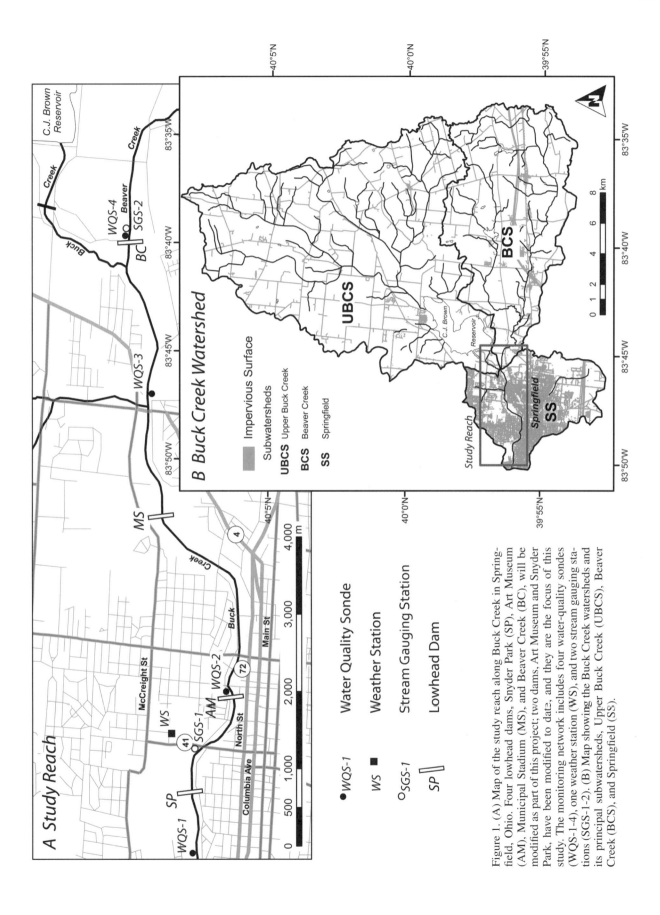

Figure 1. (A) Map of the study reach along Buck Creek in Springfield, Ohio. Four lowhead dams, Snyder Park (SP), Art Museum (AM), Municipal Stadium (MS), and Beaver Creek (BC), will be modified as part of this project; two dams, Art Museum and Snyder Park, have been modified to date, and they are the focus of this study. The monitoring network includes four water-quality sondes (WQS-1-4), one weather station (WS), and two stream gauging stations (SGS-1-2). (B) Map showing the Buck Creek watersheds and its principal subwatersheds, Upper Buck Creek (UBCS), Beaver Creek (BCS), and Springfield (SS).

TABLE 1. MORPHOLOGIC AND LAND-USE AND LAND-COVER CHARACTERISTICS OF THE BUCK CREEK WATERSHED AND ITS SUBWATERSHEDS

		Buck Creek Watershed	Subwatersheds		
			Upper Buck Creek	Beaver Creek	Springfield
Morphology	Area (km^2)	362.9	214.8	102.4	45.7
	% of total area	100	59	28	13
	Average relief (m)	129	114	104	72
	Average slope (%)	3.4	3.5	3.2	3.3
Land use or cover	Urban or built (%)	8.7	1.5	3.2	54.6
	Park, wetland, or water (%)	11.2	9.3	6.9	29.8
	Forest (%)	8.0	8.7	6.8	7.0
	Agriculture (%)	70.0	78.6	79.7	8.4
	Other (%)	2.1	1.9	3.4	0.2

Creek subwatershed is the largest of these subwatersheds, constituting 59% of the total watershed area. It is dominated by agricultural land use and cover (Table 1). The C.J. Brown Reservoir is at the downstream end of this subwatershed. Reservoir outflow is controlled by the U.S. Army Corps of Engineers (USACE) and rarely exceeds 22.6 m^3/s (800 ft^3/s). The USACE adjusts the release from the reservoir between 6:30–10:30 a.m., and the change in release is easily recognized at the downstream gauging station as a step function, up or down, ~30–45 min later (Fig. 2). The Beaver Creek subwatershed is 28% of the total watershed area and is dominated by agricultural land use and cover (Table 1). A gauging station and water-quality sonde on Beaver Creek are located ~150 m upstream of the lowhead dam on Beaver Creek and ~365 m above its confluence with Buck Creek (Fig. 1). Flow from the Beaver Creek subwatershed is not regulated and generally dominates the downstream hydrograph (Fig. 2). The 2 yr discharge for Beaver Creek subwatershed is ~41.7 m^3/s (1472 f^3/s) based on the historical record (1943–1976) for a U.S. Geological Survey gauging station on Beaver Creek upstream of its confluence with Buck Creek (HUC 05080001).

The Springfield subwatershed includes the urbanized area of Springfield, which drains directly to Buck Creek and comprises ~13% of the total watershed area. Urban land use and cover account for 55% of the Springfield subwatershed area (Table 1). Because Springfield has a combined sewer system, 34 combined sewers discharge wastewater and runoff from impervious surfaces directly into Buck Creek during wet weather events. Hydrographs recorded at the Plum Street gauging station (Fig. 1A, SGS-1) typically have two different peaks. The first hydrograph peak is flashy, peaking as the rainfall tapers, and represents input from the Springfield subwatershed (Fig. 2). A second peak, ~5–9 h after the first peak, reflects stormflow from the Beaver Creek subwatershed (Fig. 2).

METHODOLOGY

The hydrology and water quality of the study reach are monitored with a telemetric high-frequency network of four water-quality sondes equipped with YSI 6920 V2–2 Sondes with temperature, conductivity, pH, oxidation reduction potential, optical dissolved oxygen, and optical turbidity sensors, a Vaisala Weather Transmitter WXT520 providing wind speed, wind direction, barometric pressure, relative humidity, liquid precipitation, and air temperature, and two stream gauging stations equipped with AccuStage vented water-level sensors (Fig. 1A). Two of the water-quality sondes, the weather station, and one gauging station were installed several months prior to the first dam modification. Measurements are made at 15 min intervals and downloaded hourly to the Internet (http://v4.wqdata.com/webdblink/buckcreek.php).

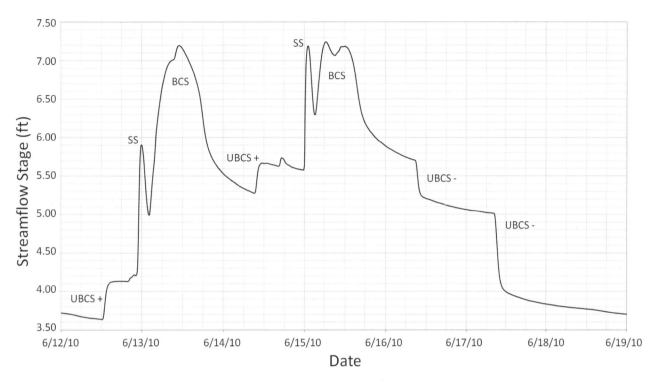

Figure 2. Stormflow hydrograph at SGS-1 (Fig. 1A) resulting from two separate rainfall events totaling as much as 114 mm (4.5 in) of rainfall. The hydrograph illustrates the relative magnitude and timing of stormwater runoff from Springfield subwatershed (SS) and Beaver Creek subwatershed (BCS). Increases (UBCS +) and decreases (UBCS –) in releases from the C.J. Brown Reservoir and the Upper Buck Creek subwatershed are also noted. Because of the changing nature of the stage-discharge rating curve for SGS-1, stage was not converted to discharge, but peak discharge for this hydrograph was estimated to be as great as 51.6 m³/s (1830 f³/s).

Each dam site was divided into three different zones for further field sampling, the upstream or undisturbed riffle, the impounded zone, and the downstream recovery riffle (Fig. 3A). The qualitative habitat evaluation index (QHEI) (Ohio EPA, 2006) was used to assess quality of stream habitat in each of the zones at the respective dam sites. The QHEI is a physical habitat index that provides a quantitative evaluation of the habitat characteristics that are important to fish communities (Ohio EPA, 2006). It is based on six channel and riparian characteristics: channel substrate, in-stream cover, riparian characteristics, channel condition, pool/riffle quality, and gradient and drainage area. The modifications made at the study sites were primarily physical changes in channel morphology and the longitudinal profile, engineered to optimize hydraulic conditions for whitewater recreation. While the modifications created potential for habitat, they were not designed for habitat. Because the relation between the QHEI metrics and the index of biotic integrity has been investigated extensively for streams in Ohio (Rankin, 1989), the QHEI methodology offered an objective evaluation of the potential impact of changes on channel and riparian characteristics and biotic quality of the sites. In addition, the Ohio Environmental Protection Agency (Ohio EPA, 2009) had recently created target values for QHEI and several of the individual metrics as part of its total maximum daily load assessment for the Mad River, of which Buck Creek is tributary. QHEI scores made prior to dam modifications were submitted to the Ohio EPA to secure the Section 401 Water Quality Certification permit for construction.

Aquatic macroinvertebrates were sampled at each dam site in June–July 2009 and 2010 using a 1400 μm mesh D-frame net that measured 30 cm along the bottom and 22 cm tall at the widest point. Three subsamples were collected from the upstream or undisturbed riffle, the impounded zone, and the downstream recovery riffle using the kick method. The pollution tolerance index (PTI) (Hippensteel, 2005), based on order-level identification of macroinvertebrates, was used to quantify the impact of dam modification on macroinvertebrate populations. The PTI method, though not as robust a method as family- or higher-level indexes such as the invertebrate community index (Ohio EPA, 1987), was used because it is a popular general tool for assessing stream health and can be replicated from year to year by biology and environmental science students with adequate training.

At the time macroinvertebrate samples were collected, the temperature, pH, conductivity, and dissolved oxygen of surface and substrate water sampled from the hyporheic zone were also measured at three sites within in each zone. Subsequent measurements were made at each dam site during

A Premodification

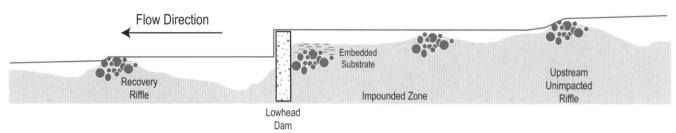

B Postmodification - with constrained vertical grade (e.g., Art Museum)

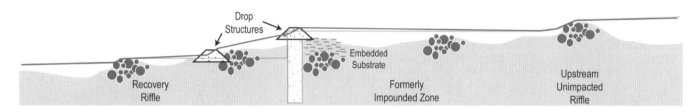

C Postmodification - with unconstrained vertical grade (e.g., Snyder Park)

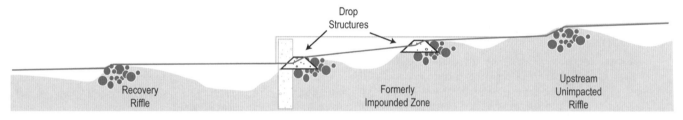

Figure 3. Schematic longitudinal profile of a stream reach showing the three different zones associated with each lowhead dam, the upstream undisturbed riffle, the impounded zone, and the downstream recovery riffle (A) prior to lowhead dam modification, (B) a situation similar to the Art Museum, where the upstream drop structure is constructed at the same location and/or downstream of the original dam, and (C) a situation similar to Snyder Park, where the upstream drop structure is constructed within the formerly impounded area.

the late summer and early fall. Temperature, pH, and conductivity were measured with a YSI Pro-Plus meter, and dissolved oxygen was measured with a YSI 600 OMS sonde with a 6150 YSI ROX optical dissolved oxygen sensor. The substrate water was sampled by driving a well point with a 2.5 cm perforated length of casing into the shallow channel substrate. The nonperforated length of casing was bailed to remove 2–3 times the volume of water in the casing to ensure that substrate water was being sampled. In order to minimize sample disturbance, especially with respect to dissolved oxygen, measurements were made in place by inserting the probes into the casing.

RESULTS

Dam Modifications

The hydraulics created on the downstream side of the Art Museum and Snyder Park dams were retentive prior to modification, and dangerous for swimmers and boaters. The modifications were engineered to eliminate the trapping nature of these hydraulics by creating hydraulics conducive to kayak play with eddies that push objects toward the banks. They distribute the hydraulic height associated with a single, broad-crested, rectangular weir over the course of two to four drop structures, essentially

Figure 4. Aerial photograph and ground view looking upstream at the lowhead dam at the Art Museum prior to (A and B) and following (C and D) modification. A second drop structure was created downstream of that shown in D.

Figure 5. Aerial photograph and ground view looking upstream at the lowhead dam at Snyder Park prior to (A and B) and following (C and D) modification. The modified site consists of four drop structures.

triangular weirs, spaced 36.5–45.7 m apart. Each drop structure is a constructed channel constriction composed of a hard step in the long stream profile immediately upstream of a scour pool, forming a morphologic sequence of constriction, step, and pool.

In-stream work on the Art Museum site began in September 2009 and was completed by November (Fig. 4). The 1.2 m hydraulic height of the Art Museum Dam was replaced by two drop structures. The base of the original dam was modified as the upstream-most drop structure and, though reshaped to form the v-shaped constriction, remains in the same location and at the same elevation (Figs. 4B and 4D). One additional drop structure was constructed downstream of the original lowhead dam (Fig. 4C). A schematic illustration of this site is shown in Figure 3B. In-stream work on Snyder Park began in December 2009 and was completed in April 2010 (Fig. 5). The 2 m hydraulic height of the Snyder Park dam was replaced by four drop structures (Fig. 5C). The downstream-most drop structure was constructed just upstream of the original lowhead dam and three additional drop structures were constructed upstream of it, within the former impounded area (Figs. 5C and 5D). A schematic illustration of the drop structures relative to the impounded area is shown in Figure 3C.

In plan view, the drop structures create multiple constrictions in stream width. This is illustrated well at Snyder Park, where the width of the channel was between 34.7 and 39.6 m in the impounded area upstream of the lowhead dam (Fig. 5C). With the addition of drop structures, it varies between 7.3 m at the constrictions to 42.7 m in the pools. As described already, the drop structures distribute the hydraulic height of the dam over a finite reach of the river either downstream (Art Museum) or upstream (Snyder Park) of the lowhead dam. Upstream of the upper drop structure, stream stage has increased in part because the triangular design of the drop structure raises the upstream water level relative to the rectangular nature of the lowhead dam (Fig. 6). At the Plum Street gauging station, located between the Snyder Park and Art Museum sites (SGS-1 in Fig. 1A), located ~619 m upstream of the Snyder Park dam and now only 475 m upstream of the upstream-most drop structure at the Snyder Park site, the stage discharge rating curve is currently being reestablished by field measurements of discharge (Fig. 6). The difference in stage

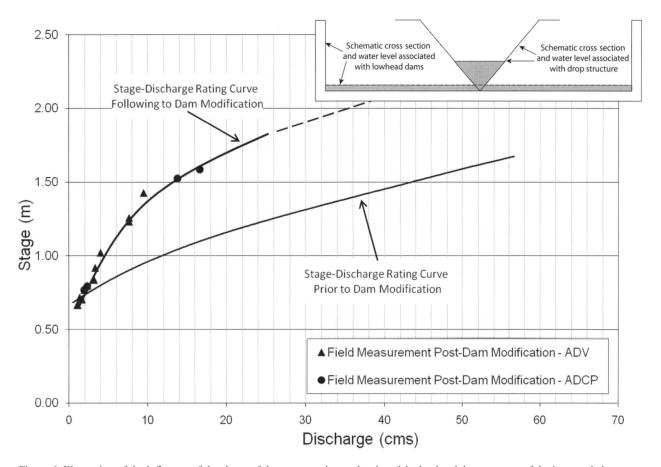

Figure 6. Illustration of the influence of the shape of the cross section at the site of the lowhead dam on stage of the impounded water upstream of the dam. The change from a rectangular to triangular cross section causes stage of the impounded water to rise. The Plum Street gauging station is just upstream from the Snyder Park site (SGS-1 in Fig. 1A). ADV—acoustic Doppler velocimeter; ADCP—acoustic Doppler current profiler.

resulting from the modifications is negligible for lower base-flow discharges of late summer and early fall, but it increases with greater discharge. For example, the stage related to a discharge of 13.6 m^3/s increased by ~0.5 m following modification (Fig. 6). At the Art Museum, the stage increase at the upstream constriction is apparent even at base flow, but it has not been measured.

Physical Characteristics of the Stream Reaches and QHEI Analysis

QHEI was measured in the undisturbed upstream riffle, the impounded zone, and the downstream recovery riffle prior to and following modifications (Fig. 7; Table 2). Individual metrics

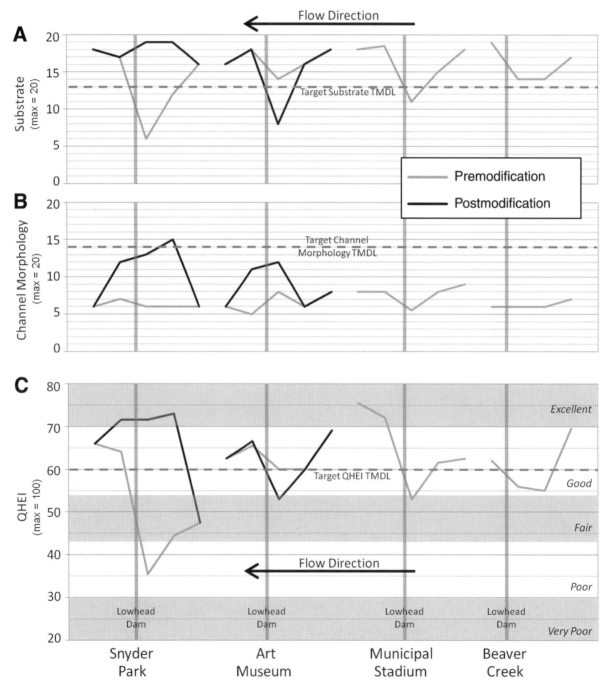

Figure 7. Individual qualitative habitat evaluation index (QHEI) metrics for (A) substrate and (B) channel morphology and (C) summary QHEI score for the lowhead dams investigated in this study. For the Art Museum and Snyder Park sites, scores are shown prior to and following dam modification. The qualitative description of habitat quality for the summary QHEI score is based on Ohio EPA (2006). TMDL—total maximum daily load.

for substrate and channel morphology are included in Figure 7 (A and B) because they, along with the bank erosion and riparian zone metric, influence the degree to which siltation affects a stream, but substrate and channel morphology were also most impacted by dam modifications. The most notable changes occurred in the impounded zone. Prior to modification, summary QHEI scores decreased in the impounded zone of each lowhead dam site (Fig. 7; Table 2). Substrates in the impounded reaches are finer grained, in the very fine to medium gravel range, compared to the undisturbed upstream and recovery riffles, and they are moderately to heavily embedded (Ohio EPA, 2006) with fine sand and silt; as a result, substrate metrics decrease in impounded areas (Fig. 7A). At the Art Museum, the dip in the substrate metric in the impounded area was less pronounced than at Snyder Park (Fig. 7A). Although the summary score decreased in the impounded zone prior to modification, it stayed within the good range (Ohio EPA, 2006). At the Snyder Park site, the summary score decreased in the impounded zone, which rated as poor (Fig. 7C) due to consistently low scores for channel morphology (Fig. 7B) and a substantial dip in the substrate score (Fig. 7A) in the impounded zone.

Following modification, the summary QHEI score decreased at the Art Museum and increased at Snyder Park (Fig. 7; Table 2). The overall quality of the impounded zone at the Art Museum decreased from good (60) to fair (53) (Table 2; Fig. 7C). In the impounded area at the Art Museum, the amount of detritus and muck increased following modification. The sand and gravel substrate was moderately to extensively embedded with heavy silt. The substrate score decreased from 14 (out of a maximum score of 20) prior to modification to 8 following modification. At Snyder Park, the summary score increased from poor (35–44) to excellent (71–73) following modification (Table 2; Fig. 7C). The substrate score for the impounded area increased from 6–12 to 17–16 following modification (Fig. 7A). Prior to modification, the sand and gravel substrate in the impounded area exhibited extensive embeddedness. Following modification, this same reach exhibited an increased range of substrate (cobble, gravel, sand, and bedrock) and normal embeddedness, defined as a

TABLE 2. QUALITATIVE HABITAT EVALUATION INDEX (QHEI) AND DISSOLVED OXYGEN VALUES FOR SAMPLES COLLECTED FOR THE ART MUSEUM AND SNYDER PARK SITES PRIOR TO (2009) AND FOLLOWING (2010) DAM MODIFICATION

			Art Museum						Snyder Park					
			2009			2010			2009			2010		
			U	I	R	U	I	R	U	I	R	U	I	R
QHEI metrics		Substrate	16–18	14	16–18	16–18	8	16–18	16	6–12	17–18	16	19	17–18
		Channel morphology	6–8	8	5–6	6–8	12	6–11	6	6	6–7	6	13–15	6–12
		Riparian zone	4	4	4	4	5	4	4	4	5	4	6	5
		Total	60–69	60	62–65	60–69	53	62–66	47	35–44	64–66	47	71–73	66–71
Dissolved oxygen	Surface water	Percent saturation	95	99	98	88	88	92	90	83	94	117	111	107
		Concentration (mg/L)	8.4	9.0	8.9	7.4	7.5	7.8	7.8	7.2	8.1	9.9	9.4	9.1
		Quality index	98	99	99	93	94	97	95	90	98	92	95	97
	Substrate water	Percent saturation	83	79	95	88	44	87	78	51	75	112	106	71
		Concentration (mg/L)	7.3	7.1	8.6	7.5	3.8	7.3	6.7	4.4	6.5	9.5	9.0	6.1
		Quality index	89	86	98	94	36	92	84	45	82	94	98	77

Note: The column headings represent sample sites (U—upstream undisturbed riffle, I—impounded zone, and R—recovery riffle). If more than one QHEI determination was made within a site, the range of values is shown. The dissolved oxygen values represent averages based on multiple samples collected within a site and on multiple days.

normal dusting without functional significance (Gordon, 2001). Sand and gravel deposits formed upstream of each drop structure and were largely free of fine sand, silt, and organics. The structures themselves, made of large blocks of natural stone grouted together for stability, mimic a bedrock channel over short reaches of the former impounded area. The target QHEI score is 60, and the target substrate score is 13 (Ohio EPA, 2009) for this reach of Buck Creek.

Channel morphology prior to modification was characterized by straight, stable channels, dominated by pools and glides. The channel morphology score at the Art Museum increased from 8 to 12 (out of a maximum of 20) as a result of modification and from 6 to 13–15 at Snyder Park (Fig. 7B). The target channel morphology score is 14 (Ohio EPA, 2009). At both sites, the channelization score improved from "no recovery" to "recovered" (after Ohio EPA, 2006) as a function of drop structure construction. In addition, the development of constriction-step-pool sequences influenced the pool/glide and riffle/run quality (Ohio EPA, 2006). Depth of the pool and the variation in channel width increased this metric. If sand and gravel deposits were present above the constriction, as was the case at Snyder Park, it was scored as a pool-riffle complex, meaning deep riffles and runs were present and pools had a maximum depth of more than 1 m (Gordon, 2001). At the Art Museum, deeper pools are present as a result of excavation during construction and the increased stage upstream of the uppermost drop structure, but sand and gravel deposits are not present. Changes in the bank erosion and riparian zone metric were minimal, but increased at both sites as a result of stabilization and terracing of the banks at each drop structure.

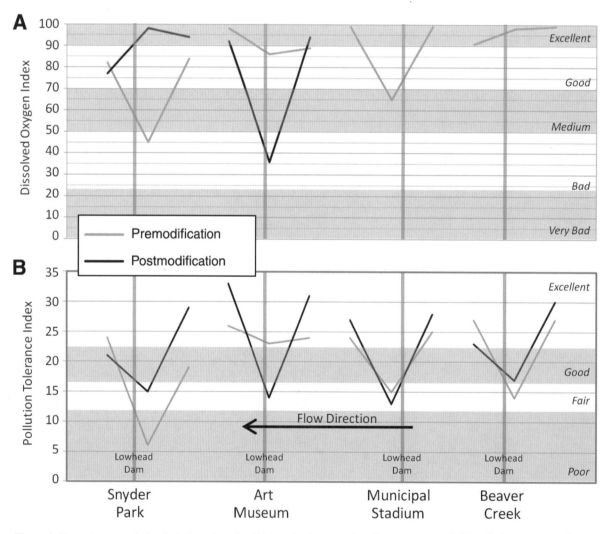

Figure 8. Downstream variation in index values for (A) dissolved oxygen in substrate water and (B) pollution tolerance for taxa at the Art Museum and Snyder Park sites prior to and following dam modification. The index values and qualitative description of quality for dissolved oxygen are based on a water-quality index created by the National Sanitation Foundation (Brown et al., 1970). The qualitative description of quality based on pollution tolerance is from Hippensteel (2005).

Quality of Surface and Substrate Water

Temperature, pH, and conductivity data from surface and substrate water did not vary significantly or consistently within or between sites and, therefore, will not be discussed here. Dissolved oxygen did show variation. In the surface water column, dissolved oxygen does not vary dramatically within a site, both increasing and decreasing downstream from the upstream undisturbed riffle through the impounded zone and into the downstream recovery riffle (Table 2). However, dissolved oxygen content of substrate water sampled from the hyporheic zone decreases in the impounded area but generally recovers in the downstream riffle (Table 2; Fig. 8A). These trends in dissolved oxygen are illustrated in Figure 8A as an index value based on the National Sanitation Foundation's (Brown et al., 1970) water-quality index. The benefit of the index is that it provides a relative sense of water quality, although as used here, it is based only on dissolved oxygen. Whereas the quality of surface water is consistently excellent relative to dissolved oxygen, the quality of the substrate water is variable between areas at a given dam site and changes over time. At the Art Museum site, the average quality of substrate water from the upstream undisturbed and recovery riffles was good to excellent both prior to and following dam modification; the same is true for substrate at the upstream undisturbed and recovery riffles at Snyder Park (Fig. 8A). Within the impounded area at the Art Museum, water quality decreased from good to bad following modification, whereas, at Snyder Park water quality increased from bad to excellent in the formerly impounded area following dam modification (Fig. 8A).

Macroinvertebrates and the PTI

Aquatic macroinvertebrates were collected at three representative sites in the undisturbed upstream riffle, the impounded zone, and the downstream recovery riffle at both dam sites prior to and following modification. Organisms were classified to the family level in all cases and genus level in many cases. Compared to premodification, the total number of individuals and pollution-intolerant taxa decreased in the impounded area at each site (Table 3). At the Art Museum, the total number of individuals

TABLE 3. MACROINVERTEBRATE POPULATIONS COLLECTED FOR THE ART MUSEUM AND SNYDER PARK SITES PRIOR TO (2009) AND FOLLOWING (2010) DAM MODIFICATION, GROUPED BY COMMON NAME

	Pollution tolerance	Art Museum						Snyder Park					
		2009			2010			2009			2010		
		U	I	R	U	I	R	U	I	R	U	I	R
Stonefly nymph	Intolerant	0	0	0	0	0	0	0	0	0	0	0	0
Mayfly nymph		44	3	20	37	0	57	35	0	46	82	24	221
Caddis fly larvae		56	6	54	2	1	6	5	0	55	7	3	178
Dobsonfly larvae		0	0	0	0	0	0	0	0	0	0	0	0
Riffle beetle		17	8	24	62	0	34	8	0	19	48	14	14
Water penny		13	0	1	3	0	1	2	0	3	2	0	0
Damselfly nymph	Moderately tolerant	0	0	0	0	0	0	0	0	0	0	0	0
Dragonfly nymph		0	0	0	0	0	0	0	0	0	0	0	0
Sowbug		0	1	0	0	0	0	0	0	0	0	0	0
Scud		3	0	1	1	1	0	0	0	3	1	0	0
Crane fly larvae		0	0	0	1	4	0	0	0	0	0	0	0
Clams/mussels		0	0	0	0	0	0	0	0	0	0	0	0
Midges	Fairly tolerant	0	0	0	39	4	17	0	4	0	17	25	141
Black fly larvae		9	2	5	0	0	0	0	0	7	0	0	37
Planaria		0	0	0	0	0	0	0	0	0	0	0	0
Leech		0	0	0	0	0	0	0	1	0	2	0	0
Aquatic worms	Very tolerant	5	1	0	0	0	0	1	5	3	15	0	1
Blood midge		4	10	7	40	7	11	4	19	15	42	25	8
Rat-tailed maggot		0	0	0	0	0	0	0	0	0	0	0	0
Total individuals		151	31	112	185	17	126	55	29	151	216	91	600
Total taxa		8	7	7	8	5	6	6	4	8	9	5	7
Pollution tolerance index		24	23	26	31	14	33	19	6	24	29	15	21

Note: The column headings represent sample sites (U—upstream undisturbed riffle, I—impounded zone, and R—recovery riffle).

and pollution-intolerant taxa also decreased in the impounded area following dam modification. The PTI, based on the taxa present weighted relative to their pollution tolerance, dips in the impounded zone at the Art Museum relative to the upstream undisturbed and recovery riffles, but within the impounded zone, it decreased from 23 (excellent) to 14 (fair) following modification (Table 3; Fig. 8B). In contrast, both the total number of individuals and pollution-intolerant taxa parameters increased in the impounded zone at Snyder Park following dam modification (Table 3; Fig. 8B). Some of those taxa collected postmodification include such pollution-intolerant taxa as caddis fly and mayfly larvae and riffle beetles, none of which was present prior to modification. Though the PTI dips at the Art Museum in the impounded zone relative to the upstream undisturbed and recovery riffles, it increased from 6 (poor) to 15 (fair) following dam modification (Table 3; Fig. 8B).

DISCUSSION

Urban stream restoration, especially through the modification and removal of lowhead dams, may be limited, even impracticable, because of urban infrastructure. Integrating restoration plans with structures designed for in-stream recreation may provide a viable alternative. The data we have collected from the Art Museum and Snyder Park locations illustrate two different and opposite ecosystem responses to dam modification. Habitat quality, dissolved oxygen in the substrate, and the pollution tolerance index all decreased in the former impounded zone at the Art Museum following dam modification, whereas they all increased at Snyder Park (Figs. 7 and 8). These results provide insight for the engineering design of the drop structure that could optimize both their recreational and restoration potential.

The recreational design of the drop structure effectively changes the channel cross section from a single rectangular weir to a series of one or more triangular weirs. While the change in shape creates the type of hydraulics desired for whitewater recreation, the increased stage associated with the triangular shape of the drop structure may maintain or even exacerbate the impact of the structure on the stream ecosystem depending on the location of the upstream structure (Fig. 6). The upstream drop structure controls stream grade and the ability of the stream to incise and redistribute sediments in the former impoundment (Fig. 3). The constriction and circulation within the downstream pool also maintain flow velocity and turbulence required to keep finer particle sizes in suspension, so that sand and gravel deposits do not become embedded.

The most critical impact of dam removal is on the longitudinal and planform evolution of the channel as it incises and redistributes sediment accumulated in the impoundment (Kondolf, 1997; Bednarek, 2001; Doyle et al., 2003b). At the Art Museum site, the location and minimum elevation of the drop structure were not changed (Fig. 3B), resulting in a rise in stage for a given discharge and an increase in the impounded area upstream of the upstream-most structure. This was by design because the dam still protects utilities within its core, but the result is that at high flow, it still very much behaves like a dam. Although flow is concentrated through the structure, resulting in increased flow velocity locally, incision of the impounded sediment is not possible because the grade of the channel bed is tied to the upstream edge of the drop structure. As a result, fine sediments continue to accumulate over areas that previously had exposed sands and gravels, degrading the habitat potential (Fig. 7) at this site. Less than a year after modification, the impounded area upstream of the drop structure behaves as an impoundment, more so than prior to modification. Dissolved oxygen in the substrate water upstream of drop structure has decreased (Fig. 8A). The difference in dissolved oxygen levels between surface and substrate flow is likely a result of decreased hyporheic exchange as the sand and gravel substrate has become embedded with fine sand, silt, and organic matter as observed by Orr et al. (2009). The macroinvertebrate population is less diverse and more pollution tolerant (Fig. 8B), perhaps as a direct result of the change in dissolved oxygen of the substrate water. A downstream drop structure at the Art Museum slightly improves the habitat potential downstream of the original dam (Fig. 7C) for reasons that may be more similar to those for structures at the Snyder Park site.

At Snyder Park, utilities associated with the lowhead dam were no longer functional. Constraints on the grade further upstream did not inhibit the recreational design, and, as a result, multiple drop structures were constructed upstream of the original dam within the former impounded area (Fig. 3C). By extending the hydraulic height of the dam over several drop structures successively located within the former impoundment, the impounded fine material has been excavated or eroded from the system. The quality of the stream ecosystem along this reach has measurably increased since modification as function of the variable channel width and depth associated with the drop structures as well as decreased embeddedness of the channel bed (Fig. 7). Sand and gravel have been deposited in bars upstream of each drop structure, and finer sediment is moved through the reach. Dissolved oxygen in the substrate water in the former impoundment has increased (Fig. 8A), and the macroinvertebrate population is more diverse and contains more pollution-intolerant taxa (Fig. 8B). The similarity of dissolved oxygen in surface and substrate water (Table 2) following modification suggests active exchange of water within the hyporheic zone. If this is indeed occurring, there may also be increased exchange of nutrients and organic matter between the surface and substrate flow as the functional significance of the hyporheic zone is restored (Boulton et al., 1998). Stage has increased upstream of the uppermost drop structure (Fig. 6), increasing the impounded area upstream slightly, but the impact has been minimal in the first year following modification. The increase in PTI in the former impoundment was modest in the first year following modification, moving from poor to fair (Fig. 8B), but its relation to increased substrate quality (Fig. 7A), QHEI (Fig. 7C), and dissolved oxygen (Fig. 8A) is encouraging.

It is clear from this work that recreational improvements associated with lowhead dams can play a significant role in stream restoration efforts in an urban setting. The morphology of successive drop structures creates expansions and contractions of channel width that contribute to variation in flow velocity and depth that are representative of a more natural condition. According to the QHEI metrics and their respective weights, these changes in morphology should increase the habitat potential of the recreational reach, but this depends on the quality and embeddedness of the substrate. If the minimal elevation of the uppermost drop structure is somewhat lower than the original elevation of the lowhead dam, our work suggests that the quality of the substrate improves, as does both water and macroinvertebrate quality. Bottom sediment restoration is an in-stream restoration technique that has been applied with some success (Kasahara and Hill, 2007; Sarriquet et al., 2007), and this type of restoration upstream of the constriction may be beneficial where the upstream grade is constrained. The difference between the rectangular lowhead dam and the triangular, or v-shaped, modification at the drop structure is that there is concentration of flow through the constriction. Concentration of flow and increased flow velocity immediately upstream and through the constriction maintain suspension of fine-grained sediment that would otherwise embed in sediments upstream of the uppermost structure. In addition, the impermeable nature of the grouted bedrock associated with the drop structure may also cause upwelling of substrate water upstream of the constriction.

SUMMARY AND RECOMMENDATIONS

Our assessment of habitat potential, substrate water chemistry, and macroinvertebrates prior to and following lowhead dam modification along an urban stream indicates that in-stream recreational structures provide opportunities for stream restoration. Recreational drop structures can be used to replace and relocate the hydraulic height of a lowhead dam for urban stream reaches requiring a stable longitudinal profile, and they can be used to create variations in channel width, depth, and velocity that act as compressed pool-riffle sequences in a formerly impounded area. Sand and gravel substrates form upstream of the structure, and fine sediments move through the reach. The coarse substrates have higher dissolved oxygen either because of the lack of embedded organics and fine sediment or a more active exchange of water in the hyporheic zone. A more diverse and higher-quality population of macroinvertebrates develops in these sediments within a year of modification.

Placement of the drop structures is especially critical. Placement of the upstream drop structure should be within the former impounded zone or at an elevation lower than the original lowhead dam (Fig. 3C) if possible. Erosion of impounded sediment is possible in this case and, more importantly, trapping of fine sediment that might otherwise embed in exposed sand and gravel is decreased. In this study, the Snyder Park modifications provide an example of this type of placement and restoration benefits. The Art Museum modifications and attendant impacts represent the opposite. The placement of drop structures at the original dam site, with a minimum elevation equal to the original dam elevation, and downstream of the dam have increased the impounded area and exacerbated the degraded conditions there. In low flow or stagnant areas near the structure, this might be mitigated by narrowing the channel by promoting deposition or constructing a low surface (i.e., a two-stage channel) that would constrain flow to a central channel during low-flow conditions but provide for water storage or conveyance and sediment deposition during higher flow.

ACKNOWLEDGMENTS

We would like to thank John Loftis, Kevin Loftis, Friends of the Buck Creek Corridor, and the Springfield Conservancy District for inviting us to partner with them in their efforts to improve safety, navigability, and recreation on Buck Creek. The monitoring network was made possible by a grant from the Springfield Conservancy District. Summer support for Evelsizor, Minter, Rigsby, Shearer, and Shaw was provided by the Floyd R. Nave Endowment Award in Geology, Wittenberg's Student Development Board Summer Research Grants, and Wittenberg's Center for Civic and Urban Engagement. The manuscript benefited greatly from constructive reviews by Jeff Clark, James Evans, and Jerry Miller.

REFERENCES CITED

American Rivers and Trout Unlimited, 2002, Exploring Dam Removal: A Decision-Making Guide: http://act.americanrivers.org/site/DocServer/Exploring_Dam_Removal-A_Decision-Making_Guide.pdf?docID=3641 (accessed 21 October 2010), 85 p.

Bednarek, A.T., 2001, Undamming rivers: A review of the ecological impacts of dam removal: Environmental Management, v. 27, p. 803–814, doi:10.1007/s002670010189.

Born, S.M., Genskow, K.D., Filbert, T.L., Hernandez-Mora, N., Keefer, M.I., and White, K.A., 1998, Socioeconomic and institutional dimensions of dam removals: The Wisconsin experience: Environmental Management, v. 22, p. 359–370, doi:10.1007/s002679900111.

Boulton, A.J., Findlay, S., Marmonier, P., Stanley, E., and Valett, H.V., 1998, The functional significance of the hyporheic zone in streams and rivers: Annual Review of Ecology and Systematics, v. 29, p. 59–81, doi:10.1146/annurev.ecolsys.29.1.59.

Brown, R.M., McClelland, N.I., Deininger, R.A., and Tozer, R.G., 1970, A water quality index—Do we dare?: Water & Sewage Works, v. 117, p. 339–343.

Doyle, M.W., Stanley, E.H., Harbor, J.M., and Grant, G.S., 2003a, Dam removal in the United States: Emerging needs for science and policy: Eos (Transactions, American Geophysical Union), v. 84, p. 29–33, doi:10.1029/2003EO040001.

Doyle, M.W., Stanley, E.H., and Harbor, J.M., 2003b, Channel adjustments following two dam removals in Wisconsin: Water Resources Research, v. 39, p. 1011–1026, doi:10.1029/2002WR001714.

Gordon, S.I., 2001, Ohio Watersheds Homepage—QHEI: Ohio State University, http://tycho.knowlton.ohio-state.edu/qhei.html (accessed 20 December 2010).

Harnik, P., 2010, Urban Green: Innovative Parks for Resurgent Cities: Washington, D.C., Island Press, 184 p.

Hippensteel, S., 2005, Volunteer Stream Monitoring Training Manual: Miami Conservancy District, http://www.miamiconservancy.org/water/pdfs/mvstm.pdf (accessed 20 December 2010), 94 p.

I.C.F. Consulting, 2005, A Summary of Existing Research on Low-Head Dam Removal Projects: National Cooperative Highway Research Program (NCHRP) Project 25-25, Task 14, Transportation Research Board, 77 p.

Kasahara, T., and Hill, A.R., 2007, Instream restoration: Its effects on lateral stream-subsurface water exchange in urban and agricultural streams in southern Ontario: River Research and Applications, v. 23, p. 801–814, doi:10.1002/rra.1010.

Kondolf, G.M., 1997, Hungry water: Effects of dams and gravel mining on river channels: Environmental Management, v. 21, p. 533–551, doi:10.1007/s002679900048.

Ohio Environmental Protection Agency (EPA), 1987, Biological Criteria for the Protection of Aquatic Life: Volume II. Users Manual for Biological Field Assessment of Ohio Surface Waters: Columbus, Ohio, Division of Water Quality Monitoring and Assessment, Surface Water Section, http://www.epa.state.oh.us/dsw/bioassess/BioCriteriaProtAqLife.aspx (accessed 21 December 2010).

Ohio EPA, 2006, Methods for Assessing Habitat in Flowing Waters: Using the Qualitative Habitat Evaluation Index (QHEI): Ohio Environmental Protection Agency Technical Bulletin EAS/2006-06-1, http://www.epa.state.oh.us/portals/35/documents/QHEIManualJune2006.pdf (accessed 9 December 2010), 26 p.

Ohio EPA, 2009, Total Maximum Daily Loads for the Mad River Watershed: http://www.epa.state.oh.us/portals/35/tmdl/MadRiverTMDL_final_dec09.pdf (accessed 20 December 2010), 115 p.

Orr, C.H., Clark, J.J., Wilcock, P.R., Finlay, J.C., and Doyle, M.W., 2009, Comparison of morphologic and biological control of exchange with transient storage zones in a field-scale flume: Journal of Geophysical Research–Biogeosciences, v. 114, G02019, 10 p.

Prince, B.F., 1922, A Standard History of Springfield and Clark County, Ohio: Chicago, American Historical Society, 562 p.

Rankin, E.T., 1989, The Qualitative Habitat Evaluation Index (QHEI): Rationale, Methods, and Application: Ohio Environmental Protection Agency, Division of Water Quality, Planning and Assessment, Ecological Assessment Section, http://www.epa.ohio.gov/portals/35/documents/BioCrit88_QHEIIntro.pdf (accessed 9 December 2010), 54 p.

Rosgen, D.L., 1996, Applied River Morphology: Pagosa Springs, Colorado, Wildland Hydrology, 390 p.

Sarriquet, P.E., Bordenave, P., and Marmonier, P., 2007, Effects of bottom sediment restoration on interstitial habitat characteristics and benthic macroinvertebrate assemblages in a headwater stream: River Research and Applications, v. 23, p. 815–828, doi:10.1002/rra.1013.

MANUSCRIPT ACCEPTED BY THE SOCIETY 29 JUNE 2012

Effects of multiple small stock-pond dams in a coastal watershed in central California: Implications for removing small dams for restoration

J.L. Florsheim
University of California, Earth Research Institute, Santa Barbara, California 93106, USA

A. Chin
University of Colorado, 1250 14th Street, Denver, Colorado 80217, USA

A. Nichols
University of California, Davis, One Shields Avenue, Davis, California 95616, USA

ABSTRACT

Small dams are often situated on low-order tributaries that drain grazed hillslopes in dry regions of the western United States. In this paper, we use remote-sensing techniques in a case study to explore the effects of multiple small stock-pond dams on tributaries to Chileno Creek, a coastal watershed in central California. Dam density, or number of dams per drainage area, is 0.76 dams per km^2, with most of the tributaries containing one or more dams. The total eroding channel length downstream of dams is ~11% greater than total eroding length upstream. The relatively high density of the small stock-pond dams leads to cumulative effects that elevate the magnitude of (1) increased downstream erosion and (2) fragmentation of longitudinal connectivity between tributary headwaters and the main channel. The total headwater area producing sediment blocked by small dams equals 30% of the Chileno watershed. From these data, we infer that reduced sediment load due to the presence of the dams slows downstream riparian recovery. Results of the Chileno Creek case study emphasize that basin-scale management approaches and restoration strategies to restore connectivity are imperative in watersheds with high dam density. Uncertainty related to the biophysical effects of small dams and their removal may be investigated through analysis of baseline and long-term monitoring data, and adaptive assessment and management.

INTRODUCTION

Removal of riparian infrastructure such as dams is a promising approach toward restoring river systems. As restoration, dam removal may be considered "passive," meaning that the constraint (the dam) is removed, and natural processes are relied on to reestablish connectivity and ecological functions. Significant management challenges exist, however, related to removal of dams that have already transformed riparian processes or societal perceptions.

In this paper, we consider removal of small dams intended to restore connectivity, hydrologic and geomorphic processes, and ecology. We first define small dams and review the state of the science regarding their biophysical effects. Second, we report results of a case study within a small coastal watershed in central California where numerous small dams exist on tributaries. This work is significant because it considers cumulative effects of small dams in a region where climate, topography, and land use differ from that documented in previous investigations. Third, we use results of this work to develop a strategy for dam removal on tributaries with multiple dams. Results of this work are critical toward providing a linkage between restoration science and policy related to small dam removal in watersheds with cumulative land-use effects.

Relevance of Dam Size

Small dams cause different physical and biological effects as compared to large dams; thus, distinguishing the size of dam is significant in developing restoration goals, designs, monitoring, and adaptive assessment and management strategies. A complicating factor is that numerous definitions of small dams currently exist. For example, the National Inventory of Dams (NID, 2009) includes small dams greater than 2.0 m in height and 61,000 m^3 of storage or 8 m in height and 18,500 m^3 of storage, or that present a significant hazard (Renwick et al., 2005). Small dams have also been defined as those having reservoir storage of less than ~100,000 m^3 (Heinz Center, 2002; Graf, 2005; Chin et al., 2008), or with heights generally less than ~4 m (Skalak et al., 2009). In the case study reported in this paper, reservoir storage is relatively small compared to these definitions, as discussed in the results section of the study.

The influence of small dams is becoming increasingly recognized (Chin et al., 2008) through regional- and local-scale studies (e.g., Stanley and Doyle, 2002; Graf, 2005; Magilligan and Nislow, 2005; Chin et al., 2008; Walter and Merritts, 2008; Skalak et al., 2009). Stanley and Doyle (2002) noted that numerous small dams greater than 80 years old exist in various areas. There is significant concern that as small dams fill with sediment, lessening their usefulness, they may become safety liabilities if lack of maintenance leads to future uncontrolled breaching.

In the past decade, the effects and pervasiveness of dams smaller than those included in national assessments have gained recognition, although in most areas, small dams are still undocumented (Graf, 2005; Magilligan and Nislow, 2005; Chin et al., 2008). In contrast, effects of large dams have been well documented. Significant inter-relations among hydrologic (Dynesius and Nilsson, 1994; Graf, 1999; Richter et al., 1996; Poff et al., 1997; Singer, 2007; Schmidt and Wilcock, 2008), geomorphic (Williams and Wolman, 1984; Graf, 2006), ecologic (Ward and Stanford, 1995; Ligon et al., 1995; Power et al., 1996), and socio-economic (Williams, 1991; Magilligan and Nislow, 2005; Graf, 2005) effects of large dams along major river systems have been documented globally. For example, effects of flow regulation downstream of such large dams may include alteration of rivers over time scales lasting from years to decades (Church, 1995; Gilvear, 2004).

Biophysical Effects of Small Dams

Small dams disrupt the longitudinal flow of water, sediment, and nutrients from upstream to downstream and may cause upstream, local, and downstream changes in physical conditions and biota. Changes upstream of dams may extend kilometers upstream of dams and include adjustments in channel slope and sediment transport capacity (Evans et al., 2007). Local changes include reduction in flow velocity as water is impounded or slowed as it flows through the reservoir. Thus, the residence time and temperature of water increase in the reservoir due to both the slower flow and increased surface area upstream of the dam (Poff and Hart, 2002). Coarse sediment contributed from upstream areas is usually trapped upstream of the dam, whereas finer material may either be trapped or transported farther downstream, depending on flow magnitude and velocity within the reservoir and dam characteristics. The increase in temperature may lead to algae proliferation and, together with sedimentation in the reservoir upstream of the dam, promote a change from lotic to lentic wetland plants. Poff and Hart (2002) noted that dam size is one factor influencing interactions between sedimentation and chemical transformation in water impounded upstream of a dam.

The disruption of sediment and water transferred from uplands to lowlands decreases downstream sediment loads (Grant et al., 2003; Graf, 2006), thus altering downstream channel capacity (Chin et al., 2002). Decreased sediment loads resulting from the trapping of sediment upstream of the dam may cause channel erosion and a change in substrate size in downstream reaches (Pizzuto, 2002). Recent work addressing geomorphic effects of small dams documented variable effects of geomorphic processes related to channel erosion and alteration of channel morphology (Walter and Merritts, 2008; Skalak et al., 2009).

Multiple dams on tributaries may be present, with effects on hydrology depending on each dam's hydrologic contributing area and the size of storms relative to that area (Dunne and Leopold, 1978). The hydrologic effects of small dams (Chin et al., 2008) as well as the effects of dams on ephemeral streams (Neave et al., 2009) are poorly understood. Small dams are often "run-of-the-river structures," a term given to dams without water-control capability, such that once filled, drainage through a spillway occurs, and the rate of discharge into the reservoir is equivalent to the rate of discharge out of the reservoir (Csiki and Rhoads, 2010). Thus, during wet periods, after the reservoir fills, small dams may not greatly influence the magnitude of peak flood flows. However, the influence of small dams on any flows occurring during relatively dry years may be significantly impacted by their widespread presence, because the entire flow generated by storms during dry years may be impounded upstream of the reservoir. Further, many river systems are already altered to such a degree that their ability to absorb disturbances, such as drought, is limited (Palmer et al., 2008). For example, impoundment of flow from upland tributaries in dry climates changes timing, magnitude, and frequency of ecologically sustainable flows (Magilligan and Nislow, 2005) and adds to cumulative land-use effects.

A primary effect of small dams is their influence on landscape connectivity and fragmentation (Graf, 2006). Biophysical effects of small dams include fragmentation in landscapes and its influence on hydrology and sediment and nutrient transfer and deposition, and associated downstream ecological effects. Ward and Stanford (1983) suggested that dams disrupt the longitudinal natural resource gradient from headwaters to the ocean. Even small dams disrupt and fragment riparian ecosystems (Renwick et al., 2005). For example, fragmentation is the major impact of small and medium dams investigated in Texas (Chin et al., 2008) and in the Connecticut River basin (Magilligan et al., 2008). Small dams act as barriers that disrupt migration pathways for biota—both from upstream to downstream, and from downstream to upstream. Barriers that fragment tributaries are important in Pacific coastal streams where seasonal anadromous fish migration from downstream to upstream and out-migration from upstream to downstream are highly dependent on tributary–main channel connectivity.

CASE STUDY ON CHILENO CREEK, COASTAL CALIFORNIA

Scope of Case Study on Chileno Creek

The case study focuses on small dams in Chileno Creek, a tributary to Walker Creek, a coastal watershed that drains to Tomales Bay in Marin County in Central California (Fig. 1). In the Chileno Creek watershed, dams are present because water is needed during the dry season to support cattle grazing. This investigation provides an example of the spatial distribution and geomorphic effects of small dams constructed to create stock ponds in a dry region where the pervasiveness and potential effects of small dams have not previously been documented. The scope of the investigation includes a remote-sensing analysis using a geographic information system (GIS) to document: (1) spatial distribution of dams present on tributary channels; (2) dam density, or the number of dams per watershed area; and (3) lengths of eroding channels upstream and downstream of dams. Our results are significant toward understanding the cumulative effects of small dams that are pervasive in dry regions where grazing is the dominant land use. We highlight restoration potential for tributary–main channel connectivity that has application to dry regions with similar climate, environmental characteristics, and land uses.

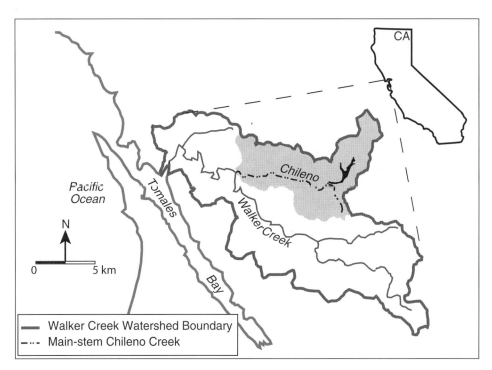

Figure 1. Location of Chileno Creek, a tributary to Walker Creek, a coastal watershed that drains to Tomales Bay in Marin County, Central California. Shaded portion highlights Chileno watershed and is the same area as shown in Figures 3 and 7. Mouth of Walker Creek is 38°13′17.25″N, 122°55′21.01″W.

Figure 2. Photograph of tributary drainage in Chileno Creek illustrating relatively low-relief, rounded, soil-mantled hills with grassland vegetation.

Study Area

Geology and Topography

Chileno Creek has a drainage area of ~52.4 km². The majority of the headwaters of the basin drain relatively low-relief, rounded, soil-mantled hills with grassland vegetation (Fig. 2). However, the tectonically active watershed also includes a low-gradient headwater area occupied by the seminatural Laguna Lake; the lake and its tributaries encompass an area of ~14.4 km², or 28% of the total watershed area (Fig. 1). Our investigation focuses on dams on tributaries joining Chileno Creek downstream of Laguna Lake. The majority of the watershed is underlain by Jurassic–Cretaceous Franciscan Complex mélange, with steeper areas underlain by metamorphic rocks, and portions of the main-stem valley filled with Quaternary alluvium (Fig. 3).

Climate and Hydrology

The watershed is located in an area of coastal California with a Mediterranean climate characterized by variable annual precipitation that ranges from ~460 mm to 2850 mm (MMWD, 2007). The annual rainfall distribution is strongly seasonal, with the majority of the rainfall occurring between November and March—characteristic of California where the majority of rain falls during relatively few days per year (Dettinger et al., 2011). The largest storms prior to the 1942 and 2005 images presented here occurred in 1941 and 1998, respectively (Fig. 4). During both these wet years, annual rainfall exceeded 2000 mm. Flow discharge in Chileno Creek is ungauged; however, field reconnaissance in 2007–2009 suggests that tributaries to Chileno Creek are seasonally dry, with runoff flowing for short periods following storms.

Effect of Land-Use Changes on Erosion and Sedimentation

Beginning in about the mid-1800s, changes in land use and watershed disturbances caused significant increases in rates of sediment erosion and yield from the watershed to Tomales Bay (Haible, 1980). Widespread hillslope erosion caused channel aggradation, and, along with sediment from other portions of the Walker Creek watershed, formed a delta that prograded into Tomales Bay. Rapid watershed erosion (Ellen et al., 1988) and

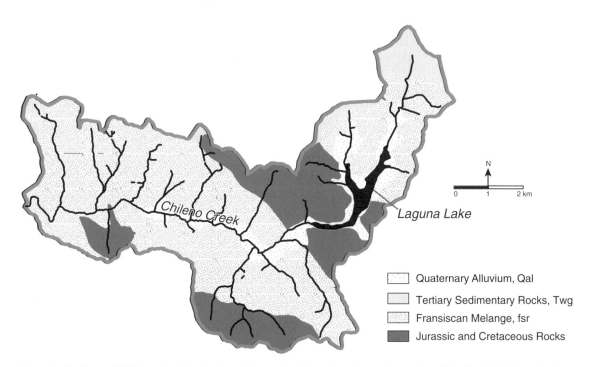

Figure 3. Geology of Chileno Creek is dominated by erosive Franciscan Formation rocks, with alluvial fill in valley bottom. Mouth of Walker Creek is 38°13′17.25″N, 122°55′21.01″W.

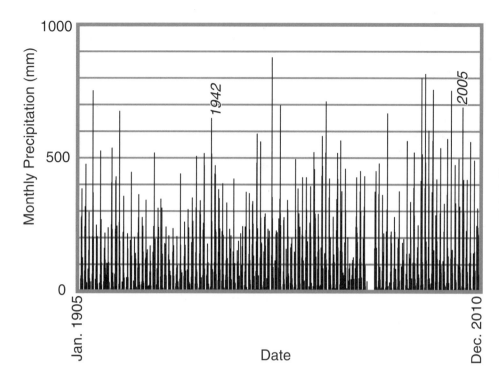

Figure 4. Monthly precipitation measured in Marin County, California, ~13 km south of Chileno Creek between 1905 and 2010.

deposition on the delta was documented during the 1982 flood (Anima et al., 1988). Proxy data including average sedimentation rates in Tomales Bay suggest that although an increase in sedimentation rates was associated with the onset of regional farming, grazing, and other activities, current rates have decreased back to average predisturbance sedimentation rates. The decrease in Tomales Bay sedimentation rates integrates long-term changes in sediment dynamics and land-use activities from several contributing watersheds. Because of the lack of hydrologic, sediment transport, and detailed topographic data in Chileno Creek tributaries, development of a sediment budget is not currently feasible.

In Chileno Creek, historical land uses primarily included vegetation denudation and conversion, potato farming, grazing, and dairy farming (Wharhaftig and Wagner, 1972; Haible, 1980; Hammack, 2005); however, the oak-grassland environment did not support intensive logging activities prevalent in nearby watersheds containing conifers. Today, the dominant land use is grazing. The need for small dams for water storage in reservoirs to support grazing arose because the dry season in coastal California often extends over half the year. A comparison of land uses in 1942 and 2005 is illustrated in Figure 5.

Methods of Case Study on Chileno Creek

Remote-sensing data and observations from field reconnaissance from publicly accessible areas were utilized in the case study on Chileno Creek. The remote-sensing approach was initiated because the entire Chileno Creek watershed is privately owned, and field data collection was not feasible.

Remote Sensing and Image Processing

We documented the presence of small dams and upstream and downstream bank erosion in Chileno Creek watershed using 1942 and 2005 aerial imagery. The 1942 historical aerial photographs were scanned at a resolution of 600 dots per inch; the orthorectification, or "rubber sheeting," was performed using ERDAS Imagine LPS image software. During this process, 10 to 15 control points were collected per image, and 40 auto-tie points were automatically generated base on the control points. Auto-tie points automatically match up adjacent images based on pixel location and color signature. The images were referenced horizontally to the National Agriculture Imagery Program (NAIP) 2005 imagery and vertically to the National Elevation Data Set 10 m digital elevation model (DEM). All imagery was projected into the Universal Transverse Mercator (UTM) Zone 10 North coordinate system. Rectified images were subsequently imported into ArcGIS, and additional error analysis was performed by quantifying the locational error (in meters) between up to five common test points on three randomly selected 1942 and 2005 aerial photographs (Hughes et al., 2006; Evans et al., 2007). Locational error ranged from 0.8 to 10.1 m, with a mean error of 5.87 ± 2.85 m.

Measurements and Analysis

Specific mapping and measurements made in the GIS included: (1) mapping the location of small dams; (2) calculating watershed area upstream of dams; and (3) measuring the length of eroding channels upstream and downstream of dams (Fig. 6). We calculated dam density—or the number of dams per watershed area—as a metric that can be used to represent the effect of

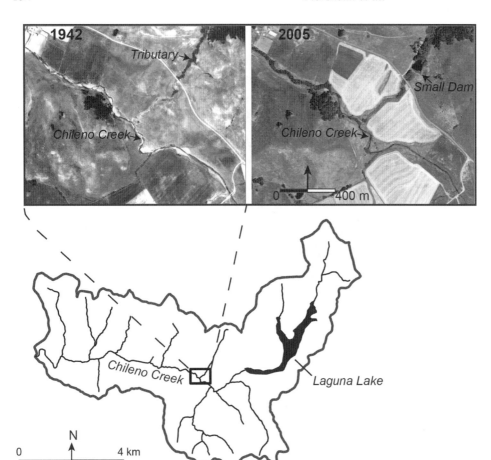

Figure 5. Comparison of land uses active in 1942 and 2005. Photographs illustrate an increase in area of cultivated land and the presence of a small dam on the tributary in 2005 that was not present in 1942.

dams on hydrology and sediment transfer in Chileno Creek in comparison to other watersheds with small dams.

To construct the drainage network, we added lower-order channel lengths visible on the photographic imagery to blue lines already present on topographic maps in ArcGIS. Eroding channel length was calculated longitudinally along one side of eroding channel banks where bare ground or erosional morphology, such as scalloped or arcuate scarps, was discerned. Riparian trees intermittently obscure the channel in the aerial photographs, introducing uncertainty in estimates of erosion length. Moreover, some erosion is not visible through aerial interpretation at the existing scale of the available imagery, and we did not include reaches obscured by trees. Thus, the remote-sensing methods likely render a conservative estimate of channel erosion.

A rationale for comparing erosion in reaches upstream of dams to reaches downstream of dams in order to assess dam effects was provided in Skalak et al. (2009). These authors suggested that the channel upstream of the hydraulic effects of reservoirs provides the best available means to define the geomorphic conditions downstream before dam construction, assuming the geomorphic character of the channel upstream is similar to that downstream. In contrast, Evans et al. (2007) emphasized upstream alterations in sediment transport processes and channel bar characteristics in response to construction of small dams, suggesting uncertainty inherent in using this method. We assume that the method of Skalak et al. (2009) provides a reasonable model for assessment of the effects of small dams on bank erosion because of the relatively steep tributary slopes in Chileno Creek watershed (average tributary slopes ≥ 0.0350). We suggest that steep tributary slopes would topographically limit the effect of the dam on upstream hydraulic and sediment transport processes, in contrast to channels where gradient was less, such as in channels that contain bars where slopes average ≤ 0.0200.

Results of Case Study on Chileno Creek

Estimate of Dam Size in Chileno Creek Watershed

Estimates of dam and reservoir size in the Chileno Creek watershed were based on field reconnaissance and mapping from aerial photographs. Height estimated during field reconnaissance suggests a range from ~2 m to 4 m. Average reservoir surface area (A) mapped from the 2005 NAIP photographs was ~4000 m². Reservoir storage volume (V) was estimated using a geometric representation as two inverted regular tetrahedrons side-by-side such that each comprises half the reservoir:

$$V = \frac{1}{3}\left(\frac{A}{2}\right)h + \frac{1}{3}\left(\frac{A}{2}\right)h, \qquad (1)$$

where h is estimated as 4 m. Thus, an average value for reservoir storage volume of small dams in Chileno Creek is ~5300 m³. Field measurements of reservoir bathymetry would refine this calculation.

Spatial Distribution, Frequency, and Dam Density

Dams were not apparent in the study area in 1942; however, by 2005, small dams were pervasive in the Chileno Creek watershed. The spatial distribution of small dams (Fig. 7) indicates a total of 78 dams present. Of the total, 40 of these small dams were on tributary channels; the others were off-channel impoundments that are not likely to influence tributary sediment transport. The frequency of dams (see Fig. 7) indicates that 10 of the 13 tributaries delineated downstream of Laguna Lake are influenced by at least one small dam, with seven of these tributaries containing multiple dams.

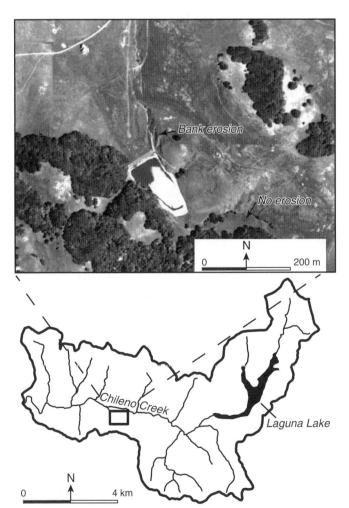

Figure 6. Photograph showing example of eroding versus noneroding channel adjacent to small dam.

Dam density defined as the number of dams per watershed area provides a quantitative metric that may be used to compare spatial distribution of small dams in Chileno Creek watershed to other watersheds regionally, or globally. We assume that Laguna Lake traps sediment from its entire contributing area but do not include it in our calculation relevant to small man-made dams because it represents a seminatural sediment trap. The density of small stock-pond dams on tributaries downstream of Laguna Lake in the Chileno watershed in 2005 was 0.76 dams per km².

Watershed Area Upstream of Dams

To document the total contributing watershed area upstream of dams, we defined the downstream-most dam on each tributary as the "final sediment trap" and summed the upstream areas. Of the total number of dams present on tributaries joining Chileno Creek downstream of Laguna Lake, 31 correspond to a final trap. On this basis, the total area upstream of the final sediment trap that blocks sediment delivery to downstream reaches of tributaries and to the main stem of Chileno Creek is 15.8 km², or 30% of the Chileno watershed. The calculation of contributing area upstream of final sediment trap dams does not differentiate the effects of a sequence of dams on a single tributary versus the effects of a single dam—rather it is intended as a cumulative measure of the effect of multiple dams on sediment and water retention in tributaries.

Length of Eroding Channels

Agricultural activities, including grazing, vegetation denudation, and farming since the 1850s, in combination with the storm flows of 1941 led to the channel erosion already evident on the 1942 imagery (Fig. 5). Trends in channel erosion between 1942 and 2005 on the main stem and tributaries of Chileno Creek are documented in Table 1. Evaluation of 1942 and 2005 aerial photographs suggests that in 2005, eroding channel length equaled ~40% of the total tributary channel length (a decrease from 1942, when almost 60% of tributary channel length was eroding). The decrease in the extent of bank erosion between 1942 and 2005, partly due to initiation of cattle-exclusion fencing, suggests a watershed-scale trend of riparian recovery; however, the proportion of actively eroding to noneroding channels in 2005 is still significant.

To assess the effect of dams on tributary channel erosion, the length of eroding channel upstream and downstream of dams was measured from the 2005 imagery and normalized by total channel length (Table 2). The percent of eroding tributary channel length to total length upstream of the final trap dams is 23%, whereas the value increases to 34% downstream of final trap dams. The greater eroding length downstream of dams suggests that the presence of the dams influences channel erosion in downstream reaches. Field reconnaissance along the road paralleling the main stem of Chileno Creek, where the lower portions of the tributaries may be observed, suggests that bank erosion is prevalent and riparian vegetation is discontinuous and relatively sparse.

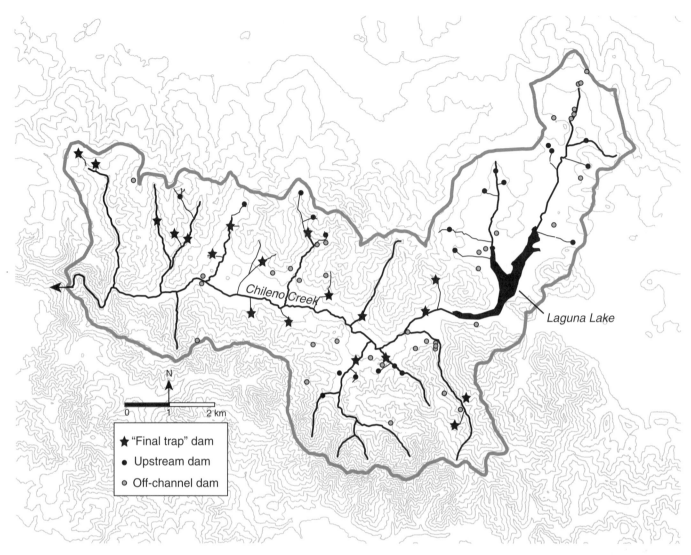

Figure 7. The spatial distribution of small dams, with 40 dams on tributaries out of a total of 78 dams present. On tributaries with multiple dams, the final trap (downstream-most dam) is likely a barrier to sediment transport, flow, and biota. Off-channel dams do not influence sediment transport. Topographic contour interval = 30 m. Mouth of Walker Creek is 38°13′17.25″N, 122°55′21.01″W.

Discussion and Conclusions of Chileno Creek Case Study

All of the small stock-pond dams on tributaries to Chileno Creek were constructed to support grazing land-use activities after 1942. These dams generally have heights less than 4 m with surface areas ~4000 m² on average. An estimate of average storage volume based on these data, of ~5300 m³, indicates that the size of reservoirs impounded upstream of small dams in Chileno Creek is about one to two orders of magnitude smaller than for small dams documented in other investigations related to the effects of small dams. Reservoir capacity in Chileno Creek is likely minimized topographically by the slope of the tributaries (average ~0.0350).

Dams are present on 10 of 13 tributaries delineated downstream of Laguna Lake, with seven of these tributaries containing multiple dams. The density of small dams on tributaries downstream of Laguna Lake in the Chileno watershed in 2005 was 0.76 dams per km². This dam density is significantly higher than that for the United States as a whole, recorded in the National Inventory of Dams (NID, 2009). Thus, the results of the remote-sensing analysis on Chileno Creek indicate that a major impact of small dams is riparian fragmentation on the majority of tributaries within the watershed.

In this work, we use total watershed area upstream of the downstream-most dam as a proxy to quantify effects of multiple small dams on sediment transfer from headwaters to main channels. Based on this rationale, we report the total area upstream of the final sediment trap that blocks sediment delivery to downstream reaches of tributaries and to the main stem of Chileno Creek as 15.8 km², or 30% of the Chileno watershed. Each small dam poses a potential barrier to transfer of water, sediment, nutrients, and biota from the headwaters to the main channel,

TABLE 1. RELATIVE PROPORTION OF ACTIVELY ERODING LENGTH TO TOTAL CHANNEL LENGTH: CHILENO CREEK MAIN STEM AND SIDE-VALLEY TRIBUTARIES

Chileno Creek	Year	Total mapped channel length (m)	Eroding channel length (m)	Eroding length/total channel length
Main stem	1942	17,240	6213	0.36
Tributaries	1942	93,346	54,365	0.58
Main stem	2005	17,240	1326	0.08
Tributaries	2005	82,718	30,579	0.37

TABLE 2. RELATIVE PROPORTION OF ACTIVELY ERODING LENGTH TO TOTAL CHANNEL LENGTH: UPSTREAM VERSUS DOWNSTREAM OF SMALL DAMS ON TRIBUTARIES

Chileno Creek	Year	Total mapped channel length upstream of dams (m)	Total mapped channel length downstream of dams (m)	Eroding channel length upstream of dams (m)	Eroding channel length downstream of dams (m)	Eroding length/total channel length upstream of dams	Eroding length/total channel length downstream of dams
Tributaries	2005	28,264	15,054	6423	5056	0.23	0.34

disrupting longitudinal connectivity and fragmenting the drainage network. On this basis, we conclude that basinwide management approaches and restoration strategies to restore connectivity are imperative. However, the likely cumulative effects of changes in hydrology and sediment caused by each individual dam are complex and require field and modeling analysis that is beyond the scope of this study.

Measurement of eroding channel length visible on the 1942 and 2005 aerial photographs suggests that in 2005, eroding channel length equaled ~40% of the total tributary channel length (a decrease from 1942, when almost 60% of tributary channel length was eroding). These data suggest that riparian recovery is taking place at the watershed scale. However, although there has been a decrease in the extent of bank erosion between 1942 and 2005, the proportion of actively eroding to noneroding channels in 2005 is still significant. Moreover, estimation of erosion upstream versus downstream of dams indicates total eroding length downstream is ~11% greater than total eroding length upstream of final trap dams. These data are significant because of the relatively high density of dams in the watershed. From these data, we infer that an increase in sediment storage upstream of dams causes a deficit of sediment in downstream reaches, and that the reduced sediment load due to the presence of the dams influences channel erosion in downstream reaches. Because channel erosion was already significant in 1942, we suggest that an important downstream effect of the dams in Chileno Creek watershed is to slow processes that account for riparian recovery.

Several uncertainties exist in this analysis. First, upstream of dams, further data acquisition and analysis are needed to discern if eroding tributary channel lengths in our analysis include knickpoint migration in gullies or if the entire eroding length may be attributed to channel bank erosion, as we assume in the preceding analysis. Alternatively, some of the channels obscured by vegetation may be eroding. Field verification would refine this estimate and aid in understanding the complexity of erosion between multiple dams on a single tributary, and in quantifying incision, substrate characteristics, hydrologic changes, and ecological attributes. Higher-resolution remote-sensing data such as light detection and ranging (LiDAR) would also help to refine estimates of erosion in channels upstream and downstream of dams.

The case study on Chileno Creek is an example of the effects of small dams in a coastal watershed that has undergone a legacy of land-use changes since the mid-1800s. The watershed's hydrology, ecology, and ability to transfer sediment have been greatly altered, and current attention is focused on potential restoration approaches to reduce sediment impacts to anadromous fish and estuarine habitat. Thus, information to support various methods to enhance riparian connectivity along with various other restoration techniques is of critical importance.

CONSIDERATIONS FOR SMALL DAM REMOVAL IN RESTORATION: SCIENCE AND POLICY

Remote-sensing data documenting the occurrence of small dams and channel erosion in the Chileno Creek watershed provide baseline information needed to begin discussions of the benefits and/or potential effects of dam removal for restoration. The first step in developing a strategy for small dam removal is to identify goals. Lejon et al. (2009) highlighted four general reasons for dam removal: safety, law and policy, economy, and ecology. Aging, maintenance costs, and liability issues ensure that removal of small dams less than 5 m in height remains a viable management option (Poff and Hart, 2002; Ashley et al., 2006). A primary goal of dam removal in Chileno Creek tributaries would be to address fragmentation of the drainage system and to restore connectivity. One long-term benefit of reestablishing connectivity in fluvial systems is the potential to increase resilience to disturbances. For example, Florsheim and Dettinger (2007) and Palmer et al. (2008) suggested removing riparian infrastructure to restore the natural capacity of riparian systems to buffer future climate change impacts.

Results provided by the Chileno Creek case study are timely, as a policy is currently being proposed for maintaining instream flows in Northern California coastal streams (North Coast

Instream Flow Policy; State Water Resources Control Board, 2010) in order to protect habitat for anadromous salmonids such as steelhead trout (threatened) and coho salmon (endangered) listed under the Federal and the California Endangered Species Acts. Because small dams act as a barrier to fish migration and disrupt the transfer of aquatic benthic macroinvertebrates, woody debris, and gravel needed for downstream fish habitat, the proposed policy would limit construction of new on-stream dams, and it addresses various hydrologic issues for existing structures (such as season of water diversion, minimum bypass flow, and maximum cumulative diversions). In Chileno Creek, numerous off-channel dams already exist, and the potential North Coast Instream Flow Policy may create an additional incentive for considering removal of small dams and/or replacement with off-channel dams for the purpose of restoration. In Chileno Creek, impounding water in stock ponds is the main purpose of the small dams; thus, mitigating their environmental impacts and potentially removing such structures would affect grazing activities. In contrast to large dams that provide hydroelectric power, flood control, or water supply, which are owned and managed by governmental agencies, removing small dams such as are present in the Chileno Creek watershed would necessitate working with the private individuals who own them.

A second step is to develop guidelines to address the cumulative effects due to the occurrence of multiple small dams on individual tributary systems. Given the high density of dams in the watershed, a basinwide management approach is needed to prioritize dam removal for restoration. First, removing one dam within a sequence of dams would have less benefit than removing the only dam present on a single tributary. Thus, an effective approach would be to remove dams on tributaries with only one dam present. Second, removing a dam in the middle of the sequence of multiple dams on a single tributary would have less benefit than starting removal with the downstream-most dam, such that connectivity with the main channel is maximized. Further, some consideration of the potential benefit of staggering dam removals over time to minimize downstream effects would be advantageous; thus, viewing potential for dam removal over the long-term would allow for temporal management of sediment releases. Finally, although there is a high density of small dams in watersheds such as Chileno Creek, and not all of the dams would need to be removed to progress toward restoration goals; instead, prioritizing removals to maximize ecological benefits as described previously herein could be effective.

Numerous uncertainties exist with respect to the geomorphic effects of dam removal, including the potential volume and size of material that would be released and the rates and mechanisms of downstream sediment transport. Quantification of the sediment size and volume trapped upstream of a dam proposed for removal would help in predicting the potential effects on upstream erosion and downstream sedimentation. After a dam is removed, transport of sediment eroded by incision into accumulated fill material upstream of the former dam influences the channel downstream of the dam. Downstream of the dam, the channel may need to adjust to a modified sediment load and grain size, and may translate as a wave or move through dispersion (Lisle et al., 2001; Pizzuto, 2002) in response to an altered postremoval flow regime. However the rate of transport of sediment downstream depends on the mechanisms of movement, grain size of sediment available for transport, channel morphology, method of dam removal, and revegetation rates. Thus, substantial uncertainty exists regarding the rate and nature of downstream channel adjustment. Further, there is concern that the release of sediment may have downstream ecological effects; thus, the removal itself should be considered a disturbance (Stanley and Doyle, 2003; Grant et al., 2003) because of increased transport of fine sediment that may raise turbidity or cause deposition of fine sediment that alters substrate in downstream reaches (Bushaw-Newton et al., 2002; Granata et al., 2008). Alternatively, effects of removal sometimes appear minor, highlighting difficulties in prediction of change (Ashley et al., 2006; Cheng and Granata, 2007; Skalak et al., 2009).

Another uncertainty following dam removal relates to the resulting form of the channel upstream and downstream of the former dam. Currently, little is understood about processes and rates of adjustment of channel form following dam removal in ephemeral streams (Neave et al., 2009). For example, in coastal California, wet winter flows may not occur every year, lengthening the time required for the system to cycle through a period of bank erosion, sediment deposition, and vegetation succession in a sequence such as outlined by Doyle et al. (2002). Most dam removal restoration projects to date have included small dams (Poff and Hart, 2002; Doyle et al., 2003). As the occurrence of small dam removal restoration projects grows, a similarly growing body of research is documenting fluvial adjustments following their removal (for example: Bednarek, 2001; Grant, 2001; Pizzuto, 2002; Stanley and Doyle, 2002; Doyle et al., 2003; Evans et al., 2007; Granata et al., 2008; Neave et al., 2009). When dam removal includes removing the structure to the elevation of the former bed of the channel, local base level is lowered. This change may initiate rapid initial incision (Cantelli et al., 2004) or headward migration of a knickpoint and a related sequence of widening (Doyle et al., 2003) into sediment formerly trapped upstream of the reservoir. Even in partially controlled dam removals, where a sill remains at a higher elevation than the original channel bed (Dalal and Florsheim, 2004), headwater migration of a knickpoint may initiate channel development in the sediment fill upstream of the former dam.

FUTURE NEEDS: MONITORING, ADAPTIVE ASSESSMENT, AND ADAPTIVE MANAGEMENT

Uncertainties related to the effects of small dams and the benefits or effects of their removal for restoration discussed in the previous sections may be reduced through long-term monitoring, adaptive assessment, and adaptive management. Hydrologic and geomorphic monitoring should focus on changes in connectivity, flows, and sediment fluxes that influence ecological recovery. A first step is to document linkages between hillslopes and channels and sediment pathways whereby sediment is transported from tributaries to downstream main channels. Understanding erosion processes that contribute sediment to downstream reaches, quantifying flow variability, rates of transport, and spatial distribution of sediment storage and erosion prior to dam removal will provide an important baseline of upstream and downstream conditions needed to assess the effects of dam removal.

A second step is to document the linkage between upstream and downstream ecological resources in order to determine riparian attributes or limiting factors that may be either improved or impacted by dam removal. In some environments, changes following dam removal might occur progressively as annual floods drive ecological change. However, in other environments, change might occur episodically. Understanding the hydrologic regime, and designing a monitoring program that captures its variability are particularly important in environments with episodic storms where no changes might occur during dry years (or decades) following dam removal, and channel alteration might only occur during infrequent floods.

Adaptive assessment and management include modifications in monitoring and analysis methods needed to document and manage changes that occur as dynamic systems evolve following restoration efforts—and using this new information to adapt management strategies. As biophysical changes become apparent through analysis and interpretation of data, new data are then used to refine management and to inform restoration science. Long-term monitoring programs that aid in understanding rates and processes of sediment removal from reservoirs, post-removal magnitudes of flows that mobilize sediment, changes in upstream and downstream erosion and deposition, and the effects of these processes on riparian ecology are essential for the success of restoration science.

ACKNOWLEDGMENTS

We thank Sheila Roberts, Andy Selle, and James Evans for comments and suggestions that improved this paper. Initial work for this project was supported by a grant from the San Francisco Bay Regional Water Quality Control Board. Photo rectification and georeferencing was conducted by the San Francisco Estuary Institute. We thank M. Napolitano, J. Marshall, N. Scolari, L. Prunske, and L. Hammack for their help and interesting discussion.

REFERENCES CITED

Anima, R., Bick, J., and Clifton H.E., 1988, Sedimentologic consequences of the storm in Tomales Bay, in Ellen, S.D., and Wieczorek, G.F., eds., Landslides, Floods, and Marine Effects of the Storm of January 3–5, 1982, in the San Francisco Bay Region, California: U.S. Geological Survey Professional Paper 1434, p. 283–300.

Ashley, J.T.F., Bushaw-Newton, K., Wilhelm, M., Boettner, A., Drames, G., and Velinsky, D.J., 2006, The effects of small dam removal on the distribution of sedimentary contaminants: Environmental Monitoring and Assessment, v. 114, p. 287–312, doi:10.1007/s10661-006-4781-3.

Bednarek, A.T., 2001, Undamming rivers: A review of the ecological impacts of dam removal: Environmental Management, v. 27, no. 6, p. 803–814, doi:10.1007/s002670010189.

Bushaw-Newton, L.L., Hart, D.D., Pizzuto, J.E., Thomson, J.R., Egan, J., Ashley, J.T., Johnson, T.E., Horwita, R.J., Keeley, M., Lawrence, J., Charles, D., Gatenby, C., Kreeger, D.A., Nitengale, T., Thomas, R.L., and Velinsky, D.J., 2002, An integrative approach towards understanding ecological responses to dam removal: The Manatawny Creek study: Journal of the American Water Resources Association, v. 38, no. 6, p. 1581–1599, doi:10.1111/j.1752-1688.2002.tb04366.x.

Cantelli, A., Paola, C., and Parker, G., 2004, Experiments on upstream-migrating erosional narrowing and widening of an incisional channel caused by dam removal: Water Resources Research, v. 40, W03304, doi:10.1029/2003WR002940.

Cheng, F., and Granata, T., 2007, Sediment transport and channel adjustments associated with dam removal: Field observations: Water Resources Research, v. 43, W03444, doi:1029/2005WR004271.

Chin, A., Harris, D.L., Trice, T., and Given, J.L., 2002, Adjustment of stream channel capacity following dam closure, Yegua Creek, Texas: Journal of the American Water Resources Association, v. 38, no. 6, p. 1521–1531, doi:10.1111/j.1752-1688.2002.tb04362.x.

Chin, A., Laurencio, L.R., and Martinez, A.E., 2008, The hydrologic importance of small- and medium-sized dams: Examples from Texas: The Professional Geographer, v. 60, no. 2, p. 238–251, doi:10.1080/00330120701836261.

Church, M., 1995, Geomorphic response to river and flow regulation: Case studies and time scales: Regulated Rivers: Research and Management, v. 11, p. 3–22, doi:10.1002/rrr.3450110103.

Csiki, S., and Rhoads, B.L., 2010, Hydraulic and geomorphological effects of run-of-river dams: Progress in Physical Geography, v. 34, no. 6, p. 755–780, doi:10.1177/0309133310369435.

Dalal, M.S., and Florsheim, J.L., 2004, Monitoring channel initiation in sediment formerly impounded upstream of a small dam removed on Murphy Creek, San Joaquin County, CA: Geological Society of America Abstracts with Programs, v. 36, no. 5, p. 285.

Dettinger, M.D., Ralph, F.M., Das, T., Neiman, P.J., and Cayan, D.R., 2011, Atmospheric rivers, floods, and the water resources of California: Water, v. 3, p. 445–478, doi:10.3390/w3020445.

Doyle, M.W., Stanley, E.H., and Harbor, J.M., 2002, Geomorphic analogies for assessing probable channel response to dam removal: Journal of the American Water Resources Association, v. 38, no. 6, p. 1567–1579, doi:10.1111/j.1752-1688.2002.tb04365.x.

Doyle, M.W., Stanley, E.H., and Harbor, J.M., 2003, Channel adjustments following two dam removals in Wisconsin: Water Resources Research, v. 39, no. 1, p. 1011, doi:10.1029/2002WR001714.

Dunne, T., and Leopold, L.B., 1978, Water in Environmental Planning: San Francisco, California, W.H. Freeman and Company, 818 p.

Dynesius, M., and Nilsson, C., 1994, Fragmentation and flow regulation of river systems in the northern third of the world: Science, v. 266, p. 753–762, doi:10.1126/science.266.5186.753.

Ellen, S., Cannon, S.H., and Reneau, S.L., 1988, Distribution of debris flows in Marin County, in Ellen, S.D., and Wieczorek, G.F., eds., Landslides, Floods, and Marine Effects of the Storm of January 3–5, 1982, in the San Francisco Bay Region, California: U.S. Geological Survey Professional Paper 1424, p. 113–131.

Evans, J.E., Huxley, J.M., and Vincent, R.K., 2007, Upstream channel changes following dam construction and removal using a GIS/remote sensing approach: Journal of the American Water Resources Association, v. 43, no. 3, p. 683–697, doi:10.1111/j.1752-1688.2007.00055.x.

Florsheim, J.L., and Dettinger, M.D., 2007, Climate and flood variability still govern levee breaks: Geophysical Research Letters, v. 34, L22403, doi:10.1029/2007GL031702.

Gilvear, D.J., 2004, Patterns of channel adjustment to impoundment of the Upper River Spey, Scotland (1942–2002): River Research and Applications, v. 20, p. 151–165, doi:10.1002/rra.741.

Graf, W.L., 1999, Dam nation: A geographic census of American dams and their large-scale hydrologic impacts: Water Resources Research, v. 35, no. 4, p. 1305–1311, doi:10.1029/1999WR900016.

Graf, W.L., 2005, Geomorphology and American dams: The scientific, social and economic context: Geomorphology, v. 71, p. 3–26, doi:10.1016/j.geomorph.2004.05.005.

Graf, W.L., 2006, Downstream hydrologic and geomorphic effects of large dams on American rivers: Geomorphology, v. 79, p. 336–360, doi:10.1016/j.geomorph.2006.06.022.

Granata, T., Cheng, F., and Nechvatal, M., 2008, Discharge and suspended sediment transport during deconstruction of a low-head dam: Journal of Hydraulic Engineering, v. 134, no. 5, p. 652–657, doi:10.1061/(ASCE)0733-9429(2008)134:5(652).

Grant, G., 2001, Dam removal: Panacea or Pandora for rivers?: Hydrological Processes, v. 15, p. 1531–1532, doi:10.1002/hyp.473.

Grant, G., Schmidt, J.C., and Lewis, S.L., 2003, A geological framework for interpreting downstream effects of dams on rivers, in O'Connor, J.E., and Grant, G.E., eds., A Peculiar River, Water Science and Application 7: Washington, D.C., American Geophysical Union, p. 203–219.

Haible, W.W., 1980, Holocene profile changes along a California coastal stream: Earth Surface Processes, v. 5, p. 249–264.

Hammack, L., 2005, Geomorphology of the Walker Creek Watershed: Sebastopol, California, Report Prepared for Marin Resource Conservation District by Prunuske Chatham, Inc., 10 August 2005, 56 p.

Heinz Center, 2002, Dam Removal: Science and Decision Making: Washington, D.C., The H. John Heinz III Center for Science, Economics, and Environment, 221 p.

Hughes, M.L., McDowell, P.F., and Marcus, W.A., 2006, Accuracy assessment of georectified aerial photographs: Implications for measuring lateral channel movement in GIS: Geomorphology, v. 74, p. 1–16, doi:10.1016/j.geomorph.2005.07.001.

Lejon, A.G.C., Renofalt, B.M., and Nilsson, C., 2009, Conflicts associated with dam removal in Sweden: Ecology and Society, v. 14, no. 2, p. 4.

Ligon, F.K., Dietrich, W.W., and Trush, W.J., 1995, Downstream ecological effects of dams: Bioscience, v. 45, no. 3, p. 183–192, doi:10.2307/1312557.

Lisle, T.E., Cui, Y., Parker, G., Pizzuto, J.E., and Dodd, A.M., 2001, The dominance of dispersion in the evolution of bed material waves in gravel bed rivers: Earth Surface Processes and Landforms, v. 26, p. 1409–1420, doi:10.1002/esp.300.

Magilligan, F.J., and Nislow, K.H., 2005, Changes in hydrologic regime by dams: Geomorphology, v. 71, p. 61–78, doi:10.1016/j.geomorph.2004.08.017.

Magilligan, F.J., Haynie, H.J., and Nislow, K.H., 2008, Channel adjustments to dams in the Connecticut River basin: Implications for forested mesic watersheds: Annals of the Association of American Geographers, v. 98, no. 2, p. 267–284, doi:10.1080/00045600801944160 (accessed 2007).

Marin Municipal Water District (MMWD), 2007, Rainfall History: http://www.marinwater.org/mmwd/documents/Rainfall_history_as_2008.pdf.

Neave, M., Rayburg, S., and Swan, A., 2009, River change following dam removal in an ephemeral stream: The Australian Geographer, v. 40, no. 2, p. 235–246, doi:10.1080/00049180902978165.

National Inventory of Dams (NID), 2009, CorpsMap National Inventory of Dams: http://geo.usace.army.mil/pgis/f?p/=397:1:2768265235071349 (accessed 2009).

Palmer, M.A., Reidy, C.A., Nilsson, C., Florke, M., Alcamo, J., Lake, P.S., and Bond, N., 2008, Climate change and the world's river basins: Anticipating management options: Frontiers in Ecology, v. 6, no. 2, p. 81–89, doi:10.1890/060148.

Pizzuto, J., 2002, Effects of dam removal on river form and process: Bioscience, v. 52, no. 8, p. 683–691, doi:10.1641/0006-3568(2002)052[0683:EODROR]2.0.CO;2.

Poff, N.L., and Hart, D.D., 2002, How dams vary and why it matters for the emerging science of dam removal: Bioscience, v. 52, no. 8, p. 659–668, doi:10.1641/0006-3568(2002)052[0659:HDVAWI]2.0.CO;2.

Poff, N.L., Allan, J.D., Bain, M.B., Karr, J.R., Prestegaard, K.L., Richter, B.D., Sparks, R.E., and Stromberg, J.C., 1997, The natural flow regime: A paradigm for river conservation and restoration: Bioscience, v. 47, p. 769–784, doi:10.2307/1313099.

Power, M.E., Dietrich, W.E., and Finlay, J.C., 1996, Dams and downstream aquatic biodiversity: Potential food web consequences of hydrologic and geomorphic change: Environmental Management, v. 20, no. 6, p. 887–895, doi:10.1007/BF01205969.

Renwick, W.H., Smith, S.V., Bartley, J.D., and Buddemeier, R.W., 2005, The role of impoundments in the sediment budget of the conterminous United States: Geomorphology, v. 71, p. 99–111, doi:10.1016/j.geomorph.2004.01.010.

Richter, B.D., Baumgartner, J.V., Powell, J., and Braun, D.P., 1996, A method for assessing hydrologic alteration: Conservation Biology, v. 10, p. 1163–1174, doi:10.1046/j.1523-1739.1996.10041163.x.

Schmidt, J.C., and Wilcock, P.R., 2008, Metrics for assessing the downstream effects of dams: Water Resources Research, v. 44, W04404, doi:10.1029/2006WR005092.

Singer, M.B., 2007, The influence of major dams on hydrology through the drainage network of the Sacramento River Basin, California: River Research and Applications, v. 23, p. 55–72, doi:10.1002/rra.968.

Skalak, K., Pizzuto, J., and Hart, D.D., 2009, Influence of small dams on downstream channel characteristics in Pennsylvania and Maryland: Implications for the long-term geomorphic effects of dam removal: Journal of the American Water Resources Association, v. 45, no. 1, p. 97–109, doi:10.1111/j.1752-1688.2008.00263.x.

Stanley, E.H., and Doyle, M.W., 2002, A geomorphic perspective on nutrient retention following dam removal: Bioscience, v. 52, no. 8, p. 693–701, doi:10.1641/0006-3568(2002)052[0693:AGPONR]2.0.CO;2.

Stanley, E.H., and Doyle, M.W., 2003, Trading off: The ecological effects of dam removal: Frontiers in Ecology, v. 1, no. 1, p. 15–22, doi:10.1890/1540-9295(2003)001[0015:TOTEEO]2.0.CO;2.

State Water Resources Control Board, 2010, Policy for Maintaining Instream Flows in Northern California Coastal Streams, Draft Revised May 2010: State Water Resources Control Board, Oakland, California Environmental Protection Agency, 34 p.

Walter, R.C., and Merritts, D.J., 2008, Natural streams and the legacy of water-powered mills: Science, v. 319, p. 299–304, doi:10.1126/science.1151716.

Ward, J.V., and Stanford, J.A., 1983, The serial discontinuity concept of lotic ecosystems, in Fontaine, T.D., and Bartell, S.M., eds., Dynamics of Lotic Ecosystems: Ann Arbor, Michigan, Ann Arbor Scientific Publishers, p. 347–356.

Ward, J.V., and Stanford, J.A., 1995, Ecological connectivity in alluvial river ecosystems and its disruption by flow regulation: Regulated Rivers: Research and Management, v. 11, p. 105–119, doi:10.1002/rrr.3450110109.

Wharhaftig, C., and Wagner, J.R., 1972. The geologic setting of Tomales Bay, in Corwin, R., ed., Tomales Bay Study Compendium of Reports: Washington, D.C., Conservation Foundation, p. 54–73.

Williams, G.P., and Wolman, M.G., 1984, Downstream Effects of Dams on Alluvial Rivers: U.S. Geological Survey Professional Paper 1286, 83 p.

Williams, P.B., 1991, The debate over large dams: Civil Engineering, August 1991, 4 p.

MANUSCRIPT ACCEPTED BY THE SOCIETY 29 JUNE 2012

The shortcomings of "passive" urban river restoration after low-head dam removal, Ottawa River (northwestern Ohio, USA): What the sedimentary record can teach us

J.E. Evans
Department of Geology, Bowling Green State University, Bowling Green, Ohio 43403, USA

N. Harris
Pennsylvania General Energy Company, 120 Market Street, Warren, Pennsylvania 16365, USA

L.D. Webb
Environmental Protection Agency, Division of Drinking and Ground Waters, P.O. Box 1049, Columbus, Ohio 43216, USA

ABSTRACT

The concept of "passive" river restoration after dam removal is to allow the river to restore itself, within constraints such as localized bank erosion defense where infrastructure or property boundaries are at risk. This restoration strategy encounters difficulties in an urban environment where virtually the entire stream corridor is spatially constrained, and stream-bank protection is widely required. This raises the question of the meaning of river restoration in urbanized settings. In such cases, the sedimentary record can document paleohydrologic or paleogeomorphic evolution of the river system to better understand long-term response to the removal of the dam. Secor Dam was a low-head weir on the Ottawa River flowing through the City of Toledo, Ohio, and its outlying suburbs. The dam was constructed in 1928 and removed in 2007 to enhance aquatic ecosystems, improve water quality, and avoid liability concerns. Predam removal feasibility studies predicted the hydrological and sedimentological responses for the dam removal and determined that reservoir sediments were not significantly contaminated. Postdam removal studies included trenching, sediment coring, geochronology, and surveying. The buried, pre-1928 channel was located and showed that watershed urbanization resulted in channel armoring. Incision in the former reservoir exhumed a woody peat layer that was subsequently shown to be a presettlement hydromorphic paleosol currently buried beneath 1.7 m of legacy sediments, mostly deposited since ca. 1959. Today, the river flows through an incised channel between fill terraces composed of legacy sediments. Additional coring and survey work documented that the channel lateral migration rates averaged 0.32 m/yr over the past ~80 yr, and that the meander wavelength is increasing in response to dam removal. Using sediment budget concepts, significant channel

bank erosion and lateral channel migration should be expected until this river system reworks and removes accumulated legacy sediments currently in floodplain storage. In this dam removal project, "active" restoration practices, such as riparian wetland restoration, would have been more in accord with scientific understandings. That did not happen in this case because of disagreements among different constituencies and because of limitations of funding mechanisms.

INTRODUCTION

Dams have been removed, or proposed for removal, for a variety of reasons, including public safety, liability issues, ecosystem restoration, recreation enhancements, and aesthetics (Evans et al., 2000a; Bednarek, 2001; Heinz Center, 2002; Doyle et al., 2003). The recognition that dam removals are tools, among other tools, in the overall goal of river restoration has been slower to emerge, and it represents a focus of this paper. Important steps in this growing understanding have included distinguishing between short-term and long-term fluvial response to dam removal (Simons and Simons, 1991; Evans et al., 2000b; Pizzuto, 2002; Doyle et al., 2002, 2003; Evans, 2007) and distinguishing between passive and active management practices for reestablishment of equilibrium channel forms, called "late successional channels" by Selle et al. (2007). In the latter context, passive management practices are characterized by allowing the channel to evolve from early successional, headcut-driven form to an anticipated late successional form utilizing little, if any, intervention (Selle et al., 2007). Active management practices, in contrast, attempt to accelerate channel development and guide it toward an anticipated late successional channel form (Selle et al., 2007).

The concept of "river restoration" presupposes both that some criterion or set of criteria about the existing conditions in the river is unsuitable, and that some change or series of changes is feasible that can ameliorate those unsuitable criteria. There are many unspoken assumptions in this attempted definition; for example, feasible changes might be interpreted by some as economically and socially feasible (which might focus on short-term results), while others might interpret feasible to mean self-sustaining (with a long-term emphasis, and possibly with significantly greater economic cost). This is not merely an exercise in semantics because river restorations are societal decisions with trade-offs (Thornton, 2003) where public perception is an important component, most especially in the case of an urban river restoration and dam removal.

This paper is a description of how a passive urban river restoration project, centered around the removal of an obsolete lowhead dam, led first to disagreements about restoration strategies, and then ultimately to additional research that documented the magnitude of recent historical change in this particular river. To summarize, this study reaffirms other recent findings that many rivers in the United States are undergoing long-term systemic changes due to human activity, resulting in channel instability (Wolman, 1967; Costa, 1975; Jacobson and Coleman, 1986; Evans et al., 2000c; Walter and Merritts, 2008; Bain et al., 2008; Wilcock, 2008; DeWet et al., 2011). Those involved in urban river restoration efforts need to be cognizant that short-term restoration strategies will have temporary and possibly mixed results, and that long-term strategies should be based on sediment budget concepts. In addition, this particular restoration project was complicated by the absence of a clearly articulated restoration plan at the onset, involvement of multiple agencies with different restoration goals, and an emphasis on short-term strategies driven by requirements of funding sources. None of these issues should detract from the fact that in this project the dam was successfully removed and ecosystem improvements have been observed.

GEOLOGIC AND HYDROLOGIC BACKGROUND

The Ottawa River is located in northwestern Ohio and southeastern Michigan (Fig. 1). The low-gradient, 446 km² watershed flows into Maumee Bay at the western edge of Lake Erie (Roberts et al., 2007). The watershed is highly urbanized within its lower reaches up to river kilometer (RK) 18 within the City of Toledo, Ohio (2010 urban population of 287,208 people in an area of 218 km² or a population density of 1317 individuals/km²). Further upstream (RK18 to RK40), the watershed is part of the outlying suburbs surrounding the City of Toledo (2010 greater metropolitan area population of 651,429 people in an area of 4193 km² or a population density of 155 individuals/km²). Upstream of RK40, the watershed is predominantly used for agricultural use (corn and soybean row crops) or for mixed use (pasture, parkland, or sand and gravel quarries) (Gottgens et al., 2004; Roberts et al., 2007; Mannik-Smith Group, Inc., 2008; Gerwin, 2003). For the purpose of this paper, the watershed up to RK40 will be referred to as urban (making no distinction between urban and suburban).

The watershed geology consists of Devonian carbonate bedrock (Coogan, 1996) overlain by multiple late Cenozoic till sheets (Forsyth, 1973), ice-contact sand bodies such as glaciolacustrine deltas (Anderhalt et al., 1984), and lacustrine sediments representing multiple stages in the early history of Lake Erie (Forsyth, 1973; Larson and Schaetzl, 2001). Ottawa River sediments are primarily silt and clay, sourced from the fine-grained tills and glaciolacustrine sediments, but there is also a moderate supply of sand, sourced from glaciolacustrine deltas, postglacial beach ridges, and eolian dune fields.

Hydrologic data are available from U.S. Geological Survey (USGS) gauge station no. 04177000 located at RK17.3 at the University of Toledo campus. There are continuous stage-discharge records between 1945 and 1948 and from 1977

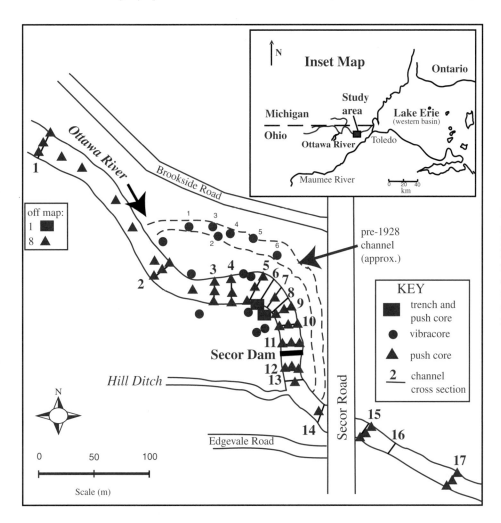

Figure 1. Location map of the study area in northwest Ohio (inset map) and locations of channel cross sections, vibracores, push cores, and trenches. Note the inferred position of the pre-1928 channel (the river was diverted when the Secor Dam was constructed). The trench and cores listed as "off map" are from 500 m upstream.

to present (Fig. 2). The mean daily flow for this interval is 5.3 $m^3\,s^{-1}$, and discharges for the 10, 25, 50, and 100 yr floods have been calculated as 91, 127, 170, and 219 $m^3\,s^{-1}$, respectively (Harris, 2008). The Ottawa River is characterized by "flashy" discharges with relatively low base flows, high peak stormflows, and very short lags-to-peak (Harris, 2008). Similar behavior in other rivers in the region has been attributed to human activities such as draining wetlands, installing agricultural field tile drains, ditching, or channelizing tributaries (Baker et al., 2004). For the Ottawa River, flashy discharge responses to rainfall events are enhanced by the combination of agricultural impacts in the upper parts of the drainage basin and urbanization of the lower parts of the drainage basin (Webb, 2010).

Previous studies on the Ottawa River have shown a good correspondence between discharge and suspended load (r^2 = 0.87), and estimate that bed-load volumes are 10%–35% of suspended load volumes (Gallagher, 1978). Field bed-load measurements over short intervals of time using a variety of bed-load traps have documented highly variable bed-load transport, ranging from zero to 6.3 kg/h, with a maximum (Q_s) of 7×10^{-7} $m^3\,s^{-1}$ (Harris, 2008). Between Lake Erie and approximately RK8, the Ottawa River is an extension of Lake Erie, with slopes approximately zero, and the channel substrate is dominantly silt and clay (Fig. 3). Between approximately RK8 and RK26, the gradient of the Ottawa River is ~0.7 m/km, and the channel substrate consists of moderately sorted, fine- to medium-grained sand that is armored with fine gravel (mostly bivalve shells and anthropogenic materials) and abundant particulate wood debris (Gottgens et al., 2004; Harris, 2008). In the headwaters above RK26, the gradient of the Ottawa River varies from 1 to 6 m/km, and the channel substrates are fine- to medium-grained sand with abundant particulate wood debris.

Within the urbanized portion of the drainage basin, the stream banks are engineered up to approximately RK14. Between approximately RK14 and RK34, the modern Ottawa River channel is incised ~2 m beneath its floodplain surface, creating the appearance of fill terraces (Fig. 4). Because these fill terraces are inundated at regular intervals, and because of other data about the sediment ages, presented in this paper and previously (Evans and Harris, 2008; Webb, 2010), the fill terraces are interpreted as anthropogenic in origin. Within this reach of the Ottawa River, evidence for active vertical and lateral bank erosion includes

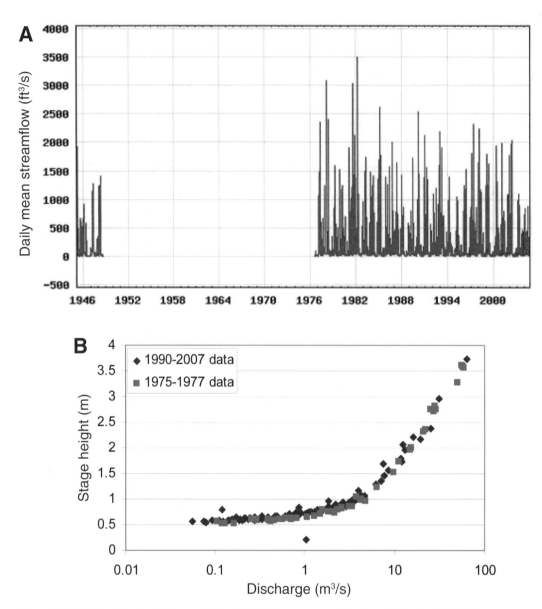

Figure 2. Hydrology data for the Ottawa River. (A) Daily mean streamflow from U.S. Geological Survey gauge station #04177000 between 1945 and 1948 and 1977 and 2006. (B) Rating curve for the gauge station.

undercut banks, rotational slumps, toe-of-slope deposits, small colluvial fans, small-scale soil avalanche or soil-fall deposits, and loss of riparian zone trees through undercutting, tree-lean, and tree-fall into the channel.

The recent land-use history can be summarized as follows. Prior to arrival of European settlers in the early 1700s, this part of northwestern Ohio was ~90% wetlands, wetland forest, and wet prairie known as the "Great Black Swamp." The technology to initiate large-scale drainage modifications did not exist until invention of steam-powered dredgers in the late 1800s (Wilhelm, 1984). By 1879, approximately half of the region had been deforested, and by 1900, almost all of the original forests were gone, thousands of kilometers of drainage ditches had been constructed, and networks of tile drains had been installed in farm fields (Black Swamp Conservancy, 2009). Today, the region retains only ~5% wetlands or wetland forests. The cleared and drained wetlands became highly productive agricultural land, and today the larger watersheds (Maumee River and Sandusky River) are ~85% agricultural land, primarily corn and soybean row crops. Soil erosion from farm fields is a major environmental problem in NW Ohio and adjacent areas; for example, the Maumee River watershed contributes ~1,000,000 m^3 of mostly fine-grained sediment to the City of Toledo harbor each year. Based on this evidence, it can be interpreted that the Ottawa River transitioned from a "blackwater" (organic-rich sediment) stream prior to European settlement to a

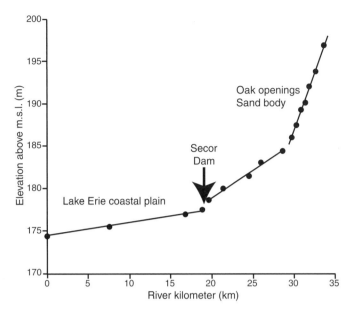

Figure 3. Longitudinal profile for the Ottawa River between Lake Erie (RK0) and the junction with Tenmile Creek and Schreiber Ditch at RK34.5. The Secor Dam was located at RK18.3. m.s.l.—mean sea level.

"brownwater" (mineral-rich sediment) stream as a consequence of land clearance activities.

The Ottawa River watershed differs somewhat from other regional watersheds because its lower portion (below RK40) has been affected by urbanization of the City of Toledo and surrounding suburbs. An analysis by Webb (2010) has shown that the population of outlying suburbs (Village of Ottawa Hills, City of Sylvania, Sylvania Township) increased dramatically in the suburbanization boom that occurred after the end of the Second World War. For example, between 1950 and 1970, the population of the Village of Ottawa Hills increased 86%, the population of the City of Sylvania increased 400%, and the population of Sylvania Township increased 136% (Webb, 2010). Between 1970 and 2000, subsequent population increases have been appreciably less (7%, 56%, and 55%, respectively). These suburban population increases were accompanied by plating and development of new subdivisions; for example, the total number of housing units in the City of Sylvania increased 369% between 1950 and 1970, compared to a subsequent (1970–2000) increase of ~113% (Webb, 2010).

METHODS

Fieldwork involved surveying, sediment coring using various techniques, trenching, bed-load and suspended load sediment sampling, global positioning system (GPS) tracking of bed forms, and specialized sampling for geochronology purposes. Surveying involved pre- and post-dam-removal channel cross sections at 17 locations (Fig. 1), using a Topcon GPT-3003W total station. Differential GPS was used to locate each cross-section pin location with a Trimble Pathfinder Pro® XRS base station unit. The location of each of the cross-section survey pins was established by repeated surveying over a 10 mo period with the Total Station and by direct measurement using differential GPS. The Total Station survey error was ±1 cm, both vertically and horizontally. From repeated measurements over 10 mo, the UTM position uncertainty of any point in a channel cross section was determined to be 0.44 ± 0.11 m (Harris, 2008). Each of the 17 channel cross sections was surveyed at least once prior to dam removal and then approximately monthly during the 6 mo following dam removal, after which time the survey pins had to be removed because of bank remediation efforts.

The types of sediment cores collected included 14 vibracores using a 7.5-cm-diameter aluminum core barrel, and 52 push cores using either 5.1 cm polyvinyl chloride (PVC) pipe or 7.5-cm-diameter aluminum pipe (Fig. 1). The maximum recovered sediment core length was 2.7 m. Trenching involved clearing slumped material from the channel bank at three locations (Fig. 1) and recording stratigraphic information and sampling layers in the field. The maximum trenched interval was 2.6 m depth. The base of each trench was extended downward by collecting an additional sediment core. Given stratigraphic overlap between trenches and cores, the composite stratigraphic interval examined in this study was ~4.5 m thick. Finally, samples were collected for ^{14}C and optically stimulated luminescence (OSL) dating, as described in the following.

In the laboratory, each of the 66 sediment cores was split lengthwise, and one half of the core was archived for future use. The working half was cleaned, photographed, described stratigraphically, and subsampled for grain-size analysis. Grain-size analysis involved removing particulate organic materials and shell debris, and then wet sieving to split sand and mud samples, if necessary. Sand samples were washed and dried, and then sieved through a nest of sieves using a sieve shaker apparatus. Mud samples were dispersed in 5% sodium hexametaphosphate solution, diluted to appropriate sediment concentrations, and then evaluated using a Spectrex PC-2300 laser particle-size analyzer. Grain-size statistics were calculated following the methodology of Folk and Ward (1957). Additional details are given elsewhere (Gottgens et al., 2004; Roberts et al., 2007; Harris, 2008; Webb, 2010).

Geochronology methods included ^{14}C dating, OSL dating, and identifying the age of anthropogenic materials found in trenches. Four peat samples were collected by sampling stratigraphic horizons observed in trenching operations. These samples were sent to Geochron Laboratories for conventional ^{14}C dating, including $\delta^{13}C$ corrections. Details of the sample treatment are given in Webb (2010). The ages were converted to calendar yr before present (cal. yr B.P.) using radiocarbon calibration program Calib revision 5.0.2® (Stuiver and Reamer, 1993). This program produces a probability distribution, and from this, a mean and standard deviation (1σ) are herein reported. The OSL samples were collected in the field from quartz-rich sandy layers exposed in trenches. Six samples were collected by pounding metal pipes 20 cm in length and 2.5 cm in diameter into the stratigraphic layer

Figure 4. Field photographs from before the Secor Dam removal. (A) Secor Dam at high-flow stage, looking upstream. (B) Secor Dam reservoir looking upstream at high-flow stage. (C) Secor Dam at low-flow stage, looking upstream across the riffle-and-plunge pool. (D) Secor Dam reservoir at low-flow stage, looking upstream. (E) Trapped debris and algae growth upstream of the dam. Water-quality issues played a role in spurring efforts to remove the dam. (F) Fill terrace flanking the incised channel (AT—anthropogenic terrace), near cross section 5 (Fig. 1).

of interest. The pipes were then capped and excavated from the trench exposure. Sample preparation involved working removal of carbonates and organics, sieving the sediment, and several heavy liquid separations to obtain the quartz fraction, etching the surface of the grains, and then evaluating the luminescence using a Riso TL/OSL-DA-15A/B luminescence reader at the USGS facility in Denver, Colorado, using the methodology of Murray and Wintle (2000). The sample preparation protocol is described in detail in Webb (2010). Finally, the age of a glass bottle recovered from a stratigraphic horizon exposed in one of the trenches was obtained from its date stamp (Whitten, 2010), and several other anthropogenic materials could be generally matched to the interval of time they were commercially in use.

REMOVAL OF SECOR DAM

Background

The Secor Dam was constructed in 1928 by the Village of Ottawa Hills for recreational purposes. The Secor Dam was a run-of-river dam or weir, 2.5 m tall and 17 m wide, constructed of reinforced concrete (Fig. 4). An older structure of poorly known history was removed from the site, and the river channel was diverted at the time of construction of the Secor Dam. As part of this study, the location of the pre-1928 channel was located and cored to look at pre-urbanization channel substrates (see later section). This extended project spanned preremoval feasibility studies, an analysis of hydrological and sedimentological effects during and immediately after dam removal on 16 November 2007, and follow-up analysis that focused on historical changes in watershed hydrology. In addition, the process of obtaining approval for the removal of the dam necessitated public hearings and other public outreach and interaction. The removal was part of a larger watershed effort to improve water quality and ecosystem health, known as the Maumee River Remedial Action Plan (Maumee River RAP). The efforts involved interagency cooperation among stakeholders, including the owner of the dam (the Village of Ottawa Hills), the Toledo Metropolitan Area Council of Governments (TMACOG), the Ohio Environmental Protection Agency (OEPA), the Ohio Department of Natural Resources (ODNR), the Ohio Department of Transportation (ODOT), the U.S. Army Corps of Engineers (COE), the University of Toledo (UT), Bowling Green State University (BGSU), and a civic organization, Partners for Clean Streams (PCS).

Initial Impetus to Remove Secor Dam

The removal of Secor Dam was first suggested in 2001 by the U.S. Fish and Wildlife Service as a means of restoring natural flows, primarily for the purpose of removing a barrier to native fish migration (*Toledo Blade*, 18 July 2001). The proposal, with some public support, led the owner of the dam, the Village of Ottawa Hills, to hold several public meetings and solicit public comments. A number of other developments helped spur the possibility of the Secor Dam removal. The first was the fact that discussion began almost simultaneously about removing the only other low-head dam on the Ottawa River, which was privately owned and located further upstream (*Toledo Blade*, 6 December 2001). This second dam, the Camp Miakonda Dam, was actually breached, thus posing a clear safety hazard, and was subsequently removed in January 2003. Removing both dams raised the possibility of the Ottawa River becoming free-flowing along its entire length again. The second development was linking improvements in the water quality of the Ottawa River with the larger regional watershed improvements mandated by the Maumee River RAP (*Toledo Blade*, 30 April 2002). Both the Ottawa River and Maumee River flow through the City of Toledo and enter Lake Erie in close proximity. Improving the water quality of the Ottawa River by removing the dams would also be in accord with the ongoing remediation of contaminated sediments at a Superfund site (Hoffman Road Landfill) near the mouth of the Ottawa River. As of 2010, this remediation has grown to a $49 million effort involving dredging of ~200,000 m^3 of contaminated sediment (*Toledo Blade*, 6 May 2010). Finally, $2.5 million funding became available for river mouth dredging for recreational boaters, gaining another constituency for improving the overall health of the watershed.

The Village of Ottawa Hills was also actively interested in the removal of Secor Dam from an environmental stewardship perspective. They wanted to restore the river to unregulated flow and remove the liability aspects of retaining the dam. The Village of Ottawa Hills encouraged, but could not fund, the necessary feasibility studies for the dam removal project. Specific concerns were the extent of sediment contamination, and the hydrological and sedimentological impacts of removing the dam. A group of scientists from two state universities in the region (UT and BGSU) completed these studies (Gottgens et al., 2004; Roberts et al., 2007) supported by a grant from the ODNR Coastal Management Program.

Feasibility Studies

Pre-dam-removal feasibility studies were conducted in 2002–2003. One part of the study was a sediment-routing model based on hydrologic analysis of seven channel cross sections between RK17 and RK27 (including bankfull stage height and bankfull width, and channel slope) and based on evaluation of the channel substrate determined by grain-size analyses of surficial samples and short cores (Gottgens et al., 2004; Roberts et al., 2007). After calculating boundary shear stress and entrainment critical shear stress using the methods described elsewhere (Evans et al., 2002), the sediment-routing model determined transport modes for different grain-size class populations. The results predicted that Secor Dam had trapped between 4500 and 9000 m^3 of mostly sandy sediment in the reach up to 1 km upstream of the dam, and that the anticipated effects of the dam removal would be deposition of most of these sands in a series of pools immediately downstream (<1 km) of the dam.

Another part of the feasibility study used the U.S. Army Corps of Engineers Hydrologic Engineering Centers–River Analysis System (HEC-RAS) to model flood-stage height and lateral flood extent both prior to and subsequent to dam removal for the 10, 25, 50, and 100 yr floods. Topography adjacent to the channel was generated from light detection and ranging (LiDAR) data with 0.3 m vertical resolution. The locations of features were input from digital orthophotography with 0.6 m pixel dimensions into ArcGIS. Following this, 400 valley cross sections oriented perpendicular to the channel were generated at 100 m spacing. The Ottawa River drainage basin was subdivided into 102 subbasins for the purpose of generating runoff curves for specific flood events denoted previously. The results show relatively minor impacts for removal of the dam (Gottgens et al., 2004; Roberts et al., 2007). This is probably because certain bridges located downstream of Secor Dam appear to have a greater impact on constricting flow and creating backwaters than the role played by the dam itself, a weir that had no flood-storage capacity.

Finally, geochemical analyses of Al, As, Cd, Cu, Fe, Ni, Pb, Zn, PCBs (polychlorinated biphenyls), PAHs (polyaromatic hydrocarbons), and petroleum hydrocarbons (C_{11} to C_{31}) were conducted using standard techniques (Gottgens et al., 2004; Roberts et al., 2007). The results showed a moderate level of metals contamination, where As and Cd commonly exceeded threshold effects levels (TEL), Ni and Pb occasionally exceeded TEL, and As and Cd rarely exceeded probable effect levels (PEL). Sediment PCB contamination was low (with rare samples that exceeded TEL values). In contrast, PAH contamination commonly exceeded PEL values, and the river sediments were considered moderately contaminated by petroleum hydrocarbons (Gottgens et al., 2004; Roberts et al., 2007). The observed higher levels of PAHs and petroleum hydrocarbons in Ottawa River sediments were attributed to gasoline, oil, and tar runoff from urban streets and parking lots.

Decision to Remove Secor Dam

The feasibility studies found no significant sediment contamination issues or potential harm from removing the dam. After receipt of these feasibility studies in December 2004, the Village of Ottawa Hills approved a village council resolution to remove the dam.

Public hearings in March 2005 attracted a diverse range of concerns and interests. Articulated concerns included worries that removing the dam would make flooding worse, change water levels, impact mosquito control efforts, or remobilize contaminants (*Toledo Blade*, 31 March 2005). Most of these concerns could be satisfactorily addressed based on the results of the feasibility study. An unexpected line of questioning was whether or not the existing low-head cement dam replicated the ecosystem impact of porous wood debris dams constructed by beavers (*Castor canadensis*), which used to be native to the area. Finally, some opposed the removal project due to possible effects on property values, taxes, or simply because they viewed it as an overstep by a governmental authority. Follow-up to these meetings included exchanges of letters-to-the-editor of local newspapers and news reports in the media for a period of 2 yr, until the dam was actually removed in November 2007. Nevertheless, following these hearings, in April 2005 the Village of Ottawa Hills decided to solicit bids to remove the dam (*Toledo Blade*, 5 April 2005).

This decision to remove the dam initiated two simultaneous and linked discussions, the river restoration plan and the source of funding. After a number of false starts, the funding issue was resolved in the following way: The removal of the dam would be accomplished by an independent contractor hired, supervised, and paid for by the ODOT as part of a wetlands mitigation project in exchange for wetland loss related to widening U.S. Highway 24 adjacent to the Maumee River, in the same watershed. The habitat restoration itself would be funded through an OEPA Section 319 Non-Point Source Program grant, as administered by TMACOG in conjunction with their efforts to improve water quality in Maumee River RAP. TMACOG then consulted COE for advice regarding stream-bank erosion mitigation.

Hydrologic Response to Dam Removal

The dam was breached on 19 November 2007 (Fig. 5A). The sedimentological response to the dam removal included erosion of the sandy bed-load deposits from the former reservoir up to ~150 m upstream of the former dam, translation of sandy bed-load downstream through the site of the former dam, and deposition of most of the material in pools within ~100 m downstream of the former dam, as predicted (Evans and Harris, 2008). In comparison to other dam removals, this response was more localized because the reservoir sediments were primarily sands with fluvial pavements (see later section).

Initially, a knick zone (diffuse knickpoint) formed near the downstream end of the reservoir and migrated upstream (Fig. 5B). The knick zone had an erosional relief of ~10 cm, and it migrated upstream ~83 m within the first few hours before it stalled upon a resistant layer, becoming a broad riffle. The resistant layer was subsequently determined to be an exhumed peat layer (Fig. 5C) representing the presettlement paleosol (see later section). Upstream migration of the knick zone exposed underlying sandy sediments in the former reservoir, and resulted in the formation of several bed forms (Fig. 5D) that migrated down channel and could be tracked using GPS at rates of ≤0.5 m/h (Harris, 2008). An analysis of cross-section data allowed creation of digital elevation models (DEMs) of the channel bed surface topography prior to dam removal, 3 mo after dam removal, and 6 mo after dam removal. The difference indicates that a volume of ~500 m^3 of mostly sand was eroded from the former reservoir within about 3 mo of the dam removal (Harris, 2008), and a volume of ~800 m^3 of mostly sand was eroded within the first 6 mo.

Within the first 6 mo, sequential changes to channel cross sections upstream of the former dam were consistent with predictions from channel evolution models: (1) incision of former reservoir sediments, (2) channel widening facilitated by bank

Figure 5. Field photographs of the Secor Dam removal and impacts. (A) Dam removal on 19 November 2007. (B) Upstream channel incision and reservoir dewatering near cross section 6 (Fig. 1), showing exhumation of the peat layer (shown in C) and downstream bed-form migration (shown in D). (C) Exhumed peat layer discovered to have wide lateral extent, near cross section 6 (Fig. 1). (D) Tracking bed-form migration within the first days after the dam removal, near cross section 8 (Fig. 1). (E) Point bar at the downstream end of the inner bank immediately upstream of the dam prior to dam removal near cross section 9 (Fig. 1). (F) Point bar incision at the same location, 41 d after the dam removal.

failures during continued incision, (3) channel aggradation and incipient floodplain formation, and (4) quasi-equilibrium due to bank consolidation and revegetation (Doyle et al., 2003; Evans, 2007). The Secor Dam removal differed in some details from these models because the reservoir sediments were primarily sands and because there was a preexisting low-stage channel through most of the reservoir sediments (Harris, 2008). For example, in this case, incision began immediately throughout the former reservoir, even though the dam was removed under low-stage conditions, resulting in mobilization of sand as a series of downstream-propagating bed forms, as described earlier, and causing the channel to rapidly narrow. The initial phase of channel incision, reaching a maximum depth of ≥1.3 m, was essentially completed within approximately the first week, and channel widening began within the first several weeks (Fig. 6A).

Within ~100 m downstream of the former dam, two pools acted as sediment traps: the former foot-of-dam plunge pool and a confluence bar pool at the junction of the Ottawa River and Hill Ditch, a small tributary (Fig. 1). Both pools were significantly infilled by the sandy sediment released from the former reservoir; in addition, a gravel riffle between the two pools was significantly modified by sand infilling the gravel matrix (Fig. 6B). Further downstream, there were minimal changes between pre- and post-dam-removal channel cross sections, indicating that very little of the reservoir sand reached as far downstream as cross sections 15–17 (Fig. 1). These survey results were verified by bed-load and suspended load field measurements (Harris, 2008). In summary, the data suggest that most of the ~500 m³ volume of sand mobilized from the former reservoir in the first 3 mo was either trapped in former pools immediately below the dam or transported completely through the study area during high-stage flows when channel surveys or field measurements were not possible (Harris, 2008).

There is evidence of improvements in aquatic ecosystems due to removal of the dam that was a barrier to fish migration. Preliminary results indicate 31 fish species moving upstream of the former dam within the 24 mo after the dam removal, and at least one fish species extending its range downstream (J.F. Gottgens, 2 October 2009, 19 November 2010, written communs.).

Remediation Following Dam Removal

As previously discussed, the successful removal of Secor Dam involved multiple agency partners and funding sources. Throughout the process of designing for and removing the dam, restoration planning became complicated, with the different participants

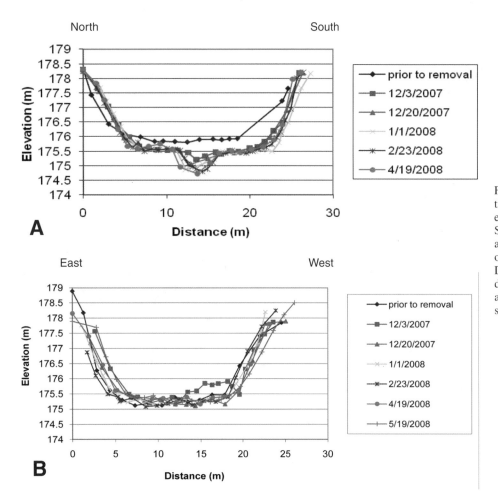

Figure 6. Surveyed channel cross sections showing sequential changes in bed elevation from prior to the removal of Secor Dam until approximately 6 mo after removal. (A) Sequential changes over 6 mo after removal of the Secor Dam at cross section 6 upstream of the dam. (B) Sequential changes over 6 mo after removal of the Secor Dam at cross section 12 downstream of the dam.

each having distinctive goals. For example, ODOT made it clear that its goal was to remove the dam by the most economical manner possible, regardless of restoration goals or concerns. In addition, COE made it clear that restoration meant stream-bank stabilization, although the COE staff members connected with the project were strong advocates for a mix of soft- and hard-stabilization structures instead of entirely hard stabilization structures. Meanwhile, the scientists involved with monitoring the hydrological, sedimentological, and ecological effects of the removal were interested in maintaining the integrity of their studies in the face of unpredictable management and construction activities. These concerns included possible construction zone impact on several species of state-listed freshwater mussels. Finally, the Village of Ottawa Hills and other government entities were concerned with protection of infrastructure components.

In retrospect, it becomes clear that the various consultants were divided into two basic philosophies. The dividing line that was articulated involved perceived threats to infrastructure, although these threats were never quantified to any degree. One infrastructure concern was potential bank failure undermining the Secor Road causeway and bridge (Fig. 1). Another was potential hillslope failure along Edgevale Road near the junction of the Ottawa River and Hill Ditch (downstream of the former dam). Finally, there was concern about buried sewer lines, the locations of which were not well documented. One group of consultants wished to anticipate all potential infrastructure problems with a combination of hard and soft bank-erosion structures. The other group of consultants wished for this river to have the opportunity to adjust to changing conditions, including lateral migration of the channel through the largely undeveloped floodplain in this stretch of the river.

Ultimately, the decisive factor in these discussions was not the science or policy, but the availability of funding resources for the restoration efforts. There were time constraints on spending the available funds, thus bolstering the argument that any potential threats to infrastructure had to be anticipated and addressed immediately. Accordingly, the meander bend upstream of the former dam was remediated as a longitudinal peaked stone-toe protection structure (Derrick and Jones, 2010), one of the former bulkhead walls of the former dam was retained, and the hillslope below Edgevale Road was extensively ripradded. Subsequent vegetative plantings were designed to help reduce any other sites of potential bank erosion (Village of Ottawa Hills, 14 August 2008, personal commun.). The process by which these decisions were made led to the follow-up studies discussed next.

POST-DAM-REMOVAL PALEOGEOMORPHIC ANALYSIS

Historical Changes in Channel Substrate

The Ottawa River was diverted in 1928 to facilitate the construction of the Secor Dam. Today, the former channel is buried beneath 0.8–1.4 m of floodplain sediments adjacent to the existing channel (Fig. 1). Historical engineering plans and existing low areas in the modern topography were used in a successful attempt to locate and vibracore the pre-1928 Ottawa River channel. In six attempts, five vibracores recovered pre-1928 channel substrates, and four of these cored entirely through the channel sediments (Fig. 7). Inspection of the vibracores reveals that the pre-1928 channel was incised into underlying proximal floodplain deposits (inundite facies; see descriptions in Harris, 2008) and that the channel was then partially infilled by fining-upward point-bar deposits, prior to the engineered infilling of the former channel in 1928.

There is a significant contrast in channel substrates between the pre-1928 channel and modern channel (Fig. 8). The pre-1928 substrates are normally distributed (i.e., nearly symmetrical skewness between $+0.10\phi$ and -0.10ϕ), while the modern channel substrates are either strongly coarsely skewed ($<-0.30\phi$), representing fluvial pavements, or strongly finely skewed ($>+0.30\phi$), representing mud-infilled pools at low-stage flow conditions (note: ϕ refers to Udden-Wentworth grain-size class intervals). The most striking contrast is that fluvial pavements (channel armoring) are common in the modern channel but entirely lacking in the pre-1928 channel (Fig. 9). Channel armoring is evidence of incision (Julien, 2002). This historical change is interpreted as one of the impact of urbanization of the watershed, which was most significant after the late 1940s, as discussed previously. Specifically, urbanized drainage systems are contributing more storm runoff to the Ottawa River, and channel degradation is the response to this increase in transport conveyance capacity.

Channel Lateral Migration Rates

A vibracore from the upstream portion of the study area cored through a complete point-bar sequence. This information was used to track the changing location of one of the upstream point bars. From the core location, it is likely that this point bar was active prior to the diversion of the Ottawa River channel in 1928, but it was not part of the diversion itself. The distance between the core location and the base of the modern bar is ≥ 25 m, representing a minimal mean channel migration rate of ≥ 0.32 m yr^{-1} over the past 79 yr (Evans and Harris, 2008). Evidence of continued lateral migration at this location, both prior to and subsequent to dam removal, includes undercut banks, slumps, toe-of-slope deposits, small colluvial fans, soil avalanches, and tree-fall at the outer bank. The outer bank is ~2.5 m tall at this location, and it exposes the stratigraphy of legacy sediments discussed in a later section.

Evidence for Post-Dam-Removal Sinuosity Changes

There were two unexpected results of the dam removal that occurred at the first meander bend upstream of the former dam. These included erosion at the upstream end of the outer bank and erosion of the downstream end of the inner bank. This is a counterintuitive pattern from the behavior of meandering

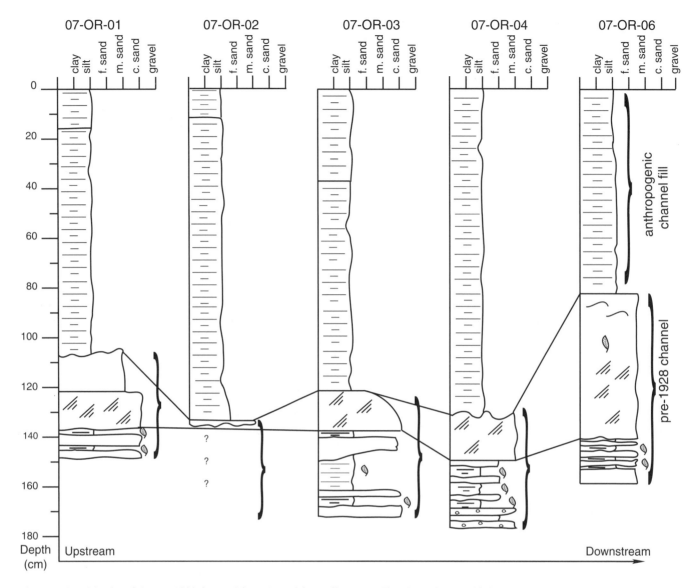

Figure 7. Stratigraphy of the pre-1928 Ottawa River channel from vibracores. The channel was artificially infilled after the river was rerouted to the Secor Dam. The cores are hung from the present ground-level datum, without showing minor variations in topography.

streams, where the zone of maximum boundary shear stress should cross from the upstream side of the point bar on the inner bank to the downstream side of the pool on the outer bank (Dietrich and Smith, 1984). In fact, there was significant erosion on the inner bank, where a point bar would be expected to form in the restored river channel (Figs. 5E and 5F). It appears that the meander wavelength was changing, adjusting the position of the point bar and cutbank.

The meander wavelength (P) is defined as the river path length divided by the valley axis path length for specific reaches of the river. Between RK18 and RK30, the meander wavelength of the Ottawa River varies from 1.2 to 2.1. Prior to dam removal, the reach encompassing Secor Dam (RK18 to RK19) had a meander wavelength of 1.3. However, an evaluation of historical documents shows that the pre-1928 Ottawa River in this reach had a much higher meander wavelength ($P = 2.0$). It is possible that the changes in height and position of the pre-1928 dam and the Secor Dam (between 1928 and 2007) affected the stability of the meander wavelength in this reach (resulting in an historical change from $P = 2.0$ to $P = 1.3$), and that the recent Secor Dam removal is driving recent changes. However, these changes were not permitted to develop naturally and were halted or at least inhibited by bank stabilization structures imposed in 2008, approximately 9 mo after the dam was removed.

Recognition of Presettlement Soil Horizon

As discussed previously, the erosional knick zone created at the time of dam removal migrated upstream until it stalled on a peat horizon exhumed in the bed of the stream (Fig. 10).

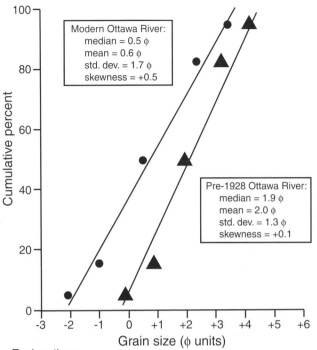

Figure 8. Comparison of average substrate grain-size distributions between the pre-1928 channel and the modern channel. The modern channel data are averaged from 84 surface samples. The pre-1928 channel data are averaged from 15 samples obtained by vibracoring the pre-1928 channel. The comparison highlights the increased grain size and skewness resulting from post-1928 channel armoring, which is attributed to an effect of urbanization.

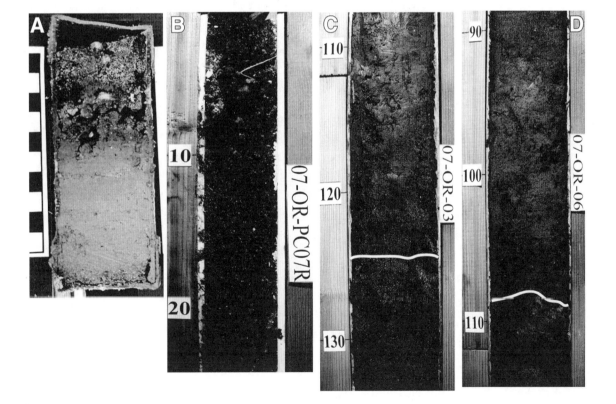

Figure 9. Photographs comparing shallow cores from the modern channel (showing fluvial pavements) with the pre-1928 channel substrate from cores (scales in centimeters). (A–B) Push cores showing modern fluvial pavements from channel armoring. (C–D) Portions of vibracores from the pre-1928 channel. The white line is the contact between the natural channel substrate pre-1928 and the artificial infilling material after the Ottawa River was rerouted to the Secor Dam. Note the lack of fluvial pavement in the pre-1928 substrates.

This peat horizon was the focus of detailed study, with geochemical profiles and textural relationships showing that it is consistent with a prehistorical wetland soil or paleo-Histosol (Webb, 2010). The peat layer is laterally continuous and could be traced throughout the study area between all three trenches and numerous cores. Based upon its lateral and vertical facies associations and geochemical profile, the peat layer can be interpreted as a riparian wetland adjacent to the Ottawa River (Evans and Harris, 2008). The peat layer has a suite of ^{14}C and OSL ages between 4889 ± 179 cal. yr B.P. to 231 ± 15 cal. yr B.P. (Fig. 11), indicating that it spans the age from mid-Holocene age until shortly before initial European land settlement in the region (Webb, 2010).

The 4.5 m composite stratigraphic record shows that the peat horizon overlies fluvial point-bar sequences ("pre–wetland fluvial stage" in Fig. 11). The transition from clastic-rich fluvial deposits to riparian wetlands about (5000 yr) coincides chronologically with a rise to the highest lake levels of Lake Erie (Nipissing I and II stages) that lasted from ca. 5.5 ka to 3.5 ka (Coakley, 1992; Holcombe et al., 2003). Thus, this peat horizon can be interpreted as the result of rising groundwater tables in the region due to the base-level rise of Lake Erie at this time, and the appearance of these wetlands along the Ottawa River can be linked to the formation of the Great Black Swamp throughout northwestern Ohio and portions of southeastern Michigan.

Overlying the peat, there is a succession of carbonaceous muds, silts, and sands, with abundant woody debris. The carbonaceous muds with thin interbedded silt and sand layers are interpreted as deposits from overbank flooding into the riparian wetland, diluting the organic content of the wetland soils. The transition stratigraphically upward from organic-rich sediment to mineral-rich sediment suggests that the riparian wetland was sequentially buried beneath clastic sediment. Other studies interpret similar changes as the transition from "blackwater" (organic-rich) streams to "brownwater" (mineral-rich) streams (Kroes and Hupp, 2010). Although such a transition could be the result of channel migration of the Ottawa River over time, it is suggestive that the uppermost sand layer has an OSL age of 231 ± 15 cal. yr B.P., which is approximately the time of arrival of European settlers in the region. It is likely that land clearance at this time was responsible for fluvial aggradation and infilling of adjacent riparian wetlands.

Within the "channel instability stage" (Fig. 11) and stratigraphically superimposed on these older interbedded clastic and organic deposits are two prominent sand layers that can be traced laterally through all three trenches and numerous cores. The first

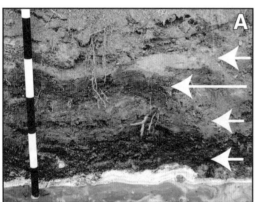

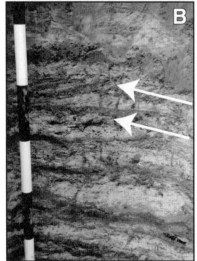

Figure 10. Field photographs of the stratigraphy of one of the trenches. (A) Peat (hydromorphic paleosol) overlain by sand with Fe-oxide grain coatings and carbonaceous muds. The upstream knick-zone migration after the dam was removed stalled on this resistant layer, and it formed a riffle. The presence of the peat is interpreted to represent the transition from blackwater (organic-sediment rich) streams prior to land clearance to brownwater (mineral-sediment rich) streams after land clearance. (B) Flood couplets representing floodplain aggradation. The cross-bedded sand layer has optically stimulated luminescence (OSL) ages indicating it is the historic 1959 flood. Overlying this sand is 1.7 m of silty sediment representing high rates of vertical accretion during watershed urbanization. Scale bars are 10 cm increments.

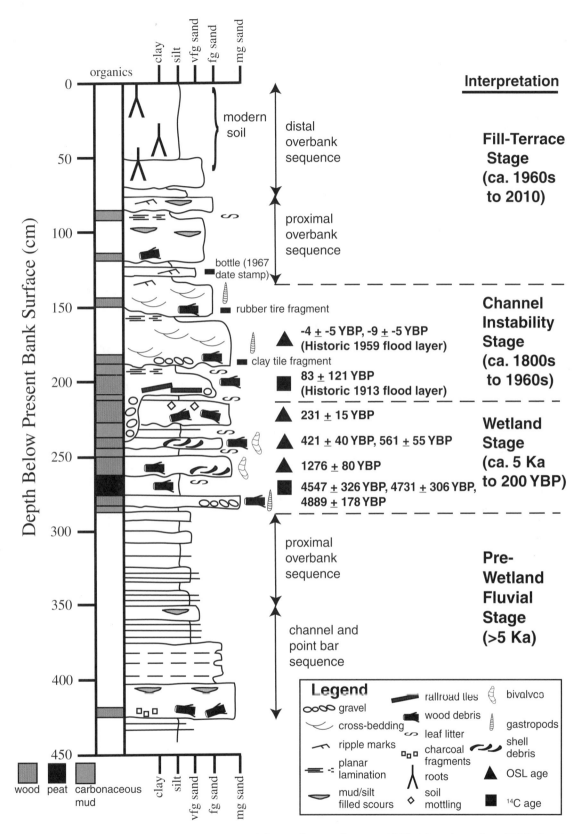

Figure 11. Composite stratigraphic section created from adjacent vibracore (09OR-11) and trenches (09OR-5 and 09OR-6). Geochronology data are matched to specific sample horizons. See text for discussion. YBP—yr B.P.; OSL—optically stimulated luminescence; vfg—very fine grained; fg—fine grained; mg—medium grained.

sand layer scours approximately 40 cm into the underlying deposits, contains pebbles, and contains anthropogenic debris including a layer of railroad ties (Fig. 11). Organic debris in this sand has a ^{14}C age of 83 ± 121 cal. yr B.P. This sand layer is interpreted as the flood layer from the historic flood of 1913, which devastated much of the region. The second sand layer has a suite of OSL dates ranging from −4 ± 5 cal. yr B.P. to −9 ± 5 cal. yr B.P., in other words between calendar years 1954 and 1959. Historical flood records point to a major flooding event that occurred during January 1959, after a storm that involved 3.7 cm rainfall on frozen ground. The historic 1959 flood submerged floodplains, roads, the decking of major highway and railroad bridges, and caused significant flood damage throughout the region (*Toledo Blade*, 21 January 1959).

Overlying the historic 1959 flood horizon, there is a 1.3–1.7-m-thick interval of mostly yellowish silts that are organized into repetitive fining-upward sequences of thin sands, silts, and thin capping muds ("fill-terrace stage" in Fig. 11). These deposits are interpreted as post-1959 historical flood horizons. The flood horizons contain anthropogenic materials, including a glass bottle with a 1967 date stamp and plastic debris consistent with post-1970s ages (Webb, 2010). We interpret this upper package of clastic floodplain sediment to represent vertical accretion of the floodplains, creating the present fill-terrace morphology. The minimum floodplain sedimentation rates for this section are between 2.85 cm yr^{-1} and 3.15 cm yr^{-1}, which are approximately three orders of magnitude higher than presettlement rates (Webb, 2010). Indirect evidence in support of these extraordinarily high floodplain aggradation rates includes the multiple sets of adventitious roots of floodplain trees that are now exposed in channel cutbanks.

DISCUSSION

Historical Changes in Rivers

Numerous previous studies have considered human impacts on rivers in eastern North America. Generally speaking, there are at least three and possibly four partly overlapping phases in the historical development of these human impacts: (1) a phase that began in the early 1600s on the East Coast and by the mid-1800s had reached the Upper Mississippi Valley region, consisting of the effects of land clearance for agriculture and other development activities such as constructing milldams by European settlers; (2) a phase that began in the mid-1900s consisting of the effects of soil conservation practices on stream sediment loads; and (3) a phase that saw its greatest effects after the mid-1900s, consisting of the effects of urbanization on watershed hydrology and stream sediment budgets. The fourth component, which remains poorly understood, would have preceded all of these, and consists of the way in which the "presettlement" land management practices of Native Americans affected rivers and/or how variations in the populations of beavers affected watershed hydrology. Each will be considered in the following.

The arrival of European settlers in eastern North America starting in the 1600s had two effects: (1) significant increases in watershed sediment supply due to upland deforestation and land clearance for agriculture (Jacobson and Coleman, 1986), and (2) the concurrent increase in intrabasinal sediment storage (alluvium and colluvium) from a variety of processes including the relative base-level rise from construction of large numbers of milldams (Walter and Merritts, 2008) and/or the accelerated subsidence and infilling of impacted riparian wetlands (Kroes and Hupp, 2010). It has been argued that the result was inundation, burial, and sequestration of presettlement riparian wetlands, leading to the development of silt-rich, broad floodplains with incised, laterally migrating channels, i.e., the classical meandering form (Walter and Merritts, 2008), although disagreements exist (Bain et al., 2008; Wilcock, 2008). From a sediment budget viewpoint, sediment input overwhelmed the transport conveyance capacity of the fluvial system, causing the accumulation of sediment volumes within the drainage basin (i.e., as intrabasinal storage in the form of alluvium or colluvium). Intrabasinal storage due to such anthropogenic causes has also been termed "legacy sediments."

Sediment budgets have been further impacted by recent (generally <50 yr) land-use changes, driven by both economic and social factors. In some regions, watershed sediment inputs have significantly declined due to abandonment of agricultural fields and subsequent reforestation (Wolman, 1967; Costa, 1975; Jacobson and Coleman, 1986). In other areas, agricultural soil erosion has declined as a result of improved soil conservation practices (Trimble and Lund, 1982; Kuhnle et al., 1996). While these changes in agricultural land use have profound impacts on sediment supply, recent studies have shown that these have been accompanied by minimal changes in streamflow (Cruise et al., 2010). However, these reductions in watershed-scale sediment inputs are not necessarily matched by reductions in stream sediment loads or sediment yield (Knox, 1987). The implication is that these streams are maintaining sediment loads and compensating for reduced upland soil erosion inputs by eroding or reworking intrabasinal storage. Examples include the decoupling of reservoir sedimentation rates from current land-use practices (Evans et al., 2000c) and the persistence of high sediment yields despite significant soil conservation efforts (Faulkner and McIntyre, 1996). These studies highlight the need to apply sediment budget concepts to understanding watershed changes (Evans et al., 2000c; Allmendinger et al., 2007).

Finally, urbanization represents a special case for sediment budgets. During the urbanization of a drainage basin, the initial soil erosion effects of land clearance and housing construction can produce sediment input rates that are orders of magnitude higher than the equivalent effect from agricultural fields (Wolman and Schink, 1967; Gellis et al., 1996; Allmendinger et al., 2007). Coinciding with increased sediment inputs in urbanizing watersheds are hydrologic modifications such as construction of impermeable surfaces and routing of runoff through urban storm-drain networks that increase runoff (Carter, 1961; Wolman, 1967; Booth and Jackson, 1997), increase peak streamflow discharge (Beighley and Moglen, 2002; Meierdiercks et al., 2010), and change channel morphology (Pizzuto et al., 2000; Segura and Booth, 2010). However, unlike the continued, long-term impact of agriculture (as long as the fields are farmed), sediment inputs in urbanized areas should decline over time as disturbed ground such as new housing developments becomes revegetated. Thus, unlike the long-term continuous sediment inputs from agricultural drainages, sediment inputs from urbanized drainages can be expected to be more episodic (related to historical intervals of peak urbanization or suburbanization in a region), and may be masked to some extent by increases in transport capacity (related to increased storm runoff in urbanized drainages). Thus, after the initial pulse of sediment from land clearance, the long-term effect of urbanization may be an increase in transport conveyance capacity, which will be manifested by bank erosion, incision, and reworking of sediment held in intrabasinal storage.

In summary, in any particular drainage basin in eastern North America, human impacts on fluvial systems generally include: (1) initially high sediment input rates related to land clearance, followed by (2) reductions in sediment input due to the abandonment and revegetation of farm fields, or improved soil conservation practices on currently farmed fields, or the revegetation of disturbed urbanized areas. During the initial phase, sediment input rates tended to vastly exceed conveyance capacity and lead to increased intrabasinal storage of legacy sediment as alluvium or colluvium. During the later phase, reduced sediment input rates lead to erosion (remobilization and reworking of previously stored legacy sediment). In other words, according to sediment budget concepts, reductions in watershed sediment input rates (often a consequence of public policy to improve soil conservation practices) *inevitably* leads to enhanced rates of incision or bank erosion. This will continue until such time as the fluvial system has adjusted sediment supply to transport conveyance capacity, which in all likelihood means until the legacy sediments are reworked and removed from the drainage basin—regardless of whether or not this was part of public policy, or an intended or unintended consequence of such policy.

Historical Changes of the Ottawa River

Historical changes in the Ottawa River, in northwestern Ohio, are an important link between studies in the Piedmont area of the East Coast of North America (Leopold, 1956; Wolman, 1967; Costa, 1975; Jacobson and Coleman, 1986; Allmendinger et al., 2007; Walter and Merritts, 2008; deWet et al., 2011) and studies in the Upper Mississippi Valley region (Knox, 1977, 1987; Trimble 1981, 1983). This study confirms earlier findings about an initial phase of increased sediment inputs resulting in burial of presettlement soils, sediment storage in presettlement wetlands, and vertical aggradation of floodplains, followed by a phase of reduced sediment inputs resulting in incision and bank erosion. These observations are interpreted conceptually using sediment budgets.

Today, the Ottawa River is a meandering stream with typical features such as point bars, confluence bars, cutbanks, and meander-loop cutoffs. However, for much of its length, the stream is deeply incised, with ≤2.0-m-tall banks hydraulically separating the channel from its floodplain except under very limited high-stage flow conditions. The high banks are sites of erosional undercutting, slumping, and tree-fall, creating erosion problems faced by land managers. It was a natural thought progression to try to address these problems as part of the river restoration effort that included removal of the Ottawa Dam.

The problem is that certain assumptions underlying this restoration approach are incorrect. The presently existing river is not in equilibrium, is demonstrably unstable, and looked fundamentally different in presettlement times. The picture that emerges is that this was a blackwater stream flanked by extensive riparian wetlands (sites of peat formation), with relatively small sediment loads, and apparently low-relief channel banks (allowing floodwaters to carry in silt and particulate organic matter). Among other values, these wetlands provided extensive flood storage capacity. After arrival of European settlers, high sediment loads associated with land clearance changed the Ottawa River to a brownwater stream. At first, the wetlands provided significant accommodation space for storage of excess sediment supply, until the wetlands were infilled, buried, and evolved into clastic floodplains. The evidence also shows that change (sediment vertical accretion in floodplains) accelerated during urbanization after World War II. Thus, the resulting clastic floodplains and fill-terrace morphology presently flanking the modern incised channel are relatively recent features. Whether or not those banks should be defended is both questionable and possibly counterproductive. Using a sediment budget approach to this situation, one could argue that this river is going to continue to attempt to erode and rework legacy sediments until they are either removed from the fluvial system or at least until the sediment supply and transport conveyance capacity are balanced. It may or may not be practical to allow this to happen, depending upon the need to defend or move infrastructure or the effect on property boundaries. However, these issues should at least be openly debated as part of any "river restoration" plan.

Setting aside the historical changes caused by land clearance and urbanization, there are still the more recent hydrological changes that are very directly related to urbanization of the watershed. These include: (1) the fairly recent transition from natural substrates to fluvial pavements (armored surfaces), (2) the measured rate of lateral channel migration of 0.32 m yr^{-1} subsequent to diverting the channel of the river during the 1928 construction of the dam, and (3) the evident attempt of the river to adjust meander wavelength in response to the dam removal in 2007. All of these are evidence of ongoing channel instability and should also be addressed in formulating any river restoration plan. Certainly, it should be a last-resort option to stabilize the channel in place when it is so clearly in the process of adjusting to changes operating over both decadal and centennial time scales.

Public Policy Aspects of This Case Study

The Secor Dam removal and related river restoration efforts for the Ottawa River in northwestern Ohio are an example of a "passive" river restoration project in an urban watershed, and they provide some important insights about these types of projects and about river restoration efforts in general. The first is the importance of understanding the magnitude and rate of historical changes in any project river. This study confirms other findings showing that there has been a two-phase impact of human activity on stream sediment budgets: (1) There was an initial phase of significant intrabasinal storage of legacy sediments due to high sediment input rates from land clearance, disturbance, and soil erosion from agriculture or housing construction in urbanized areas, and (2) there was a more recent phase of incision, bank erosion, lateral channel migration, and riparian tree-fall into channels due to subsequent reductions in sediment inputs (from farm field abandonment, or improved soil conservation practices, or from revegetation of urban construction sites, and combinations thereof) and reworking of legacy sediments. An understanding of sediment budget concepts is critical. For example, these occurrences of localized bank erosion issues are better explained to the general public not as "problems" but as "manifestations of a problem," where that broader underlying problem consists of ongoing, long-term adjustments to sediment budgets.

Second, urban river restoration projects are likely, as this project was, to involve numerous agencies and various constituencies, and might, as this project did, have a complicated funding mechanism reliant on multiple funding sources with different criteria and expectations. In these instances, communication is a key factor, and a unified restoration plan agreed upon by all parties should be the most important first task. The alternative is a more ad hoc approach of decision making, where decisions made by one party, and not necessarily agreed to by others, inevitably limit available choices down the road. An important component of this is the realities of actions and decisions permitted by the different funding agencies. In retrospect, there was a gap in understanding the difference between issues directly controlled by the removal of the dam and more systemic issues that transcended the spatial extent of the dam's influence.

Third, the easiest choice is not necessarily the best choice or even the least expensive choice (in terms of time, money, or effort). Particularly in a project with multiple agencies and constituencies, it was easier to arrive at some operational level of agreement with a "passive" restoration approach (remove the dam and let the river adjust) while targeting specific "hotspots" of bank erosion. There are multiple problems with this approach. First, it effectively ignored all of the advances in scientific understandings of historical changes in rivers. Second, it will require scrambling to constantly repair local problems. Third, it may eventually lead to an engineered solution where much of the banks are protected, raising the question of what constitutes success in river restoration projects (Florsheim et al., 2008). Certainly, it would be hard to advocate that a riprapped ditch is a successful restoration. Countering the arguments for engineered solutions requires hard data to document the rates and magnitude of change in a particular river. This should not be oversimplified to mean there was some pristine predisturbance state that can be reattained, a statement that would not only be theoretically incorrect but practically impossible to achieve (Bain et al., 2008; Wilcox, 2008).

What could have been done in this particular project? Given the evidence that this river transitioned from a blackwater stream flowing through low banks adjacent to riparian wetlands, an interesting active river restoration effort in this case might have been to scale back the anthropogenic fill terraces, confine the low-stage flow between low berms, and restore riparian wetlands outboard of those low berms. The restored riparian wetlands would have provided important habitat, acted to increase floodwater storage, and improved water quality due to natural filtration. Boardwalks or other access points with explanatory signage could have provided a public education function.

That did not happen in this project, but it could have. We propose that hydrologists have a particular role to play in such projects of educating public policy decision makers, the general public, and possibly other involved scientists or engineers about the evidence for, and implications of, long-term anthropogenic impacts on a river system using a sediment budget approach. In projects such as this one, focusing on the removal of a low-head dam, the resulting incision, bank erosion, substrate changes, and lateral migration of the channel are inevitable consequences of manipulations of the stream sediment budget. Failure to understand these key concepts will result in restored rivers that are highly engineered and do not serve the functions of natural river systems.

SUMMARY AND CONCLUSIONS

For a project focusing on removal of a low-head dam, the removal of the Secor Dam on the Ottawa River was relatively unusual because of the wealth of both pre- and post-dam-removal scientific studies. Such studies predicted and then measured the hydrological and sedimentological response of the dam removal (Evans and Harris, 2008; Harris, 2008; Roberts et al., 2007), and looked at synoptic issues of historical changes in the Ottawa River due to urbanization and land clearance (Webb, 2010). The paleohydrological evidence, including data from the pre-1928 paleochannel, showed that the effect of urbanization of the watershed was creation of fluvial pavements in the channels, lateral channel migration rates of 0.32 m yr^{-1}, and rapid floodplain aggradation rates (>1.7 m of aggradation since 1959). The removal of the dam resulted in incision, channel widening, and the start of channel aggradation (and channel narrowing), as anticipated from models and previous case studies (Doyle et al., 2003; Cui et al., 2006; Evans, 2007; Harris, 2008). Incipient changes in meander sinuosity (erosion at the upstream end of the outer bank and erosion at the downstream end of the inner bank) were not anticipated, but they are consistent with overall channel instability. In other words, the conclusion, well supported by the science, is that the Ottawa River has been in a prolonged (both decade- and century-scale) state of channel instability linked to drainage-basin scale anthropogenic activities.

Geomorphic changes related to channel instability have been the focus of numerous studies. One approach is to contrast sediment supply (ratio of available sediment below the dam to available sediment above the dam) with transport conveyance capacity (ratio of the frequency of sediment-transporting flows after dam removal to before dam removal) (Grant et al., 2003). Prior to dam removal, the Ottawa River showed evidence for sediment supply less than transport capacity, such as incision, armoring, and bank erosion. The Ottawa River returned to this condition approximately 3 mo after the dam removal. The intervening ~3 mo interval immediately after dam removal was characterized by sediment supply greater than transport capacity, such as pool infilling by fine-grained sediment, infiltration of fines into gravel matrices, bar construction and migration, and overall channel aggradation. These changes indicate channel instability over decadal time scales that preceded removal of the dam and may continue. Already, the Ottawa River is attempting to outflank or undermine the imposed bank stabilization structures that were the outcome of this river restoration project.

Looking further back in time, paleohydrologic analysis makes it clear that the Ottawa River has been in a status of channel instability over at least the past 60 yr (post-1950), specifically related to urbanization of the drainage basin. These changes included episodic inputs of sediment supply in excess of transport capacity, leading to extraordinarily high rates of floodplain aggradation, lateral channel migration, and flashy discharge associated with higher runoff volumes and frequency related to urbanization of drainage networks. There is no indication that these driving causes of channel instability will not persist for many decades to come. Accordingly, significant incision, armoring, and bank erosion are highly probable outcomes. Even further back in time, operating over a time scale of several centuries, land clearance associated with the arrival of European settlers in this region changed blackwater (organic-sediment rich) streams flowing through riparian wetlands to brownwater streams where wetlands were infilled by clastic sediment and evolved to silt-rich floodplains.

None of this is surprising. However, the evidence for prolonged channel instability raises significant hurdles for any river restoration project. First, there is a need to evaluate whatever baseline model is to be used to guide river restoration, given the likelihood the river has been in a long-term state of channel instability. This is the key step where paleohydrologic and paleogeomorphic analyses can provide necessary data. Second, there is a need to recognize that removing a dam is not synonymous with river restoration. Removing a dam is better stated as a societal decision to replace one aquatic ecosystem with another (Thornton, 2003), or (in our words) to replace one case of channel instability with another. The ramifications are significant—management of a restoration project must be understood as a long-term commitment.

Third, rivers should be evaluated as systems and given sufficient lateral space to erode, deposit, and adjust sediment loads, in contrast to the prevailing policy of defending the arbitrary location at which an unstable channel happened to find itself at a particular time. Even in this highly urbanized setting, there was room for active and creative solutions. For example, the fill terraces could have been scaled back, a low berm could have been constructed to channel the river at low-flow stage, and the floodplain could have been replaced with reconstructed riparian wetlands. In this particular case, there was too much emphasis on removing the dam (which was the easiest and least complicated part of the entire project) and neither the patience nor the funding viability to deal with the complex fluvial response that followed. In other words, the true goal of river restoration was lost in the decision-making process.

ACKNOWLEDGMENTS

We wish to thank the Village of Ottawa Hills, particularly Marc Thompson, for their continued support and encouragement of scientific studies on their property. We thank our coworkers, Hans Gottgens, Todd Crail, and Patrick Lawrence (University of Toledo) and Sheila J. Roberts and Enrique Gomezdelcampo (Bowling Green State University), for sharing information and ideas. This study received significant support and advice from Matt Horvatt (Toledo Metropolitan Area Council of Governments [TMACOG]), Cherie Blair (Ohio Environmental Protection Agency), and David Derrick (U.S. Army Corps of Engineers). Numerous graduate and undergraduate students at Bowling Green State University contributed to the collection of data; in particular, we wish to thank Andrew Clark, Allan Adams, Colleen O'Shea, Zach Mueller, Matt Bradford, Meghan Castles, Will Emery, Steven King, Jessica Lawrence, Chris Pepple, Mary Scanlan, and Thor Zednik. We thank the Ohio Geological Survey, particularly Mike Angle, for equipment support. This study benefited from funding support from the Ohio Department of Natural Resources (Gottgens and Evans), Ohio Environmental Protection Agency (TMACOG), Geological Society of America (Harris and Webb), and Bowling Green State University. Finally, this manuscript has been significantly improved by reviews from Jerome V. De Graff, Andrew Wilcox, and Andrew Selle.

REFERENCES CITED

Allmendinger, N.E., Pizzuto, J.E., Moglen, G.E., and Lewicki, M., 2007, A sediment budget for an urbanizing watershed 1951–1996, Montgomery County, Maryland, U.S.A.: Journal of the American Water Resources Association, v. 43, p. 1483–1498.

Anderhalt, R., Kahle, C.F., and Sturgis, D., 1984, The sedimentology of a Pleistocene glaciolacustrine delta near Toledo, Ohio, in Carter, C.H., and Guy, D.E., Jr., eds., Field Guidebook to the Geomorphology and Sedimentology of Late Quaternary Lake Deposits, Southwestern Lake Erie: 14th Annual Field Conference of the Great Lakes Section of SEPM: Tulsa, Oklahoma, SEPM, p. 59–90.

Bain, D.J., Smith, S.M.C., and Nagle, G.N., 2008, Reservations about dam findings: Science, v. 321, p. 910, doi:10.1126/science.321.5891.910a.

Baker, D.B., Richards, R.P., Loftus, T.T., and Kramer, J.W., 2004, A new flashiness index: Characteristics and applications to Midwestern rivers and streams: Journal of the American Water Resources Association, v. 40, p. 503–522, doi:10.1111/j.1752-1688.2004.tb01046.x.

Bednarek, A.T., 2001, Undamming rivers: A review of the ecological impacts of dam removal: Environmental Management, v. 27, p. 803–814, doi:10.1007/s002670010189.

Beighley, R.E., and Moglen, G.E., 2002, Assessment of stationarity in rainfall-runoff behavior in urbanizing watershed: Journal of Hydrologic Engineering, v. 7, p. 27–34, doi:10.1061/(ASCE)1084-0699(2002)7:1(27).

Black Swamp Conservancy, 2009, Black Swamp Conservancy: http://www.blackswamp.org/ (accessed 14 April 2009).

Booth, D.B., and Jackson, C.R., 1997, Urbanization of aquatic systems: Degradation thresholds, stormwater detection, and the limits of mitigation: Journal of the American Water Resources Association, v. 33, p. 1077–1090, doi:10.1111/j.1752-1688.1997.tb04126.x.

Carter, R.W., 1961, Magnitude and Frequency of Floods in Suburban Areas: Washington, D.C.: U.S. Geological Survey Professional Paper 424-B, p. 9–11.

Coakley, J.P., 1992, Holocene transgression and coastal-landform evolution in northeastern Lake Erie, Canada, *in* Quaternary Coasts of the United States, Marine and Lacustrine Systems: Tulsa, Oklahoma, SEPM (Society for Sedimentary Geology) Special Publication 48, p. 416–426.

Coogan, A.H., 1996, Ohio's surface rocks and sediments, *in* Feldman, R.M., and Hackathorn, M., eds., Fossils of Ohio: Ohio Division of Geological Survey Bulletin 70, p. 31–50.

Costa, J.E., 1975, Effects of agriculture on erosion and sedimentation in the Piedmont province, Maryland: Geological Society of America Bulletin, v. 86, p. 1281–1286, doi:10.1130/0016-7606(1975)86<1281:EOAOEA>2.0.CO;2.

Cruise, J.F., Layman, C.A., and Al-Hamdan, O.Z., 2010, Impact of 20 years of land-cover change on the hydrology of streams in the southeastern United States: Journal of the American Water Resources Association, v. 46, no. 6, p. 1159–1170, doi:10.1111/j.1752-1688.2010.00483.x.

Cui, Y., Parker, G., Braudrick, C., Dietrich, W.E., and Cluer, B., 2006, Dam removal express models (DREAM): Part I. Model development and validation: Journal of Hydraulic Research, v. 44, p. 291–307, doi:10.1080/00221686.2006.9521683.

Derrick, D.L., and Jones, M.M., 2010, Innovative Streambank Protection in an Urban Setting, Accotink Creek, VA: U.S. Army Corps of Engineers Report ERDC/CHL CHETN-VII-10, August 2010, 22 p.

deWet, A., Williams, C.J., Tomlinson, J., and Loy, E.C., 2011, Stream and sediment dynamics in response to Holocene landscape changes in Lancaster County, Pennsylvania, *in* LePage, B.A., ed., Wetlands, Integrating Multidisciplinary Concepts: New York, Springer Science + Business Media B.V., doi:10.1007/978-94-007-0551-7_3.

Dietrich, W.E., and Smith, J.D., 1984, Bed load transport in a river meander: Water Resources Research, v. 20, p. 1355–1380, doi:10.1029/WR020i010p01355.

Doyle, M.W., Stanley, E.H., and Harbor, J.M., 2002, Geomorphic analogies for assessing probable channel response to dam removal: Journal of the American Water Resources Association, v. 38, p. 1567–1579, doi:10.1111/j.1752-1688.2002.tb04365.x.

Doyle, M.W., Stanley, E.H., and Harbor, J.M., 2003, Channel adjustments following two dam removals in Wisconsin: Water Resources Research, v. 39, p. 1011–1026, doi:10.1029/2002WR001714.

Evans, J.E., 2007, Sediment impacts of the 1994 failure of IVEX dam (Chagrin River, NE Ohio): A test of channel evolution models: Journal of Great Lakes Research, v. 33, Special issue 2, p. 90–102.

Evans, J.E., and Harris, N., 2008, Preliminary study of the sediment impacts of the 2007 removal of the Secor Dam (Ottawa River, Ohio), *in* Davenport, T., Gibson, R., Bonnell, J., and D'Ambrosio, J., eds., Conference Proceedings of the 16th National Nonpoint Source Monitoring Workshop, 14–18 September 2008: Columbus, Ohio, U.S. Environmental Protection Agency/Ohio Environmental Protection Agency/Ohio State University Extension Service, p. 57–58.

Evans, J.E., Mackey, S.D., Gottgens, J.F., and Gill, W.M., 2000a, From reservoir to wetland: The rise and fall of an Ohio dam, *in* Schneiderman, J.A., ed., The Earth around Us: Maintaining a Livable Planet: San Francisco, W.H. Freeman, p. 256–267.

Evans, J.E., Mackey, S.D., Gottgens, J.F., and Gill, W.M., 2000b, Lessons from a dam failure: The Ohio Journal of Science, v. 100, p. 121–131.

Evans, J.E., Gottgens, J.F., Gill, W.M., and Mackey, S.D., 2000c, Sediment yields controlled by intrabasinal storage and sediment conveyance over the interval 1842–1994, Chagrin River, northeast Ohio: Journal of Soil and Water Conservation, v. 55, p. 263–269.

Evans, J.E., Levine, N.S., Roberts, S.J., Gottgens, J.F., and Newman, D.M., 2002, Assessment using GIS and sediment routing of the proposed removal of Ballville Dam, Sandusky River, Ohio: Journal of the American Water Resources Association, v. 38, p. 1549–1565, doi:10.1111/j.1752-1688.2002.tb04364.x.

Faulkner, D., and McIntyre, S., 1996, Persisting sediment yields and sediment delivery changes: Water Resources Bulletin, v. 32, p. 817–829, doi:10.1111/j.1752-1688.1996.tb03479.x.

Florsheim, J.L., Mount, J.F., and Chin, A., 2008, Bank erosion as a desirable attribute of rivers: Bioscience, v. 58, p. 519–529, doi:10.1641/B580608.

Folk, R.L., and Ward, W.C., 1957, Brazos River bar: A study on the significance of grain-size parameters: Journal of Sedimentary Petrology, v. 27, p. 3–26.

Forsyth, J.L., 1973, Late-glacial and postglacial history of western Lake Erie: Compass, v. 51, p. 16–26.

Gallagher, R.E., 1978, Ottawa River (Tenmile Creek) Hydrology: Streamflow Quantity and Erosion at Toledo, Ohio [M.S. thesis]: Toledo, Ohio, University of Toledo, 187 p.

Gellis, A.C., Webb, R.M.T., Wolfe, W.J., and McIntyre, S.C.I., 1996, Land use, upland erosion, and reservoir sedimentation, Lago Loiza, Puerto Rico: Geological Society of America Abstracts with Programs, v. 28, no. 7, p. 79.

Gerwin, K., 2003, Land-Use Changes in the Ottawa River Watershed: http://www.utoledo.edu/as/geography/pdfs/research_gerwin2.pdf (accessed 14 April 2009).

Gottgens, J.F., Evans, J.E., Levine, N.S., Roberts, S.J., and Spongberg, A.L., 2004, Dam Removal in the Ottawa River, Ohio: A Feasibility Study: Toledo, Report to the Ohio Department of Natural Resources, Office of Coastal Management, 72 p.

Grant, E.G., Schmidt, J.C., and Lewis, S.L., 2003, A geological framework for interpreting the downstream effect of dams on rivers: American Geophysical Union, Water Science and Application, v. 7, p. 203–219.

Harris, N., 2008, Sedimentological Response of the 2007 Removal of a Low-Head Dam, Ottawa River, Toledo, Ohio [M.S. thesis]: Bowling Green, Ohio, Bowling Green State University, 228 p.

Heinz Center, 2002, Dam Removal: Science and Decision Making: Washington, D.C., The Heinz Center for Science, Economics, and the Environment, 221 p.

Holcombe, T.L., Taylor, L.A., Reid, D.F., Warren, J.S., Vincent, P.A., and Herdendorf, C.E., 2003, Revised Lake Erie postglacial lake level history based on new detailed bathymetry: Journal of Great Lakes Research, v. 29, p. 681–704, doi:10.1016/S0380-1330(03)70470-5.

Jacobson, R.B., and Coleman, D.J., 1986, Stratigraphy and recent evolution of Maryland Piedmont floodplains: American Journal of Science, v. 286, p. 617–637, doi:10.2475/ajs.286.8.617.

Julien, P.Y., 2002, River Mechanics: Cambridge, UK, Cambridge University Press, 434 p.

Knox, J.C., 1977, Human impacts on Wisconsin stream channels: Annals of the Association of American Geographers, v. 67, p. 323–342, doi:10.1111/j.1467-8306.1977.tb01145.x.

Knox, J.C., 1987, Historical valley floor sedimentation in the Upper Mississippi Valley: Annals of the Association of American Geographers, v. 77, p. 224–244, doi:10.1111/j.1467-8306.1987.tb00155.x.

Kroes, D.E., and Hupp, C.R., 2010, The effect of channelization on floodplain sediment deposition and subsidence along the Pocomoke River, Maryland: Journal of the American Water Resources Association, v. 46, p. 686–699, doi:10.1111/j.1752-1688.2010.00440.x.

Kuhnle, R.A., Binger, R.L., Foster, G.R., and Grissinger, E.H., 1996, The effect of land use changes on sediment transport in Goodwin Creek: Water Resources Research, v. 32, p. 3189–3196.

Larson, G., and Schaetzl, R., 2001, Origin and evolution of the Great Lakes: Journal of Great Lakes Research, v. 27, p. 518–546, doi:10.1016/S0380-1330(01)70665-X.

Leopold, L.B., 1956, Land use and sediment yield, *in* Thomas, W.L., Jr., ed., Man's Role in Changing the Face of the Earth: Chicago, University of Chicago Press, p. 639–647.

Mannik-Smith Group, Inc., 2008, Ottawa River Habitat Restoration Inventory: http://www.tmacog.org/Environment/Ottawa_River_Habitat/Ottawa_River_Habitat_Restoration_Inventory.pdf (accessed 5 May 2009).

Meierdiercks, K.L., Smith, J.A., Baeck, M.L., and Miller, A.J., 2010, Heterogeneity of hydrologic response in urban watershed: Journal of the American Water Resources Association, v. 46, no. 6, p. 1221–1237, doi:10.1111/j.1752-1688.2010.00487.x.

Murray, A.S., and Wintle, A.G., 2000, Luminescence dating on quartz using an improved single-aliquot regenerative-dose protocol: Radiation Measurements, v. 32, p. 57–73, doi:10.1016/S1350-4487(99)00253-X.

Pizzuto, J.E., 2002, Effects of dam removal on river form and process: Bioscience, v. 52, p. 683–692, doi:10.1641/0006-3568(2002)052[0683:EODROR]2.0.CO;2.

Pizzuto, J.E., Hession, W.C., and McBride, M., 2000, Comparing gravel-bedded rivers in paired urban and rural catchments of southeastern Pennsylvania: Geology, v. 28, p. 79–82, doi:10.1130/0091-7613(2000)028<0079:CGRIPU>2.0.CO;2.

Roberts, S.J., Gottgens, J.F., Spongberg, A.L., Evans, J.E., and Levine, N.S., 2007, Assessing potential removal of low-head dams in urban settings: An example from the Ottawa River, NW Ohio: Environmental Management, v. 39, p. 113–124, doi:10.1007/s00267-005-0091-8.

Segura, C., and Booth, D.B., 2010, Effects of geomorphic setting and urbanization on wood, pools, sediment storage, and bank erosion in Puget Sound streams: Journal of the American Water Resources Association, v. 46, p. 972–986, doi:10.1111/j.1752-1688.2010.00470.x.

Selle, A., Burke, M., Melchior, M., and Koonce, G., 2007, Active versus passive channel recovery following dam removal: A comparison of approaches, in Kabbes, K.C., ed., Proceedings of the 2007 World Environmental and Water Resources Congress, Restoring Our Natural Habitat, May 15–19, 2007, Tampa, Florida: Washington, D.C., American Society of Civil Engineering, Environmental and Water Resources Institute, p. 3856–3865.

Simons, R.K., and Simons, D.B., 1991, Sediment problems associated with dam removal—Muskegon River, Michigan, in Shane, R.M., ed., Proceedings of the 1991 National Conference of the American Society of Civil Engineers: New York, American Society of Civil Engineers, p. 680–685.

Stuiver, M., and Reamer, P.J., 1993, Radiocarbon calibration program—Calib rev5.0.2: Radiocarbon, v. 35, p. 215–230.

Thornton, J.A., 2003, Discussion of "Geomorphic analogies for assessing probable channel response to dam removals," by Martin W. Doyle, Emily H. Stanley, and Jon M. Harbor: Journal of the American Water Resources Association, v. 38, p. 1567–1579.

Trimble, S.W., 1981, Changes in sediment storage in the Coon Creek basin, Driftless Area, Wisconsin, 1853–1975: Science, v. 214, p. 181–183, doi:10.1126/science.214.4517.181.

Trimble, S.W., 1983, A sediment budget for Coon Creek basin in the Driftless Area, Wisconsin, 1853–1977: American Journal of Science, v. 283, p. 454–474, doi:10.2475/ajs.283.5.454.

Trimble, S.W., and Lund, S.W., 1982, Soil Conservation and the Reduction of Erosion and Sedimentation in the Coon Creek Basin, Wisconsin: U.S. Geological Survey Professional Paper 1234, 35 p.

Walter, R.C., and Merritts, D.J., 2008, Natural streams and the legacy of water-powered mills: Science, v. 319, p. 299–304, doi:10.1126/science.1151716.

Webb, L.D., 2010, Historical Changes in the Geomorphology of the Ottawa River (NW Ohio, U.S.A.) due to Urbanization and Land Clearance [M.S. thesis]: Bowling Green, Ohio, Bowling Green State University, 169 p.

Whitten, D., 2010, Glass Factory Marks on Bottles: http://www.myinsulators.com/glass-factories/bottlemarks.html (accessed 12 May 2010).

Wilcock, P., 2008, What to do about those dammed streams: Science, v. 321, p. 910–912, doi:10.1126/science.321.5891.910b.

Wilhelm, P.W., 1984, Draining the Black Swamp: Henry and Wood Counties, Ohio, 1870–1920: Northwest Ohio Quarterly, v. 56, p. 79–95.

Wolman, M.G., 1967, A cycle of sedimentation and erosion in urban river channels: Geografiska Annaler, v. 49A, p. 45–62.

Wolman, M.G., and Schink, A.P., 1967, Effects of construction on fluvial sediment: Water Resources Research, v. 3, p. 451–464, doi:10.1029/WR003i002p00451.

Manuscript Accepted by the Society 29 June 2012

The rise and fall of Mid-Atlantic streams: Millpond sedimentation, milldam breaching, channel incision, and stream bank erosion

Dorothy Merritts*
Robert Walter
Michael Rahnis
Department of Earth and Environment, Franklin and Marshall College, P.O. Box 3003, Lancaster, Pennsylvania 17604-3003, USA

Scott Cox
Jeffrey Hartranft
Pennsylvania Department of Environmental Protection, P.O. Box 8460, Harrisburg, Pennsylvania 17105-8460, USA

Chris Scheid
Department of Earth and Environment, Franklin and Marshall College, P.O. Box 3003, Lancaster, Pennsylvania 17604-3003, USA

Noel Potter
Department of Geology, P.O. Box 1773, 28 N. College Street, Dickinson College, Carlisle, Pennsylvania 17013-2896, USA

Matthew Jenschke
Austin Reed
Derek Matuszewski
Laura Kratz
Lauren Manion
Andrea Shilling
Katherine Datin
Department of Earth and Environment, Franklin and Marshall College, P.O. Box 3003, Lancaster, Pennsylvania 17604-3003, USA

ABSTRACT

For safety and environmental reasons, removal of aging dams is an increasingly common practice, but it also can lead to channel incision, bank erosion, and increased sediment loads downstream. The morphological and sedimentological effects of dam removal are not well understood, and few studies have tracked a reservoir for more than a year or two after dam breaching. Breaching and removal of obsolete milldams over the last century have caused widespread channel entrenchment and stream bank erosion in the Mid-Atlantic region, even along un-urbanized, forested stream reaches. We document here that rates of stream bank erosion in breached millponds

*dorothy.merritts@fandm.edu

remain relatively high for at least several decades after dam breaching. Cohesive, fine-grained banks remain near vertical and retreat laterally across the coarse-grained pre-reservoir substrate, leading to an increased channel width-to-depth ratio for high-stage flow in the stream corridor with time. Bank erosion rates in breached reservoirs decelerate with time, similar to recent observations of sediment flushing after the Marmot Dam removal in Oregon. Whereas mass movement plays an important role in bank failure, particularly immediately after dam breaching, we find that freeze-thaw processes play a major role in bank retreat during winter months for decades after dam removal. The implication of these findings is that this newly recognized source of sediment stored behind breached historic dams is sufficient to account for much of the high loads of fine-grained sediment carried in suspension in Mid-Atlantic Piedmont streams and contributed to the Chesapeake Bay.

INTRODUCTION

Dam removal, particularly of small dams, has become increasingly common since the 1980s (cf. Heinz Center, 2002). The reasons commonly cited for dam removal include safety, aquatic and riparian habitat improvement, and economics. Low-head dams (<7 m in hydraulic height) have been dubbed "drowning machines" because submerged hydraulic jumps downstream of the dams can trap and drown victims (Tschantz and Wright, 2011). Dams fragment fluvial systems and associated aquatic and riparian ecosystems (Graf, 1999). Removing dams eliminates safety hazards, restores variable hydrologic flows, and allows for unimpeded passage of fish and other aquatic organisms. For tens of thousands of obsolete low-head dams built to power mills, forges, and other industries in the eighteenth to early twentieth centuries, removal can be more cost effective than continued maintenance.

Despite safety, ecologic, and economic advantages, however, dam removal also can lead to channel incision, bank erosion, and increased sediment loads downstream. The morphological and sedimentological effects of dam removal are not well understood, and few studies have tracked a reservoir for more than a year or two after dam breaching (Csiki and Rhoads, 2010). As a result, considerable uncertainty exists regarding channel evolution trajectories and rates of stream bank erosion over a period of decades following dam breaching.

In this paper, we examine rates of erosion of fine-grained sediment upstream of three breached low-head dams in Pennsylvania for which prebreach conditions and time of breach are known (two cases) or are constrained to within several years (one case) (Fig. 1). These dams, which were breached 10, 26, and ~39 yr ago, are used to quantify rates of sediment production from breached reservoirs over decadal time scales. The three dams are ≤4 m in height and extended across the entire valley width. Because water flowed freely over their crests before removal, they are referred to as run-of-river dams.

In addition to rates of erosion, we examine the processes by which incised stream banks retreat laterally across the coarse-grained floors of the breached reservoirs. Our primary concern is to determine rates of erosion of sediment from banks of incised streams that are in different stages of post-dam-breach condition, and to assess how these rates change with time. Fine-grained sediment and nutrients are the leading pollutants in the Chesapeake Bay, the largest estuary in the United States and an

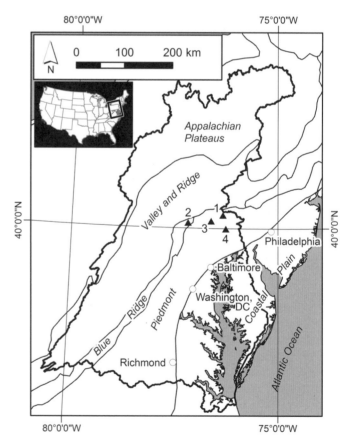

Figure 1. Locations of Mid-Atlantic sites discussed in text, with physiographic provinces and Chesapeake Bay watershed. Sites are as follows: 1—Hammer Creek, Pennsylvania; 2—Mountain Creek, Pennsylvania; 3—Conoy Creek, Pennsylvania; 4—Little Falls, Maryland.

impaired water body under the Clean Water Act (Phillips, 2002). Understanding the sources of sediment in streams is critical to developing successful strategies to reduce erosion and sediment flux to the bay.

Channel Evolution Models

Conceptual channel evolution models (CEM) and geomorphic studies of dam failure provide guidelines to predict the morphological and sedimentological effects of dam removal (Simon and Hupp, 1986; Evans et al., 2000a, 2000b; Evans, 2007; Doyle et al., 2003). When a dam is removed, local base level for upstream reaches is lowered (cf. Schumm et al., 2001; Simon and Darby, 1997). The stream cuts into unconsolidated sediment at the breach site immediately after dam breaching, forming a knickpoint in the stream profile. Vertical incision generally ceases once the stream reaches the base of the dam and the bottom of the original valley. Across this zone of increased grade, the stream has greater scouring capacity than upstream along the stream profile, where it remains perched in reservoir sediment. The knickpoint propagates up the valley through the reservoir sediment as the stream scours its bed. If the sediment is noncohesive and fine grained, the stream is able to erode and transport sediment easily, so the knickpoint propagates rapidly. With continued incision and erosion of the bed, mass movement commonly occurs along incised channel banks near the dam as a result of loss of lateral support (confining pressure) in wet reservoir sediment with high pore pressure (Simon and Darby, 1997; Evans et al., 2000a, 2000b; Evans, 2007; Doyle et al., 2003; Cantelli et al., 2004). Water slope decreases as the stream incises throughout the reservoir and upstream reaches become graded to the new local base level.

Simon and Hupp (1986) and later Doyle et al. (2002, 2003) ascribed these temporal patterns of channel adjustment to stages within a CEM, as follows: A—preremoval; B—lowered water surface; C—bed degradation; D—bed degradation and channel widening; E—bed aggradation and channel widening; and F—quasi-equilibrium. Doyle et al. (2002, 2003) tested this CEM by monitoring two dam removal sites in Wisconsin for a period of 1–2 yr after dam removal (Doyle et al., 2003). After removal of the Rockdale milldam on the Koshkonong River in Wisconsin, Doyle et al. (2003) documented that a headcut migrated upstream at a rate of ~10 m/h for 24 h, but decelerated to an average rate of 40 m/mo over the next 11 mo. Downstream of the headcut, a deep, narrow channel had high boundary shear stresses (up to 20–30 N/m^2) capable of eroding bed and bank material. Upstream of the headcut, however, low boundary shear stresses (<5 N/m^2) were insufficient to erode the bed or banks, and the reservoir sediment surface remained largely undisturbed (see figures 9 and 12c *in* Doyle et al., 2003).

Evans (2007) provided further empirical evidence for the development and duration of each stage of channel evolution by evaluating the response of the Chagrin River to seepage piping failure of the IVEX milldam in 1994. Their 12 yr study found that the general progression of stages of channel evolution was similar to the CEM of Doyle et al. (2003), but stage E was dominated by lateral migration of a single meandering channel rather than by overall bed aggradation and channel widening. Both incision and aggradation occurred as the Chagrin channel migrated through former millpond sediment, with undercutting and slumping at one bank coincident with point-bar deposition on the opposite bank. Furthermore, during stage F, the quasi-equilibrium stage, some bank erosion persisted locally.

Breached Millponds and Bank Erosion

The response of streams to dam breaching became a more prominent problem when Walter and Merritts (2008) documented that late seventeenth to early twentieth-century valley sedimentation in the unglaciated Mid-Atlantic region resulted not only from accelerated upland erosion during post–European settlement land clearing and agriculture, but also from contemporaneous, widespread valley-bottom damming for water power. For centuries, valley damming trapped immense amounts of fine sediment in extensive backwater areas upstream of tens of thousands of low-head milldams. Furthermore, Walter and Merritts (2008) proposed that local drops in base level caused widespread incision into historic reservoir sediment as aging dams breached or were removed during the last century.

For the Mid-Atlantic Piedmont region, Merritts et al. (2011) reported that modern stream-channel entrenchment largely is decoupled from current upland land use. Their case studies demonstrated that a breached dam can lead to incision, stream bank erosion, and increased loads of suspended sediment for streams in forested, rural areas as well as agricultural and urban areas, regardless of whether or not upland land use has altered stormwater runoff or sediment supply.

Pizzuto and O'Neal (2009) concurred with Walter and Merritts (2008) that dam breaching leads to higher rates of stream bank erosion. Of eight millpond reaches along 30 km of the South River, a tributary to the Potomac River in Virginia, all but one of the eighteenth- to nineteenth-century milldams were breached in the 1950s, and the last was breached by 1976. Studying changing bank lines on aerial photos, Pizzuto and O'Neal (2009) found a statistically significant, strong correlation between accelerated rates of bank erosion and dam breach conditions, with normalized estimates of mean bank erosion rates increasing by more than a factor of 3 in the first two decades after dam breaching. Furthermore, they concluded that accelerated erosion could not be explained by climatic factors (e.g., storm intensity or frequency of freeze-thaw cycles) or changes in the density of riparian trees along stream banks. Although the South River study showed that breached reservoirs have higher rates of bank erosion than unbreached reservoirs, it did not provide information on changing rates of postbreach bank erosion over decadal time scales.

Stream bank erosion is the detachment and removal of particles from the surface of the bank. It occurs through three main types of processes (Hooke, 1979; Lawler, 1995; Lawler et al., 1997; Couper and Maddock, 2001; Wynn and Mostaghimi, 2006a; Wynn et al., 2008):

(1) subaerial processes—freezing and thawing or wetting and drying of the bank surface, leading to weakening and erosion;
(2) mass wasting—instability of bank material and failure via collapse, calving, toppling, or other mass failure; and
(3) fluvial entrainment—detachment and entrainment of particles by hydraulic forces on stream banks from flowing water.

The combination of processes of freeze-thaw, wetting and drying, weakening of bank material, mass wasting, undercutting, bank collapse, and removal of material by stream flow causes banks to retreat laterally. At any one site, all three of these processes might occur and contribute to cumulative erosion with time. Freeze-thaw is more frequent at higher elevations and/or higher latitudes, and wetting-drying is more common where precipitation is highly seasonal or where streams are incised and hydrographs are strongly peaked (i.e., flashy) due to high channel banks. Mass wasting is promoted by scour and undercutting, which depend on the nature and erodibility of material at the base of the bank. Fluvial entrainment is directly proportional to shear stress of flowing water, which is proportional to flow depth and water surface slope (see Eq. 1 later herein).

An incised stream with high banks of sediment leads to greater bank instability in a breached reservoir. As sediment is dewatered, gravitational forces and rapidly changing pore pressures result in settling (compaction) and mass wasting. The interaction of gravitational and hydraulic forces acting on bank sediment maintains oversteepened, unstable banks and controls rates of bank erosion (Simon et al., 2000). We have observed rotational slumping, calving, and other types of mass wasting failure at more than 100 breached dam sites in Pennsylvania and Maryland, including sites where dams were breached more than 50 yr ago (Fig. 2).

Erodibility of Stream Bank Sediment and Freeze-Thaw Processes

Stream bank erodibility, k_d (in m³/N-s), is the rate per unit area at which mass (sediment) is removed from the bank face once it begins to erode. The lateral erosion rate of a stream bank, E_r (in m/s) is proportional to its erodibility and the amount of available excess shear stress (in Pa, or N/m²). The excess shear stress is the difference between the shear stress, τ, acting on the

Figure 3. (A) Apron of debris forming on the right bank of a breached millpond on Little Falls, Maryland (see Fig. 1 for location), near the town of Whitehall. Bank height is ~2 m, and downstream is to left. Dark soil at midbank level is the presettlement land surface. The site is close to the valley wall, and the millpond sediment (overlying brown sediment) is thinning toward the original reservoir margin. (B) Freeze-thaw has produced large needles of ice from pore water in bank sediment. As ice needles grew in pores perpendicularly to the bank face (horizontal), a vertical fracture opened parallel to the bank face, causing a large slab of bank to collapse onto the debris apron in the winter of 2008–2009 (photos taken in February 2009). Trowel for scale in lower image.

Figure 2. Mass movement occurred along the left bank of the incised Hammer Creek just downstream of XS-5 as a result of wetting and drying of the banks by high flow from a late June 2006 storm. Double arrow indicates person for scale along tape measure at section. Flat surface at top of bank is the level of sedimentation in the millpond. See Figure 5 for location of cross section.

bank and the critical shear stress, τ_c (in Pa, or N/m^2), needed to entrain material from the bank (Osman and Thorne, 1988; Darby and Thorne, 1996), as follows:

$$E_r = k_d (\tau - \tau_c), \qquad (1)$$

Previous work has shown that k_d and τ_c can vary up to four to six orders of magnitude along a given stream reach, and both vary seasonally as a result of soil desiccation (during dry and/or vegetation growth periods) and winter freeze-thaw cycling (Wynn and Mostaghimi, 2006a, 2006b; Wynn et al., 2008). Detailed monitoring of sites along the Ilston River, South Wales, by Lawler (1986, 1993) and along Strouble Creek, Virginia, by Wynn et al. (2008) established that freeze-thaw processes significantly lower the critical shear stress and increase the erodibility of cohesive stream bank sediment. Examination of Equation 1 indicates that lowered τ_c and increased k_d would result in greater rates of bank erosion.

The action of freeze-thaw directly results in bank erosion by the action of needle ice, as observed by us at numerous sites in the Mid-Atlantic region (Fig. 3) and described by Wolman (1959, p. 215) from his observations along banks of sand, silt, and clay at Watts Branch, Maryland, in December 1955: "Particles are heaved out from the bank by ice crystals and upon melting of the crystals the sediment drops into the stream ... slabs of sediment perhaps one foot square containing thin ice lenses have been observed. The action of frost appears to be one of preparation of a veneer of sediment for erosion..."

Wolman observed at Watts Branch that most bank erosion occurred during winter months and was associated with freeze-thaw processes (discussed later herein). Bank pins (metal bars), surveyed channel cross sections, and two baselines parallel to the retreating bank edge were used to document up to 0.2 m/yr of bank erosion over a period of several years. Wolman (1959) observed that rises in water stage were more effective at removing bank material after frost-related processes had increased its susceptibility to fluid erosion. In contrast, "little or no erosion was observed" during the highest flood on record at the time in July 1956 (Wolman, 1959, p. 204).

Wolman (1959) concluded that there is an obvious "lack of erosion in summer and marked erosion in winter" and determined that 85% of bank erosion during his 2 yr study occurred during the winter months of December through March. He further noted that lateral channel migration of Watts Branch by bank erosion takes place primarily during the winter (Wolman, 1959, p. 208 and 216). Unbeknownst to Wolman, this reach of Watts Branch was immediately upstream of a recently breached mill dam from a former nineteenth-century (and possibly earlier) grist mill (see supporting online material for Walter and Merritts, 2008).

Lawler's (1986) statistical analysis of data from 230 erosion pins in stream banks consisting of sand, silt, and clay at two meander bends on the Ilston River in South Wales over a 2 yr period (1977–1979) indicated a strong seasonality to stream bank erosion. Nearly all bank erosion took place in the winter, from December to February. The strongest control on average and maximum rates of bank erosion was frost action, and in particular the number of days for which minimum temperatures were below freezing (0 °C). Lawler (1986, p. 230) observed that a "skin of friable, cohesionless" sediment formed on stream banks by needle ice growth and was easily removed by a subsequent rise in stage. Stepwise multiple regression analysis revealed that air frost frequency, the variable most strongly associated with erosion rate, explained 94.2% of the variation in bank erosion rate.

Bank erosion processes are highly dependent upon the nature of the bank material (cf. Julian and Torres, 2006). In all three of the studies cited here (Watts Branch, Maryland; Strouble Creek, Virginia; and the Ilston River, South Wales), the banks varied from 1 to 2 m in height and consisted primarily of sand, silt, and clay. Cohesive silt and clay—with moderate to high critical shear stress when moist to dry, respectively—are particularly susceptible to freeze-thaw, wetting-drying, and mass failure. In contrast, noncohesive material such as sand—with a low to moderate critical shear stress—is more prone to erosion by hydraulic forces (Julian and Torres, 2006). Cohesive sediment (e.g., silt loam) commonly forms vertical banks from which slabs have been observed to slake and calve to the toe of slope or into the stream. Much less cohesive sand and gravel, on the other hand, forms banks that are closer to the angle of repose, generally 35°–40°. Undercutting and bank collapse can reset the slopes of the banks as the stream erodes material from the toe, or base, of the bank.

Vegetation on banks also plays a role, with tree roots and grasses adding various degrees of mechanical reinforcement. For nonplastic stream bank sediment in Virginia, increases in root volume were correlated with reduced K_d (Wynn and Mostaghimi, 2006b). On the other hand, forested stream banks experienced greater diurnal temperature ranges and up to eight times more freeze-thaw cycles than banks with dense herbaceous cover (Wynn and Mostaghimi, 2006b). In addition, fallen trees from bank erosion can obstruct flow and trap debris within incised channels, leading to localized scour and accelerated bank erosion around the obstruction.

Changing Rates of Bank Erosion after Dam Breaching

It is probable that rates of stream bank erosion and mass wasting are highest immediately after a dam breaches. This supposition is supported by the flume experiments of Cantelli et al. (2004, 2007),[1] as well as by post-dam-breach monitoring of Doyle et al. (2003) in Wisconsin and of others in Oregon after removal of the Marmot Dam on the Sandy River (Major et al., 2008, 2012). Based on the previous discussion of the role of freeze-thaw processes in bank erosion, however, it is possible that bank erosion will continue long after dam breaching, provided that banks of sand, silt, and clay are exposed to air temperatures that drop to freezing.

---

[1]Movies of Cantelli's flume experiments can be downloaded at https://repository.nced.umn.edu/browse.php?dataset_id=28.

Shear stresses at the stream bed are much higher downstream of a knickpoint that is propagating through a breached reservoir than upstream of the knickpoint (e.g., Doyle et al., 2003). Basal shear stress, τ, acting on the channel bed is the product of fluid density, flow depth, and water surface slope:

$$\tau = \gamma RS, \qquad (2)$$

where γ is the specific weight of water (9800 N/m³), R is hydraulic radius, calculated as A (channel area) divided by P (wetted perimeter), and S is the energy slope. Both hydraulic radius and slope increase as a result of post-dam-breach channel incision.

We posit several scenarios for the relation between stream bank erosion and time since dam breach. Erosion rates might remain constant with time or diminish gradually until the majority of reservoir sediment is eroded. It is more likely, however, that the rate decelerates with time. About 9%–13% of the reservoir sediment of the IVEX dam was eroded during and immediately after dam breaching in 1994, and a laterally migrating stream channel has continued to erode reservoir sediment since then (Evans et al., 2000a, 2000b; Evans, 2007). Average monthly rates of sediment removal from two millponds in southern Wisconsin indicate a rapid decline in volume of sediment removed, from 0.7% to 1.7% per month within the first 8–10 mo after dam removal, to 0.2%–0.5% per month from 8 to 13 mo after dam removal (Doyle et al., 2003).

Upstream of the breached Marmot Dam, the rate of removal of sediment decelerated rapidly during the first year after dam breach. About 17% of the volume of reservoir sediment was eroded within 3 wk, 28% after 5 wk, 39% after ~2 mo, and 51% after 11 mo (Major et al., 2008, 2012; C. Podolak, 28 April 2011, personal commun.).

Once the geometry of an incised channel is established and adjusted for upstream runoff conditions, substrate resistance, and other factors, it is possible that a lower rate of stream bank erosion will continue until most or all of the reservoir sediment is gone. The long-term trend might correspond to a negative power function, as with the rate of removal of sediment from the Marmot Dam reservoir. It also is likely, however, that sporadic, stochastic events, such as high-magnitude floods, or tree falls that lead to localized scour, could cause short-term deviations in this long-term signal. In subsequent sections, we present data that enable us to quantify the trend in long-term sediment removal from three breached reservoirs. In the discussion section, we compare the observed trend to the scenarios posited here.

BACKGROUND

Mid-Atlantic Region Streams and Milldams

Tens of thousands of grist mills, sawmills, furnaces, forges, and other industries relied upon hydropower from first- to fourth-order Mid-Atlantic streams throughout the seventeenth to early twentieth centuries (Walter and Merritts, 2008; cf. U.S. industrial censuses of 1840, 1860, 1870, and 1880 [U.S. Bureau of the Census, 1841, 1865, 1872, and 1884, respectively). Such streams comprise greater than 70% of stream length in the region, and damming them had a substantial impact on a large portion of watersheds, including upstream tributaries. Hydropower was dominant when the Mid-Atlantic region was one of the world's leading suppliers of wheat and iron, and mills were particularly abundant in areas close to major shipping ports, including Philadelphia, Pennsylvania, and Baltimore, Maryland.

Few historic milldams are included in the U.S. Army Corps of Engineers National Inventory of Dams, which lists only 1546 dams in Pennsylvania. Of these, only 833 are listed as being less than ~7.6 m in height. In contrast, the Pennsylvania Department of Environmental Protection has an inventory of ~8400 low-head milldams (generally <5 m) in Pennsylvania, of which 4100 are breached, and it estimates that 8000–10,000 more might exist (J. Hartranft, Pennsylvania Department of Environmental Protection, 19 September 2007, personal commun.). These estimates result in an average density of 1 dam per 6–7 km² for the state of Pennsylvania. Considering that mill and dam building continued throughout the nineteenth century, the possibility of 16,000–18,000 dams in Pennsylvania is consistent with the ~10,000 mills listed for Pennsylvania in the U.S. census of 1840 (U.S. Census Bureau, 1840).

Our research of township-scale maps in southeastern and central Pennsylvania indicates that at least 1200 milldams existed in Chester, Lancaster, and York Counties, 153 in Cumberland County, 205 in Huntingdon County, and 186 in Centre County (Walter and Merritts, 2008; Merritts et al., 2011). Similar to Pennsylvania, adjacent states in the Mid-Atlantic region had ubiquitous milldams, with at least 211 in Baltimore and Montgomery Counties of Maryland (Walter and Merritts, 2008; Merritts et al., 2011). From detailed historic records, we have calculated the mean height of milldams in Lancaster County, Pennsylvania, as ~2.4 m ($n = 246$). Historic dams ranged in height from as low as 1.5 m to as high as ~9 m (Lord, 1996). Nineteenth-century U.S. census reports indicate that milldams in other Mid-Atlantic counties had similar heights (U.S. Census Bureau, 1840, 1870, 1880).

Milldams commonly lined Mid-Atlantic streams in series, forming chains of slack-water pools that enabled millers to maximize the potential energy of falling water. For example, at least 13 milldams operated on the lower 21.3 km of one of the streams investigated here, Hammer Creek, during the eighteenth and nineteenth centuries (Fig. 4; Bridgens, 1858, 1864; Lord, 1996). This number yields a milldam spacing of ~6 km along Hammer Creek.

High trap efficiencies in historic millponds are corroborated by large volumes of historic sediment stored along stream corridors upstream of milldams (Walter and Merritts, 2008; Merritts et al., 2011). Previous work has shown that low-head dams built across small (first- to third-order) stream valleys have high sediment trap efficiencies of >40%–80% (Brune, 1953; Gottschalk, 1964; Dendy and Champion, 1978; Petts, 1984; Evans et al., 2000a; Doyle et al., 2003). A reservoir's trap efficiency, a measure of its ability to trap and retain sediment, is expressed as a ratio of sediment retained by settling to incoming sediment (Brune, 1953; Verstraeten and Poesen, 2000).

High-resolution topographic data from airborne laser swath mapping (LiDAR) provides a means of tracing the tops of millpond sediment fill surfaces to the crests or spillways of milldams. Such analyses indicate that wedges of sediment exist upstream of thousands of milldams in Pennsylvania and Maryland (Walter and Merritts, 2008; Merritts et al., 2011). These wedges thicken downstream toward the dams that formed the slack-water reservoirs.

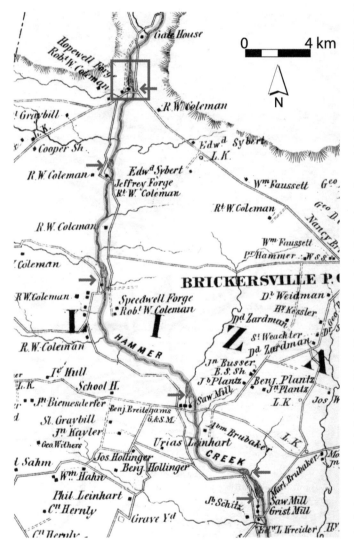

Figure 4. Historic township maps indicate that at least 13 milldams were located along a distance of ~28 km on Hammer Creek in the mid-nineteenth century (Bridgens, 1858). Inset box is area of Figure 5. Earliest milldams were built in the early seventeenth century, and many ponds were partly or nearly filled with sediment by the late nineteenth century. Milldams located in six counties in Pennsylvania and two in Maryland for the nineteenth century, as well as a number of mills per county in the eastern United States as of the 1840 U.S. census, can be viewed at the following website: http://www.fandm.edu/x17479.

Site Descriptions

Our three study sites are located within the Piedmont and the Ridge and Valley physiographic provinces of the Mid-Atlantic United States. The headwaters of Hammer and Conoy Creeks, both third-order streams in the Piedmont, consist of a low-relief undulating landscape formed in Triassic- to Cretaceous-age rift basin sedimentary rocks. Conglomerates and sandstones form the hills, whereas valley bottoms are underlain by shale. A thick cobble- to boulder-size weathered residuum with sandy matrix occurs in relict periglacial slope deposits that bury the shale, so it rarely is exposed along valley bottoms. Upstream drainage area is 47 km^2 at the Hammer Creek site and 20 km^2 at the Conoy Creek site. Hammer Creek drains southward into the Conestoga River, which, in turn, drains westward into the Susquehanna River and ultimately Chesapeake Bay. Conoy Creek flows westward directly into the Susquehanna River. Mountain Creek is located along the easternmost edge of the Ridge and Valley physiographic province (see Fig. 1). This fourth-order stream has a drainage area of 120 km^2 and flows northeastward into Yellow Breeches Creek, which, in turn, drains into the Susquehanna River and ultimately to Chesapeake Bay.

Hammer Creek

Hammer Creek was named during the Colonial period for the constant hammering of iron at forges and mills along the stream. In 1901–1902, a 2.4-m-high concrete and block dam was built within the incised stream channel of Hammer Creek at the approximate site of an older, breached milldam (40.2421°N, 76.3359°W). This older dam was associated with an iron forge on nineteenth-century maps (see Fig. 4). Slack water from the twentieth-century pump station dam extended upstream at least 500 m, but the thickness of historic sediment from the older reservoir indicates that the impact of the original dam extended even farther upstream. Assuming an upstream-thinning wedge for the reservoir and a trapezoidal valley shape over a length of 0.5 km, we estimate ~22,300 m^3 of reservoir sediment. The slack-water reservoir formed by the Hammer Creek pump station dam four decades after the dam was built is shown in an historic air photo from 1940 (Fig. 5A). A digital orthophoto acquired in 1993, just 8 yr before dam removal, shows that sedimentation had narrowed the stream channel substantially.

An 11 m section of the Hammer Creek dam was removed in September 2001. The upper 1 m of the dam was removed in September 2001, leaving a rock ledge with concrete ~0.5 m in height that forms a local base-level control. The post-dam-breach gradient of Hammer Creek in the former reservoir is 0.0015. A digital ortho-image acquired in 2005 shows an incised stream channel 3.5 yr after dam removal (Fig. 5B). Photographs taken by state officials at the time of partial dam removal show the channel during and shortly after breaching (Figs. 6A–6D). A narrow incised channel produced a knickpoint that propagated rapidly upstream more than 500 m within the first few days, and a substantial amount of fine-grained sediment (sand, silt, and clay) exposed in

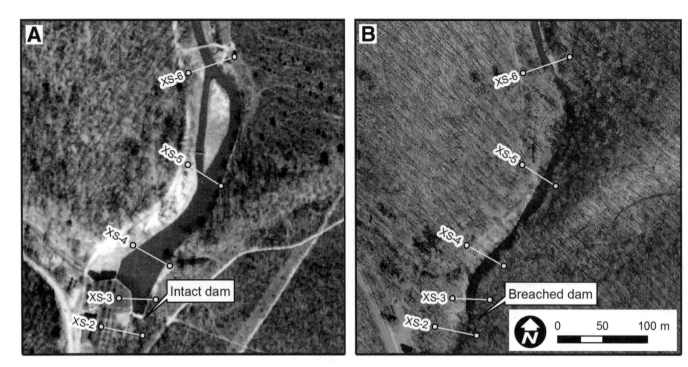

Figure 5. (A) Historic (1940) aerial photo of the Hammer Creek pump station dam showing the reservoir upstream and wing wall connecting the main-stem dam to that on a small tributary (Walnut Run) from the west. The pump station dam was built in 1901–1902 and had substantial sedimentation by the time of this photo. (B) Digital ortho-image from 2003, acquired by the state of Pennsylvania (horizontal resolution 0.6 m), showing the incised stream channel and remnant paired fill terraces from the millpond 2 yr after dam breaching. The dam on Walnut Run is not breached, so this tributary has not yet incised to adjust to the lowered base level on the main stem.

the reservoir was removed by bank erosion within several weeks. Some thin beds of pebble-sized quartz gravel derived from local Mesozoic conglomerates also occur in the uppermost part of the historic reservoir sediment.

Conoy Creek

At the Conoy Creek site, a 1.2-m-high dam (40.1327°N, 76.6212°W) was built for local water supply in 1930 near a breached 1.8-m-high dam originally built to power an eighteenth- to nineteenth-century sawmill (see Fig. 1). This second-generation dam was constructed within the older, incised millpond reservoir. A 1940 air photo shows the intact twentieth-century dam set within the valley flat (the older millpond fill surface). A state inspection report from 1959 indicates that the reservoir upstream of the dam was "silted up," with no remaining capacity, and the stream banks around the wing walls were eroded. Our field mapping indicates that the younger inset fill forms a prominent bench along the valley ~0.3 to 0.6 m lower than the larger valley flat formed by the older millpond sediment. Assuming an upstream-thinning wedge for the millpond reservoir and a trapezoidal valley shape for the ~1 km of stream impacted by the millpond, we estimate ~29,240 m^3 of historic reservoir sediment.

A 1971 air photo shows the Conoy Creek dam as intact, although erosion can be seen along the left (southeastern) bank between the masonry wall and the valley margin, and some water appears to be passing through this eroded area. Air photos from the late 1970s indicate that the channel had completely bypassed the dam along this margin, effectively causing a dam breach without breaching the actual structure. We estimate the timing of complete dam bypass as 1972, the year that Hurricane Agnes caused severe flooding in the region and damaged many old dams. Digital ortho-images acquired since the 1990s show a channel with significant meander migration and a more sinuous channel at multiple locations, in marked contrast to the limited channel migration prior to 1971. Modern channel gradient in the former millpond is 0.002.

Mountain Creek

Ridges adjacent to Mountain Creek consist of early Paleozoic quartzite, and the valley is underlain by early Paleozoic dolomite. As at Hammer Creek, hillslopes adjacent to the valley bottom are mantled with unconsolidated Pleistocene periglacial deposits. Our mapping along the valley slopes indicates that these deposits consist of thick sheets (~1 to 4 m thick) of quartzite cobbles and boulders within a sandy loam matrix. Exposures of these colluvial deposits in quarries, and light detection and ranging (LiDAR) analysis of landforms on the slopes of South Mountain indicate that many are gelifluction sheets and lobes. Periglacial slope deposits also underlie the historic millpond sediment along Mountain Creek.

The Mountain Creek study site extends from the 4-m-high Eaton-Dikeman paper mill dam (40.1015°N, 77.1834°W) to the

Figure 6. Photos, taken by staff of the Pennsylvania Department of Environmental Protection, document dam breaching and subsequent channel incision at Hammer Creek. All views are looking north, upstream. (A) 5 September 2001, just after the dam was breached. Note person standing on wing wall that attaches this dam to a small dam on a tributary from the west just out of view on the left. Note that historic millpond sediment is graded to the original dam crest. (B) 6 September 2001, showing exposed and eroding fine-grained reservoir sediment. Mesozoic rift basin sedimentary rocks in the watershed, including red shales and sandstones, produce sediment with strong red hues. (C) 27 September 2001, showing the breached reservoir after rocks were placed near the breach and the surface just upstream of the dam was graded. (D) The breached reservoir in April 2011. Note exposed banks upstream of the breach and remnant millpond surface forming paired terraces on each side of the incised stream channel. About 0.5 m of the base of the dam remains in place, as do the ends of the dam on each side of the breach, and rubble from the breached dam was used by anglers to create a pool for fishing.

upstream end of the reservoir, a distance of ~1.2 km. Built in 1855, the milldam was ~213 m long, but field evidence indicates that an older dam might have existed in the vicinity (within 10 m upstream) of this structure. Pennsylvania state dam inspections reported that the reservoir was substantially filled with sediment by 1914 and the reservoir volume was ~173,000 m³. The Eaton-Dikeman reservoir as shown in an historic air photo from 1968 reveals a deltaic lobe of sediment crossing the valley from southeast to northwest near the dam (Fig. 7A). Bathymetric surveying by Dickinson College students in 1976 determined that the greatest water depth near the dam was ~1 m, and the majority of the reservoir had water depths less than 0.3 m. By assuming a trapezoidal valley shape, we estimate that the reservoir volume might have been as large as 250,000 m³.

In 1985, an ~15 m section of the northern end of the 213-m-long Eaton-Dikeman dam was removed. An incised channel formed immediately at this breach, and the modern channel gradient is 0.003. State records and photos from 1985 to 1986 indicate that the channel incised and then widened rapidly after dam breaching. Digital ortho-images from 2003 show the incised channel of Mountain Creek 18 yr after dam removal (Fig. 7B), and our photographs of the site show the widened stream corridor 25 yr after dam breaching (Fig. 8).

At all three sites, the contact between historic reservoir sediment and the original valley bottom is marked by a thin stratum of Holocene organic-rich wetland soils (mucks) and fine-grained sediments. At Mountain Creek, tree stumps, logs, and forest soils rich in bark, nuts, and leaves are exposed by the incised channel at valley margins, indicating that the reservoir buried forested toe-of-slope as well as valley bottom landforms and soils. At Conoy Creek, weathered toe-of-slope colluvium is exposed where the incised channel has cut into the valley margins. Such

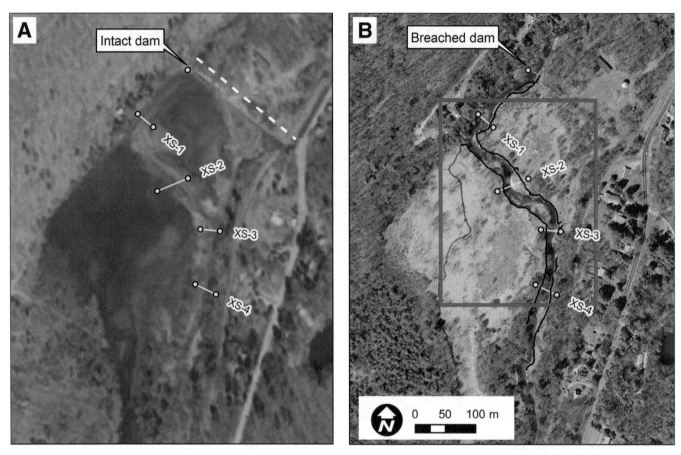

Figure 7. (A) Historic (1968) aerial photo of the Mountain Creek Eaton-Dikeman milldam (213 m long) showing the reservoir and deltaic lobes of sediment filling in the reservoir. The dam was built in 1855, and was reported as having substantial sediment infilling by the early twentieth century (see text). (B) Digital ortho-image from 2003, acquired by the state of Pennsylvania (horizontal resolution 0.6 m), showing the incised stream channel and remnant paired fill terraces from the millpond 18 yr after dam breaching. Channel cross sections for this study are shown. Solid black lines represent millpond fill-terrace edge break lines surveyed in 2008; note substantial retreat of the terrace edge from 2003 to 2008 in many places. Red box is area of Figure 13.

Figure 8. (A) Downstream view of XS-2 at the breached Eaton-Dikeman reservoir on Mountain Creek. Note the point bar on right bank prograding to the left (northwest) within paired fill terraces from the original millpond level. White arrow on left bank indicates exposure of Pleistocene periglacial gravel at base of bank, beneath millpond sediment. This sediment is interpreted as part of a toe-of-slope deposit of South Mountain in the background. (B) View of XS-1 at the breached Eaton-Dikeman reservoir looking across (southeast) the original reservoir surface, toward Piney Mountain in distance. Right bank height is ~3 m at this location, and left bank has eroded into colluvium (periglacial) at the toe of South Mountain (behind and to left of photographer).

exposures have not been observed at Hammer Creek, as it has not yet incised to the predam valley bottom nor has it eroded into the margins of the reservoir fill.

METHODS

Standard methods (cf. Wolman, 1959; Lawler, 1993) were used to estimate bank erosion rates for Hammer, Conoy, and Mountain Creeks. Erosion was measured as lateral retreat at a point (one-dimensional horizontal) with bank pins or as lateral retreat perpendicular to the stream bank face (two-dimensional vertical) with surveyed cross sections, and then converted to volume removed (three dimensions) by multiplying erosion from the incised stream corridor over a given length interval of stream. In addition, plan-view changes in bank edges and bar were determined for the breached Eaton-Dikeman reservoir on Mountain Creek. Stream bank and bar edges were digitized on two sets of color digital ortho-images (orthorectified) acquired by the state of Pennsylvania (PA MAP) during late spring leaf-off conditions in 2003 (0.6 m ground resolution) and 2006 (0.3 m ground resolution). Our mapping of break lines (water edge, bar edge, and terrace edge) with a Trimble GeoXH global positioning system (GPS) unit in 2008 and 2009 was interpreted in combination with high-resolution topographic data from LiDAR (PA MAP). We also determined particle size distributions for sediment of different ages within the Eaton-Dikeman reservoir and compared grain size of point-bar sediment to estimates of predicted particle size mobility based on flow depth and the Shields parameter. Each of these methods is discussed in more detail next.

Particle Size Analysis

Standard sieve methods were employed for particles greater than or equal to very fine sand, and a laser particle analyzer (Micromeritics Saturn Analyzer) was used to estimate particle sizes less than 300 μm (50 mesh). Particle size distribution was determined for four different ages of sediment at the Eaton-Dikeman reservoir on Mountain Creek. From oldest to youngest, these are (1) coarse presettlement substrate buried beneath millpond sediment, probably Pleistocene in age; (2) fine-grained presettlement substrate, probably Holocene in age; (3) historic millpond sediment probably dating from the eighteenth to early nineteenth centuries in age; and (4) sediment deposited in an actively migrating, unvegetated point bar within the incised channel corridor. Examination of digital ortho-images from 2003 and 2007 indicates that sediment on the bar at the sample site is likely to have been deposited within the past 6–8 yr. The historic and underlying presettlement sediment eroding from banks generally is much finer grained than the older coarse substrate or sediment deposited in point bars, and grain-size analysis was used to evaluate these differences.

Sediment particle size was evaluated at XS-1 and XS-2 in the breached Eaton-Dikeman reservoir (see Fig. 8). Samples were collected at ~10–40 cm increments, following stratigraphic boundaries, from top to bottom of the incised stream channel bank. Total sample depth was 280 cm at XS-1 and 245 cm at XS-2. Air-dried, lightly crushed (for disaggregation) samples were sieved to 0.6 mm grain size, and particle size for the fraction finer than 0.6 mm was analyzed with a Micromeritics Saturn laser diffraction particle size analyzer. Sieve data were merged with laser diffraction data to produce a complete grain-size distribution.

The coarse layer of sediment that underlies the historic sediment exposed in stream banks was exhumed as the bank of historic sediment retreated by lateral erosion. A fresh exposure of this pebbly-cobble substrate was provided after high flows in the vicinity of XS-2, where the average annual lateral erosion rate on left bank is ~0.3 m/yr. This coarse substrate is winnowed after exhumation, so the grain size estimate presented here represents the coarse fraction of the presettlement substrate that remained after being exposed to stream flow for several years. A pebble count was performed in the exposed bed substrate on 13 August 2008, using the standard "Wolman pebble count" method and a grain-size template (Wolman, 1954).

Sediment in the active point bar on the right bank at XS-2 was sampled on 25 September 2008. The sample was wet sieved in the field to the 2 mm fraction, and each fraction was dried and weighed. Wash water with particles finer than 2 mm was collected in a bucket and dry sieved in the laboratory. Some fine sediment <0.5 mm possibly was lost during wet sieving. Total mass sampled was 30.4 kg.

Shields Parameter and Particle Mobility in an Incised Reservoir

The Shields parameter (τ^*) for particle entrainment is the ratio of driving forces (τ_b, the basal shear stress acting on the sediment particles) to resisting forces (buoyant weight of sediment particles), given as

$$\tau^* = \tau_b / ([\rho_s - \rho]gD), \qquad (3)$$

where ρ_s is the density of sediment (2650 kg/m³), ρ is the density of water (1000 kg/m³), g is acceleration from gravity, and D is particle diameter (in m). The basal shear stress, τ_b, is estimated as shown in Equation 2. Assuming that particle entrainment occurs when $\tau^*_c = 0.03$–0.07, as reviewed in Buffington and Montgomery (1997), we estimated a range in predicted D_{50} values for given flow depths in the breached Eaton-Dikeman reservoir on Mountain Creek.

Bank and Bed Erosion Measurements

Repeat surveys of channel cross sections at all three sites and of the long profile at Hammer Creek, done with either a laser level or total geodetic station during various surveys, enable the calculation of change in cross-sectional area during intervals between surveys, as well as cumulative change in channel geometry and erosion. Channels were surveyed perpendicular to stream flow between section end points marked with concrete monuments at Hammer and Mountain Creek, and with rebar embedded 0.6 m in the ground and marked with survey caps at Conoy Creek. Measurement errors for cross-section surveys are ±0.5 cm and ±1 cm for horizontal and vertical dimensions, respectively. Estimates of uncertainty for erosion volume from repeat cross-section surveys are on the order of ±26%–32% for typical measurements (Table 1).

Two monumented cross sections (XS-1 and XS-2) downstream and three upstream (XS-3, XS-4, and XS-5) of the dam were installed and surveyed with a total geodetic station on Hammer Creek during the summer of 2001, just prior to dam removal in September. Two more sections (XS-0 and XS-6) were added and surveyed with a laser level during the summer of 2006, with XS-0 downstream of XS-1, and XS-6 upstream of XS-5. Locations of five sections are shown in Figure 5. In addition, the long profile of the water surface and bed along the thalweg were surveyed before and after dam removal. All cross sections upstream of the dam were located in historic reservoir sediment, but the right bank at XS-3 was lined with a stone block wing wall connecting the pump station dam with a dam and gate on a small tributary from the west (see Fig. 5A).

Four cross sections were installed with a total geodetic station upstream of the Conoy Creek dam by LandStudies, Inc., an engineering firm in Lititz, Pennsylvania, in 2005. All were located within historic reservoir sediment. We resurveyed these cross sections several times with a laser level between February 2005 and July 2008, but we used only the total change during the entire time period to estimate an average rate of bank erosion.

Two cross sections were installed on Mountain Creek upstream of the breached dam in the winter of 2007–2008 (XS-1 and XS-2), and two more were added upstream of these (XS-3 and XS-4) in the summer of 2008 (see Fig. 7). The first two sections were surveyed with a total geodetic station, and the more recent two were surveyed with a laser level. At XS-1, the left channel bank has eroded into the steep colluvial slope along the south side of South Mountain (see Fig. 7B), but all other cross sections are located within historic, unconsolidated reservoir sediment.

At Mountain Creek, bank pins (1 m metal rebar rods) were inserted horizontally into stream bank faces at the top, middle, and bottom of the bank at all four cross sections. Rod exposure at different times was measured to determine the cumulative amount of linear bank erosion, the average erosion rate, and seasonal variations in rates of erosion. Pins were installed on the left bank on all but XS-1, where pins were installed on the right bank. All pins were located within historic, unconsolidated millpond sediment. Measurement error for bank pins is ±1 cm, and estimates of uncertainty for erosion volume from bank pins are on the order of 15%–18% for typical measurements (see Table 1).

Channel-Normalized Sediment Production

We quantify erosion of sediment along the stream corridor as "channel-normalized sediment production." This parameter is useful in addition to the lateral erosion rate of a specific bank because (1) bank height varies with distance upstream of a dam; (2) both erosion and deposition occur along incised channels; and (3) one or both banks can erode at a given reach. We calculated channel-normalized sediment production in $m^3/m/m/yr$, a parameter for volume of sediment eroded per unit stream length per unit bank height per year. Normalizing to volume/height/length/time provides comparative numbers with the same units for different methods of measurement as well as for different bank heights.

For two-dimensional channel cross sections, we measured net area removed in m^2 and multiplied by one unit of stream length to get m^3, which then is presented as m^3 per meter of height per meter of stream length. This method accounts for changes in bed elevation as well as erosion and deposition within the stream corridor.

For the one-dimensional bank pin method, we measured lateral retreat at a point and converted this value to volume by

TABLE 1. UNCERTAINTY ESTIMATES FOR DIFFERENT METHODS, CALCULATED FROM TYPICAL MEASUREMENT ERRORS FOR TYPICAL MEASURED VALUES

Method-ology	Dimension	Half-range (cm)	Source	Typical measured values			Uncertainty		
				Measured value (m)	Period (yr)	Change value (m^3)	Limits (m^3)	1σ, assumed triangular distribution (%)	1σ, assumed normal distribution (%)
Pins	Horizontal	2	Pin measurement	0.30					
	Vertical	4	Bank height measurement	1.8	1	0.17	±0.073	±18.0	±14.6
	Vertical, longitudinal	4	Extrapolation to unit stream length on irregular surface	0.30					
Cross sections	Horizontal	1	Kinematic survey horizontal RMSE	15.2					
	Vertical	2	Kinematic survey vertical RMSE	1.8	1 to 2	0.85	±0.66	±32.0	±25.8
	Vertical, longitudinal	4	Extrapolation to unit stream length on irregular surface	0.30					

Note: RMSE—root mean square error.

multiplying lateral retreat and bank height (m) for one unit length of stream (m). This value is presented as m³ of sediment eroded per meter of bank height per meter of stream length. It does not account for deposition at point bars that form opposite of eroding banks, or for bed aggradation or degradation.

The following example illustrates the concept of a sediment production unit (SPU) for a given stream corridor using the two methods described here. Consider a stream reach of 100 m with left and right bank heights of 2.4 m. One of the two banks is eroding at a rate of 0.3 m/yr, and the other is not eroding. Bank pins and repeat channel cross sections could be used to quantify rates of bank erosion and net channel change. Over the 100 m length of channel, 0.3 m/yr of bank erosion would produce 72 m³/yr of sediment. This volume yields 72 m³/100 m/2.4 m/yr = 0.3 SPU. If both banks were eroding at 0.3 m/yr, sediment production from the 100 m reach would be 144 m³/yr of sediment, or 144 m³/100 m/2.4 m/yr, or 0.6 SPU. If the banks were 1.2 m high instead of 2.4 m, then for a bank retreat rate of 0.3 m/yr at only one bank, the rate of sediment eroded would be 0.3 m³/m/m/yr, or 0.3 SPU, and the total annual amount of sediment produced would be 36 m³/yr. Note that these estimates could be presented in units of m/yr, but SPU indicates the procedure by which we estimated the rate, as it is not merely a lateral rate of retreat measured at a point.

Temporal variability also is captured in the different methods of measuring bank erosion rates. A measure of the total volume removed along a channel corridor 25 yr after a dam breach yields a long-term, 25 yr average rate of erosion. However, bank pins installed 23 yr after dam breaching and measured for 2 yr yield a post-dam-breach, short-term average rate from years 23–25. If bank pins or channel cross sections are monitored over a lengthy period, it is possible to compare short-term rates from different intervals within the longer measurement period, and to compare these short-term estimates to the long-term average rates.

RESULTS

Here, we present the results of particle size analysis of stream bed, bank, and bar sediment and of particle size mobility calculations from the Shields parameter equation for the breached Eaton-Dikeman reservoir on Mountain Creek. We compare stream bank erosion rates measured from repeat surveys of cross sections in the three breached reservoirs and evaluate changes with time since dam breach. To determine plan-view changes at Mountain Creek from 2003 to 2009, we compared bank and bar edges from digital ortho-images from 2003 and 2007 with our 2008 and 2009 surveys of bank and bar edges over a distance of 600 m upstream of the dam. Finally, we consider the role of freeze-thaw processes in bank erosion by examining seasonal variations in bank erosion from bank pin measurements at Mountain Creek, and compare these variations to those measured 50 years earlier by Wolman (1959) at a breached millpond on Watts Branch, Maryland.

Sediment Size in the Eaton-Dikeman Reservoir

Particle size data are presented for XS-2 in the Eaton-Dikeman reservoir on Mountain Creek (Fig. 9). Our particle size analysis indicates that grain sizes are similar at XS-1 and XS-2, and field observations indicate they are similar at XS-3, but the historic reservoir sediment is coarser at XS-4, the upstream-most cross section. At XS-2, the historic reservoir sediment consists of 2%–23% clay, 8%–76% silt, and 30%–55% sand, with minor (<1%–25%) amounts of fine gravel. The fine gravel occurs as low-density, porous slag in thin beds with a quartz sand matrix in the uppermost 1 m of reservoir sediment. This slag most likely is from the eighteenth–nineteenth-century Pine Grove and Laurel Forge iron workings located 12 km upstream.

With exception of the uppermost 40 cm, the upper 155 cm section of historic sediment is much sandier than the lower

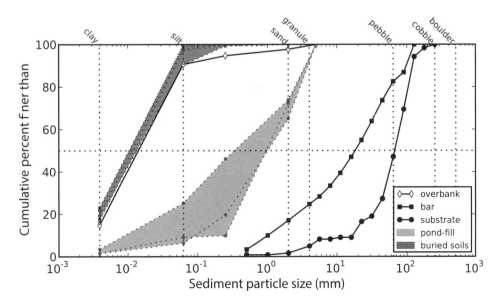

Figure 9. Cumulative grain-size distribution curves for XS-2, with cumulative percent finer on y-axis and grain size on x-axis, for the following sediments: basal gravel substrate exposed in the channel bed by bank erosion, the point bar on right bank, millpond (pond fill) sediment sampled at 10–40-cm-depth increments, and buried fine-grained presettlement sediment between gravel substrate and pond fill. Overbank sediment is uppermost 40 cm of pond fill deposited in shallow pond as overbank deposits (see Fig. 7A).

90 cm of organic-rich prehistoric sediment (from 155 to 245 cm depth, measured from top of reservoir fill). The latter is mostly silt and has a D_{50} particle size of ~0.01 mm. For the millpond sediment above 155 cm, the D_{50} particle size ranges from ~0.03 mm to 1 mm.

Coarse sediment underlying millpond strata throughout the reservoir is exposed at the base of the stream banks and in the channel bed along most of the incised channel. The uppermost part of this exhumed substrate is winnowed of finer sediment along the actively migrating stream bed. As noted earlier herein, this sediment is poorly sorted and can be traced to the hillslope of South Mountain. It is interpreted as exhumed toe-of-slope, late Pleistocene periglacial deposits (e.g., gelifluction sheets). This exhumed and winnowed substrate was sampled at the channel bed just downstream of XS-2, and yielded 2% sand, 3% granules, 65% pebbles, and 30% cobbles. The D_{50} particle size is ~68 mm. The two largest clasts were embedded, and their intermediate axes were estimated at 210 and 220 mm. Both upstream and downstream of the sample site, we measured boulders in the channel with diameters up to ~600 mm.

The inset historic bar forming on the right bank at XS-2 is much coarser than historic millpond sediment, but finer than the periglacial substrate beneath the historic sediment. It consists of 17% sand, 8% granules, 62% pebbles, and 13% cobbles. The D_{50} particle size is ~19 mm.

In sum, the different-aged sediments exposed in the breached Eaton-Dikeman reservoir range in mean grain size from pebbles and cobbles for the Pleistocene periglacial substrate exposed in the channel bed, to clay for prehistoric (Holocene) sediment, silt to sand for millpond sediment, and sand- to cobble-sized sediment for the inset point bar forming within the incised channel corridor. Note that the finest 10% of the point-bar sediment is coarser than 50% of the millpond sediment, and coarser than nearly all of the Holocene sediment between the historic millpond and lower periglacial sediment. The coarsest 70% of sediment in the channel bed is larger than all but the coarsest ~30% of sediment in the point bar.

We interpret the periglacial substrate—exhumed from beneath the eroding banks and exposed as colluvium along valley margins—as the source of most coarse sediment forming point bars within the incised stream corridor of the Eaton-Dikeman reservoir. This interpretation is consistent with the grain-size distributions shown in Figure 9. Some of the coarser parts of the historic millpond sediment probably are mixed and stored with sediment in the point bar. This interpretation is consistent with the grain-size distributions.

Shields Parameter and Particle Mobility in an Incised Reservoir

A review of particle entrainment in flume and field studies indicates that τ^*_c, the critical Shields stress for entrainment of the D_{50} size particle, generally ranges from 0.03 to 0.07 (Buffington and Montgomery, 1997). Assuming that particle entrainment occurs within this range of τ^*_c, we estimate a range in predicted D_{50} values of 34–79 mm for high flow (basal shear stress = 38 N/m²) at the point bar on the right bank at XS-2 in the breached Eaton-Dikeman reservoir on Mountain Creek. For these estimates, we use bankfull flow depth of 1.3 m and slope of 0.003 (water surface slope measured from LiDAR). High flow depths at XS-2 have been observed to be at least 1.3 m. This flow depth is consistent with the height of the active point bar, ~1 m, on the right bank at XS-2. Values of 34–79 mm are higher than the D_{50} of 19 mm measured at the point bar.

If, on the other hand, we predict flow depth based on the D_{50} of 19 mm measured at the point bar, and again assume that particle entrainment occurs when τ^*_c = 0.03–0.07, we estimate a range in predicted flow depths of 0.3–0.7 m. The point bar was sampled just after Hurricane Hanna occurred in 2008 (September 6–8), and water depth was at least 1.3 m during that event. It is possible that our measurements of grain size done on 25 September 2008 would be different if we had sampled before rather than after a large storm.

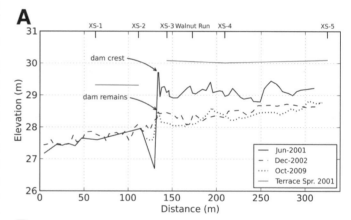

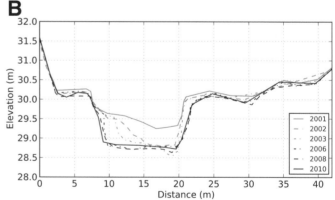

Figure 10. (A) Long profile of Hammer Creek before and after dam breaching. About 0.5 m of the base of the dam remains, forming a grade control that prevents further bed lowering. Terrace surface represents the fill terrace formed by sedimentation to the level of the original dam crest. (B) Repeat surveys of XS-5 showing post-dam-breach vertical incision (2001–2002) followed by channel corridor widening as a result of bank erosion (2002–2010).

Erosion along Incised Streams in Breached Millponds

The four cross sections upstream of the Hammer Creek dam showed similar patterns of incision during the first year after dam breaching, followed by bank erosion and channel widening for the 10 yr since breaching. In 2002, a year after dam breaching at Hammer Creek, the channel bed had degraded ~1 m just upstream of the dam; bed lowering diminished to ~0.5 m ~300 m upstream of the breach (Fig. 10A). With exception of a slightly high part of the bed 200 m upstream of the dam, the bed was lowered an additional 0.2 m between 2002 and 2009. Upstream of the dam, the surface of the historic sediment settled and subsided during the first 5 yr after dam breach, probably as a result of dewatering of the sediment-filled reservoir. Although bank retreat and channel widening occurred at all sections, no prominent point bars formed upstream of the breached pump station dam (Fig. 10B). Downstream of the dam, changes in the bed and banks were insignificant from 2002 to 2009, although the bed was slightly elevated by deposition of sandy gravel immediately after dam breaching.

A plot of cumulative net increase (erosion minus deposition) in channel area for XS-4 and XS-5 on Hammer Creek versus time since dam breach is logarithmic (Figs. 11A and 11B). Cross section 6 has only 1 yr of data (from 2006 to 2007), as it has not been resurveyed since 2007, so a long-term trend cannot be discerned. The stone wall on the right bank of XS-3 prevents its use for monitoring long-term trends in bank erosion. The majority of the increase in channel area for XS-4 and XS-5, where both banks are in historic reservoir sediment, has been the result of lateral bank erosion since 2001.

The four cross sections in the breached Eaton-Dikeman reservoir on Mountain Creek are characterized by bank erosion with little change in bed elevation (Fig. 12). The channel has incised to the level of periglacial pebbles, cobbles, and boulders along the entire length of the reservoir. As a result, most erosion is lateral rather than vertical. Post-dam-breach inset point bars are prominent in the reach of stream between XS-1 and XS-3, where the channel crosses the breached reservoir from the southern to northern sides of the valley (see Fig. 7B). Along this channel reach, both banks consist of historic fine-grained millpond sediment. Scour winnows the underlying periglacial gravel and has produced several prominent point bars within the incised stream banks (see Figs. 7B and 8A). Digitizing bank and bar edges (break lines) on repeat digital ortho-images from 2003 to 2006 and LiDAR (PA MAP) from 2007 reveal that these bars are migrating rapidly downstream at a rate of several meters per year (Fig. 13). Bar migration occurs as upstream ends of the bars are eroded, and deposition occurs on the downstream ends of the bars.

Rates of sediment production for XS-4 and XS-5 at Hammer Creek have decreased since dam breaching, from as high as 7.6 SPU in 2001–2003 to ~0.2–0.5 SPU in 2006–2008 (Table 2, Fig. 14). Sediment production rates calculated from repeat channel cross-section surveys along Mountain Creek in the breached Eaton-Dikeman reservoir varied from –0.5 to 1.0 SPU over a period of 1–1.4 yr from 2008 to 2009. Transient negative values occur when the volume of point-bar growth is greater than volume of bank eroded over short measurement intervals. As discussed earlier herein, however, the point bar consists of much coarser material than the eroded bank sediment. Comparison of digitized bank and bar edges (break lines) from digital ortho-images for 2003 and 2006, and of LiDAR for 2007, yields a reach-averaged sediment production rate of 0.3 SPU for the 2003–2007 time period. Note that comparison of digital ortho-images and LiDAR elevation models yields an average rate of bank erosion over a greater length of channel than do surveys of individual cross sections.

The four cross sections at Conoy Creek yield sediment production rates that ranged from 0.1 to 0.4 SPU during 2.4 yr of repeat surveys in 2006–2008 (Table 2). The second lowest rate of 0.1 SPU at Conoy Creek is for XS-4, located along the southeastern valley margin near the bypass that eroded behind the masonry wall of the dam. At this location, a large amount of rubble, including masonry and concrete, existed in the channel bed, probably as a result of deterioration of the dam. This cross section did not experience the deep incision and scour that occurred upstream at the other three cross sections, which were not limited by the grade control of remnants of the dam.

Bank pin measurements at Mountain Creek yield sediment production rates of 0.3–0.6 SPU, and reveal that more lateral bank erosion occurs in late winter to early spring than during other

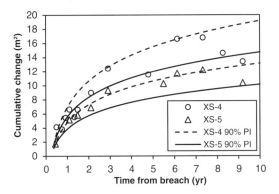

Figure 11. Net channel enlargement of XS-4 and XS-5 at Hammer Creek modeled as a function of the natural log of time. For XS-4, $n = 13$, $y = 4.48\ln(x) + 5.85$, $R^2 = 0.89$, and $p = 1.139 \times 10^{-6}$, with residuals normally distributed. For XS-5, $n = 9$, $y = 3.35\ln(x) + 4.72$, $R^2 = 0.94$, and $p = 1.826 \times 10^{-5}$, with residuals normally distributed. The 90% prediction interval (PI) is shown for each cross section and its model. Most channel enlargement is the result of bank erosion and widening, as little vertical change in the bed has occurred since 2002. Instances of recent vertical aggradation at Hammer Creek result from construction, by anglers, of a low rubble dam (~0.5 m height) immediately downstream of the former dam to create a pool for fishing.

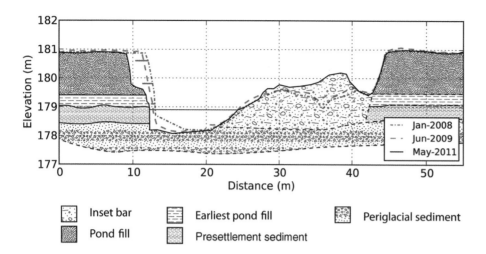

Figure 12. Repeat surveys of channel cross section 2 at Mountain Creek in 2008 (total geodetic station), 2009 (laser level), and 2011 (total geodetic station) reveal erosion of left bank and bar deposition on right bank. Red lines indicate positions of bank pins. Pins are reset periodically to keep pace with erosion. Cumulative erosion of the top pin from 2008 to 2010 was >1.6 m, as the pin was removed during bank erosion in March 2010, and replaced immediately after. Stratigraphic data indicate different age deposits, from Pleistocene periglacial gravel at the base of the section, to fine-grained, organic-rich presettlement sediment, to fine-grained historic millpond sediment, to point-bar sand and gravel on right bank within the incised stream corridor. A thin, dark, organic-rich wetland soil is found at the contact between Pleistocene gravel and millpond sediment at most sites in the Mid-Atlantic region, and it has been dated at numerous localities as Holocene in age (Walter and Merritts, 2008; Merritts et al., 2011).

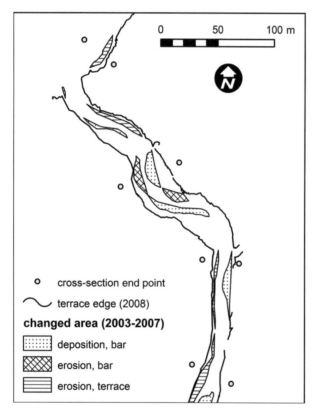

Figure 13. Ortho-images from 2003 and 2006 were mapped and compared with light detection and ranging (LiDAR) elevation data from 2007 and global positioning system (GPS) surveys in 2008 to identify areas of change (erosion and deposition) during the 4 yr interval. Point bars migrated along the channel reach between XS-2 and XS-3 at several meters per year, with upstream tips eroding and downstream tips prograding with time. In addition, incised channel banks eroded laterally, and the stream channel corridor widened with time.

seasons (Table 2; Fig. 15). The majority of bank retreat occurs in March and April. Similar observations were made by Wolman at Watts Branch in Maryland during the 1950s (Wolman, 1959), and by Lawler in his studies of stream banks in England (cf. Lawler, 1986). Data from Wolman (1959) are plotted here for comparison with our data from Mountain Creek (see Fig. 15). Remarkably, the rates and seasonal timing of bank retreat measured with bank pins and survey lines along the bank edge by Wolman (1959) on Watts Branch, Maryland, are nearly identical to those we measured at Mountain Creek with similar procedures from 2007 to 2010, half a century later. The distance between the two sites is ~140 km, with Mountain Creek due north of Watts Branch.

We observed freeze-thaw processes and needle ice in the incised banks of all three reservoirs during winter months since observations began in 2007. Winter freeze-thaw processes weaken and disaggregate bank sediment by freeze-thaw, and spring flow events are able to remove much of the apron of debris that accumulated on the banks during the preceding winter. Prominent notches form at various levels in the apron of disaggregated debris throughout the spring until it is removed completely. We observe that it takes 1 to 3 mo to remove this apron at most localities. Subsequently, warming, evaporation, and plant growth during summer months lead to drying and desiccation of banks.

DISCUSSION: POST-DAM-BREACH STREAM BANK EROSION

A primary objective of the research presented here was to evaluate decadal changes in rate of erosion of reservoir sediment for incised stream channels in breached reservoirs. Channel cross-section data for Hammer Creek, for which we have the longest record (9 yr) of repeat cross-section data, demonstrate that

TABLE 2. SITE, MEASUREMENT METHOD, TIME PERIOD, AND ESTIMATE OF SEDIMENT PRODUCTION BY BANK EROSION PER UNIT METER OF CHANNEL LENGTH

Site	Station	Method	Start date	End date	Period (yr)	Change	Unit	Length (m)	Bank height (m)	Production (SPU* or m/yr)
Hammer Creek	XS-4	Cross section	26-Jun-2001	23-Oct-2008	7.33	16.8	m²	1	0.8	2.9
Hammer Creek	XS-4	Cross section	26-Jun-2001	2-Aug-2002	1.10	6.6	m²	1	0.8	7.6
Hammer Creek	XS-4	Cross section	2-Aug-2002	8-Aug-2003	1.02	2.4	m²	1	1.0	2.2
Hammer Creek	XS-4	Cross section	8-Aug-2003	25-May-2004	0.80	3.4	m²	1	1.4	3.0
Hammer Creek	XS-4	Cross section	25-May-2004	11-Apr-2006	1.88	-0.9	m²	1	1.1	-0.4
Hammer Creek	XS-4	Cross section	11-Apr-2006	14-Aug-2007	1.34	5.1	m²	1	1.0	3.7
Hammer Creek	XS-4	Cross section	14-Aug-2007	23-Oct-2008	1.19	0.2	m²	1	0.8	0.2
Hammer Creek	XS-4	Cross section	23-Oct-2008	1-Oct-2009	0.94	-2.2	m²	1	1.2	-1.9
Hammer Creek	XS-4	Cross section	1-Oct-2009	8-Sep-2010	0.94	-1.2	m²	1	1.2	-1.1
Hammer Creek	XS-5	Cross section	26-Jun-2001	2-Aug-2002	1.10	5.2	m²	1	1.0	4.5
Hammer Creek	XS-5	Cross section	2-Aug-2002	6-Aug-2003	1.01	1.7	m²	1	1.1	1.5
Hammer Creek	XS-5	Cross section	6-Aug-2003	25-May-2004	0.80	2.5	m²	1	1.1	2.9
Hammer Creek	XS-5	Cross section	25-May-2004	26-Dec-2006	2.59	0.9	m²	1	0.9	0.4
Hammer Creek	XS-5	Cross section	26-Dec-2006	14-Aug-2007	0.63	1.5	m²	1	0.9	2.5
Hammer Creek	XS-5	Cross section	14-Aug-2007	23-Oct-2008	1.19	0.5	m²	1	0.9	0.5
Hammer Creek	XS-5	Cross section	26-Jun-2001	23-Oct-2008	7.33	12.2	m²	1	0.9	1.8
Hammer Creek	XS-5	Cross section	23-Oct-2008	8-Sep-2010	1.88	-1.8	m²	1	1.2	-0.8
Mountain Creek	XS-1	Cross section	15-Jan-2008	23-Jun-2009	1.44	2.0	m²	1	2.2	0.6
Mountain Creek	XS-2	Cross section	15-Jan-2008	23-Jun-2009	1.44	-1.4	m²	1	2.0	-0.5
Mountain Creek	XS-3	Cross section	13-Aug-2008	7-Aug-2009	0.98	1.8	m²	1	2.0	0.9
Mountain Creek	XS-4	Cross section	13-Aug-2008	7-Aug-2009	0.98	2.0	m²	1	2.1	1.0
Conoy Creek	XS-1	Cross section	15-Feb-2006	18-Jul-2008	2.42	0.2	m²	1	1.8	0.1
Conoy Creek	XS-2	Cross section	15-Feb-2006	25-Jul-2008	2.44	0.9	m²	1	1.7	0.2
Conoy Creek	XS-3	Cross section	15-Feb-2006	18-Jul-2008	2.42	1.4	m²	1	1.6	0.4
Conoy Creek	XS-4	Cross section	15-Feb-2006	25-Jul-2008	2.44	0.6	m²	1	1.6	0.1
Mountain Creek	XS-1	Bank pin	18-Dec-2007	7-Aug-2009	1.64	0.3	m	1	2.7	0.2
Mountain Creek	XS-2	Bank pin	18-Dec-2007	7-Aug-2009	1.64	0.4	m	1	2.6	0.3
Mountain Creek	XS-3	Bank pin	13-Aug-2008	7-Aug-2009	0.98	0.3	m	1	1.9	0.3
Mountain Creek	XS-4	Bank pin	13-Aug-2008	7-Aug-2009	0.98	0.4	m	1	2.1	0.4
Mountain Creek	XS-2	Bank pin	18-Dec-2007	22-May-2008	0.43	0.5	m	1	2.6	1.2
Mountain Creek	Reach	Bank edge digitization	1-Apr-2003	23-Apr-2007	4.06	1539	m³	611	2.0	0.3

*SPU—sediment production unit.

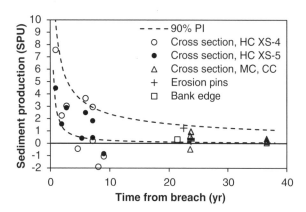

Figure 14. Channel-normalized rate of sediment production from the stream corridor with time is modeled with a negative power function for Hammer (HC), Mountain (MC), and Conoy (CC) Creeks. Data sources include channel cross sections for all three breached reservoirs ($n = 20$), and bank pins ($n = 5$) and repeat digital ortho-images ($n = 1$) for Mountain Creek. Each data point represents 1 to 2 yr measurement intervals, with the exception of the single data point from repeat digital ortho-images, which represents 3 yr. In some years and at some locations, net change is negative, indicating no change or deposition; however, these values are transient. At Hammer Creek, recent negative values result from construction, by anglers, of a low rubble dam (~0.5 m height) in 2007 and subsequent aggradation in the bed upstream from the dam. We therefore fit a power function to the positive data ($n = 21$) and indicate a 90% prediction interval [PI] for those data. The negative power function indicates that erosion rates slowly diminish with time after an early period of rapid erosion. SPU (sediment production unit): This estimate of sediment produced along a given length of stream channel by stream bank erosion takes into account bank height and erosion rate.

the cumulative increase in cross-sectional area is a function of the natural log of time (see Fig. 11A). This increase in channel area is largely an increase in width of the incised stream corridor, as most of the vertical incision occurred soon after dam breaching. Although some bed scour and deposition continue to occur, the 0.5 m of unremoved dam at Hammer Creek and the coarse periglacial substrate beneath historic sediment prevent substantial bed erosion. Similarly, the coarse periglacial substrate at Conoy and Mountain Creeks limits bed degradation (cf. Figs. 8 and 12). For the remaining discussion, we make the assumption that the majority of decadal-scale erosion from the breached millponds studied here is due to lateral bank retreat and that the rate of enlargement of the incised stream corridor width is proportional to the rate of bank erosion, E_r.

We proposed earlier that the rate of erosion in a breached reservoir might decelerate with time after dam breach, as was observed with the Marmot Dam removal on the Sandy River in Oregon (Major et al., 2008, 2012). All cross section ($n = 25$), bank pin ($n = 5$), and bank edge digitization ($n = 1$) data from Hammer, Mountain, and Conoy Creeks indicate that decadal rates of sediment production, proportional to linear bank retreat rates, do indeed diminish with time (Fig. 14). We model this decrease in rate of bank erosion as a power function, with time as the independent variable and sediment production rate as the dependent y variable.

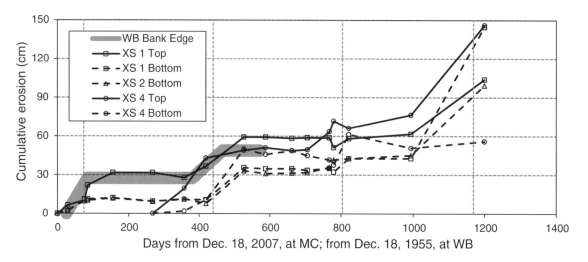

Figure 15. A record of erosion for ~1220 d (3.3 yr) for bank erosion pins at XS-1, XS-2, and XS-4 on Mountain Creek reveals that the majority of bank retreat occurs during mid- to late winter and early spring (December through early April). This same phenomenon was observed by Wolman (1959) for Watts Branch (WB), Maryland, also upstream of a breached milldam and in historic reservoir sediment. Watts Branch data from Wolman (1959) are shown as black squares. Wolman collected data from 5 December 1955 to early 1957. For comparison here, the Watts Branch and Mountain Creek (MC) records are plotted versus time since 18 December 2007, the start date for our measurements at XS-1 and XS-2. Pins at the middle of the bank commonly were buried by an apron of debris from above, and were not as frequently swiped clean as the lower bank, and they do not yield as clear a seasonal signal of erosion as top and bottom pins. Vertical lines indicate 1 March for each year.

A best-fit power function for all positive values of sediment production ($y = 5.1356x^{-0.832}$, $n = 26$, $R^2 = 0.6008$; $p = 3.326 \times 10^{-6}$) can be used to predict that erosion rates will be 0.2 SPU some 50 yr after dam breaching, and 0.1 SPU after 100 yr. These numbers, though seemingly small, can produce ~400–220 m³/yr of sediment, respectively, per kilometer of incised channel length from a breached reservoir with 2-m-high banks, yielding tens of thousands of cubic meters of sediment over decadal time spans.

We propose that the rate of increase in channel width subsequent to dam breach might decelerate more slowly, leading to a smaller exponent in the power function, in regions where freeze-thaw is an important process compared to in warmer climes. Freeze-thaw processes occur winter after winter regardless of land-use change or increased width of the stream corridor. An exposed bank is weakened and disintegrated each year and prone to removal by flow depths sufficient to reach the debris that accumulates in a freeze-thaw apron. Greatest rates of erosion occur where stream flow has access to this debris, as on the outside banks opposite point bars. Wetting and drying from variable flow depths are likely to accelerate freeze-thaw processes by pumping more water into stream banks.

CONCLUSIONS

An important implication of the results presented here is that incised stream banks in breached milldam reservoirs continue to be sources of fine-grained sediment for decades after dam breaching. Within the first 10 yr of dam breaching, rates of sediment production from breached reservoirs are highest, but they decelerate with time. Even 50–100 yr after dam breaching, however, millponds with typical bank heights of 2 m can produce hundreds of cubic meters of sediment per kilometer of stream length per year.

Freeze-thaw processes are significant in weakening the banks of incised streams and are most effective where banks have a large component of silt, as is the case in Mid-Atlantic region millpond reservoirs. Lawler (1986) determined that bank erosion rates from freeze-thaw processes are proportional to the number of days with air frost (air temperature ≤0.0 °C). Furthermore, bank erosion rate corresponds more strongly with this parameter than any of the other 16 meteorological and hydrologic variables examined for the Ilston River by Lawler (1986). For southeastern Pennsylvania, air temperature dropped below 0 °C at least 100 d in both 2008 and 2009. Lawler's statistical regression equation for erosion rate as a response to number of days of frost (see Table IV *in* Lawler, 1986) yields a predicted value of bank erosion of 0.6 m/yr, similar to rates measured at the sites discussed here, including Watts Branch, Maryland (see Fig. 15).

These implications and the results presented here have substantial portent for evaluating sources of fine-grained sediment to impaired water bodies, such as the Chesapeake Bay (see Fig. 1). Thousands, perhaps tens of thousands, of milldams exist throughout this large watershed in the states of New York, Pennsylvania, Maryland, and Virginia (Walter and Merritts, 2008). An unknown number, but probably thousands, were breached in the last century. Each is in different stages of post-dam-breach channel incision. We estimate that typical milldams contain 50,000–250,000 m³ of sediment, depending on dam height and the geometry and gradient of the valley upstream of the dam. Freeze-thaw processes are common throughout the Mid-Atlantic region, as are stream banks incised into historic sediment associated with post–European settlement and milling (Wohl and Merritts, 2007; Walter and Merritts, 2008; Merritts et al., 2011). Nevertheless, watershed models typically do not take into account stream bank erosion as a source of sediment. Even less recognized are the significance of time of dam breaching and the role of freeze-thaw processes in rates of bank erosion. Instead, watershed models use modern land use as a predictor of sediment loads in streams.

Linking upland soil erosion with sediment loads in streams has substantial uncertainties at present. The values of soil erosion predicted by empirical relations such as the Revised Universal Soil Loss Equation, referred to as "edge-of-field" estimates, are inadequate for predicting sediment delivery to streams (Boomer et al., 2008), despite their common use for such purposes. Widely used watershed models (e.g., HSPF [Hydrological Simulation Program–FORTRAN] and SWAT [Soil and Water Assessment Tool]) predict sediment loads in streams based on empirical relations among modern land use, land cover, and soil erosion (cf. Nasr et al., 2004). The Chesapeake Bay Watershed Model, for example, estimates the delivery of sediment and nutrients to the bay, which drains most of the Mid-Atlantic Piedmont, by simulating hydrologic and nutrient cycles for given land-use and land-cover conditions.

The limitations of models that simulate only upland sediment sources and modern land use can be illustrated with an example of a forested watershed such as Mountain Creek (nearly 100% forest cover), for which such models would predict low sediment yields. Recent breaching of milldams with reservoirs of fine-grained historic sediment, however, might result in high suspended sediment loads. In this case, causality is assumed to be a function of modern land use and upland soil erosion, rather than changes in stream channel slope due to base-level fall and the transient storage and release of historic sediment. Breached millponds with historic sediment are the source of sediment that originated from upland erosion during prior land-use conditions over a period of decades to centuries, representing decadal to centennial lag times in different components of the system. These legacy effects and inherent lag times are missing from a watershed model that relies upon current land use to estimate sediment sources.

Breached historic reservoirs are sources of fine sediment loads to Mid-Atlantic streams, and, as more obsolete dams breach and channels become incised with time, that source could grow. As a result of the deliberate, close spacing of milldams to maximize waterpower on streams even as small as first order, the potential for trapping significant amounts of fine-grained sediment has historically been substantial throughout the Mid-Atlantic region. The corollary is that the potential for releasing significant amounts of sediment after dam breaching and channel incision is likewise substantial. The results of this study indicate that fine-grained reservoirs continue to be sediment sources for at least several decades after dam breaching and most likely for at least a century.

ACKNOWLEDGMENTS

Funding for this work was provided by the Pennsylvania Department of Environmental Protection, the Chesapeake Bay Commission, Franklin and Marshall College, the U.S. Environmental Protection Agency, and the National Science Foundation (EAR-0923224). We are particularly grateful to the Masonic Villages, The Nature Conservancy, and the Pennsylvania Game Commission for granting us permission to work on their property. This work benefited from discussions with W. Oberholtzer, D. Altland, and E. Wohl. Reviews by Tess Wynn and Jason Julian were helpful in improving the presentation of various concepts and data.

REFERENCES CITED

Boomer, K.B., Weller, D.E., and Jordan, T.E., 2008, Empirical models based on the Universal Soil Loss Equation fail to predict sediment discharges from Chesapeake Bay catchments: Journal of Environmental Quality, v. 37, no. 1, p. 79–89, doi:10.2134/jeq2007.0094.

Bridgens, H.F., 1858, Bridgens' Atlas of Cumberland County, Pennsylvania: Philadelphia, Wagner and McGuigan Publishers, 27 p.

Bridgens, H.F., 1864, Bridgens' Atlas of Lancaster County, Pennsylvania: From Actual Surveys by H.F. Bridgens and Assistant: D. S. Bare, 43 p.

Brune, G.M., 1953, Trapping efficiency of reservoirs: American Geophysical Union Transactions, v. 34, no. 3, p. 407–418.

Buffington, J.M., and Montgomery, D.R., 1997, A systematic analysis of eight decades of incipient motion studies, with special reference to gravel-bedded rivers: Water Resources Research, v. 33, no. 8, p. 1993–2030, doi:10.1029/96WR03190.

Cantelli, A., Paola, C., and Parker, G., 2004, Experiments on upstream-migrating erosional narrowing and widening of an incisional channel caused by dam removal: Water Resources Research, v. 40, W03304, doi:10.1029/2003WR002940.

Cantelli, A., Wong, M., Parker, G., and Paola, C., 2007, Numerical model linking bed and bank evolution of incisional channel created by dam removal: Water Resources Research, v. 43, W07436, doi:10.1029/2006WR005621.

Couper, P.R., and Maddock, I.P., 2001, Subaerial river bank erosion processes and their interaction with other bank erosion mechanisms on the River Arrow, Warwickshire, UK: Earth Surface Processes and Landforms, v. 26, no. 6, p. 631–646, doi:10.1002/esp.212.

Csiki, S., and Rhoads, B.L., 2010, Hydraulic and geomorphological effects of run-of-river dams: Progress in Physical Geography, v. 34, no. 6, p. 755–780, doi:10.1177/0309133310369435.

Darby, S.E., and Thorne, C.R., 1996, Development and testing of riverbank-stability analysis: Journal of Hydraulic Engineering, v. 122, no. 8, p. 443–454, doi:10.1061/(ASCE)0733-9429(1996)122:8(443).

Dendy, F.E., and Champion, W.A., 1978, Sediment Deposition in U.S. Reservoirs: Summary of Data Reported through 1975: Miscellaneous Publication 1362: Washington, D.C., Agricultural Research Service, U.S. Department of Agriculture.

Doyle, M.W., Stanley, E.H., and Harbor, J.M., 2002, Geomorphic analogies for assessing probable channel response to dam removal: Journal of the American Water Resources Association, v. 38, no. 6, p. 1567–1579, doi:10.1111/j.1752-1688.2002.tb04365.x.

Doyle, M.W., Stanley, E.H., and Harbor, J.M., 2003, Channel adjustments following two dam removals in Wisconsin: Water Resources Research, v. 39, no. 1, p. 1011, doi:10.1029/2002WR001714.

Evans, J.E., 2007, Sediment impacts of the 1994 failure of IVEX Dam (Chagrin River, NE Ohio): A test of channel evolution models: Journal of Great Lakes Research, v. 33, Supplement 2, p. 90–102.

Evans, J.E., Gottgens, J.F., Gill, W.M., and Mackey, S.D., 2000a, Sediment yields controlled by intrabasinal storage and sediment conveyance over the interval 1842–1994: Chagrin river, northeast Ohio, U.S.A.: Journal of Soil and Water Conservation, v. 55, no. 3, p. 264–270.

Evans, J.E., Mackey, S.D., Gottgens, J.F., and Gill, W.M., 2000b, Lessons from a dam failure: The Ohio Journal of Science, v. 100, no. 5, p. 121–131.

Gottschalk, L.C., 1964, Reservoir sedimentation, in Chow, V.T., ed., Handbook of Applied Hydrology: New York, McGraw Hill, p. 1–33.

Graf, W.L., 1999, Dam nation: A geographic census of American dams and their large-scale hydrologic impacts: Water Resources Research, v. 35, no. 4, p. 1305–1311, doi:10.1029/1999WR900016.

Heinz Center, 2002, Dam Removal: Science and Decision Making: Washington, D.C., H. John Heinz III Center for Science, Economics, and the Environment, 221 p.

Hooke, J.M., 1979, An analysis of the processes of river bank erosion: Journal of Hydrology (Amsterdam), v. 42, no. 1–2, p. 39–62, doi:10.1016/0022-1694(79)90005-2.

Julian, J.P., and Torres, R., 2006, Hydraulic erosion of cohesive riverbanks: Geomorphology, v. 76, no. 1–2, p. 193–206, doi:10.1016/j.geomorph.2005.11.003.

Lawler, D.M., 1986, River bank erosion and the influence of frost: A statistical examination: Transactions of the Institute of British Geographers: New Series, v. 11, no. 2, p. 227–242, doi:10.2307/622008.

Lawler, D.M., 1993, The measurement of river bank erosion and lateral channel change: A review: Earth Surface Processes and Landforms, v. 18, no. 9, p. 777–821, doi:10.1002/esp.3290180905.

Lawler, D.M., 1995, The impact of scale on the processes of channel-side sediment supply: A conceptual model, in Osterkamp, W.R., ed., Effects of Scale on Interpretation and Management of Sediment and Water Quality: Boulder, Colorado, International Association of Hydrological Sciences Publication no. 226, p. 175–184; http://iahs.info/redbooks/a226/iahs_226_0175.pdf.

Lawler, D.M., Thorne, C.R., and Hooke, J.M., 1997, Bank erosion and instability, in Thorne, C.R., Hey, R.D., and Newson, M.D., eds., Applied Fluvial Geomorphology for River Engineering and Management: Chichester, UK, Wiley, p. 137–172.

Lord, A.C., 1996, Water-Powered Grist Mills, Lancaster County, Pennsylvania: Millersville, Pennsylvania, A.C. Lord, 87 p.

Major, J.J., O'Connor, J.E., Grant, G.E., Spicer, K.R., Bragg, H.M., Rhode, A., Tanner, D.Q., Anderson, C.W., and Wallick, J.R., 2008, Initial Fluvial Response to the Removal of Oregon's Marmot Dam: Eos (Transactions, American Geophysical Union), v. 89, no. 27, p. 241–252, doi:10.1029/2008EO270001.

Major, J.J., O'Connor, J.E., Podolak, C.J., Keith, M.K., Grant, G.E., Spicer, K.R., Pittman, S., Bragg, H.M., Wallick, J.R., Tanner, D.Q., Rhode, A., and Wilcock, P.R., 2012, Geomorphic Response of the Sandy River, Oregon, to Removal of Marmot Dam: U.S. Geological Survey Professional Paper 1792, 64 p. and data tables; available at http://pubs.usgs.gov/pp/1792/.

Merritts, D., Walter, R., Rahnis, M., Hartranft, J., Cox, S., Gellis, A., Potter, N., Hilgartner, W., Langland, M., Manion, L., Lippincott, C., Siddiqui, S., Rehman, Z., Scheid, C., Kratz, L., Shilling, A., Jenschke, M., Datin, K., Cranmer, E., Reed, A., Matuszewski, D., Voli, M., Ohlson, E., Neugebauer, A., Ahamed, A., Neal, C., Winter, A., and Becker, S., 2011, Anthropocene streams and base-level controls from historic dams in the unglaciated Mid-Atlantic region, USA: Royal Society of London Philosophical Transactions, ser. A, v. 369, no. 1938, p. 976–1009.

Nasr, A., Bruen, M., Jordan, P., Moles, R., Kiely, G., Byrne, P., and O'Regan, B., 2004, Physically-based, distributed, catchment modelling for estimating sediment and phosphorus loads to rivers and lakes: Issues of model complexity, spatial and temporal scales and data requirements, in The Water Framework Directive—Monitoring and Modelling Issues for River Basin Management: Tullamore, Ireland, The Irish National Committees of the International Hydrological Programme and the International Commission on Irrigation and Drainage, p. 55–63.

Osman, A.M., and Thorne, C.R., 1988, Riverbank stability analysis: I. Theory: Journal of Hydraulic Engineering, v. 114, no. 2, p. 134–150, doi:10.1061/(ASCE)0733-9429(1988)114:2(134).

Petts, G.E., 1984, Impounded Rivers: Perspectives for Ecological Management: New York, Wiley, 326 p.

Phillips, S.W., editor, 2002, The U.S. Geological Survey and the Chesapeake Bay—The role of science in environmental restoration: U.S. Geological Survey Circular 1220, 32 p.; available at http://pubs.usgs.gov/circ/c1220/.

Pizzuto, J., and O'Neal, M., 2009, Increased mid-twentieth century riverbank erosion rates related to the demise of mill dams, South River, Virginia: Geology, v. 37, no. 1, p. 19–22, doi:10.1130/G25207A.1.

Schumm, S.A., Harvey, M.D., and Watson, C.C., 2011, Incised Channels: Morphology, Dynamics, and Control: Lone Tree, Colorado, Water Resources Publications, 208 p.

Simon, A., and Darby, S.E., 1997, Disturbance, channel evolution and erosion rates: Hotophia creek, Mississippi, in Wang, S.S.Y., Langendoen, E.J., and Shields, F.D., eds., Proceedings of the Conference on Management of Landscapes Disturbed by Channel Incision: Oxford Campus, University of Mississippi, Center for Computational Hydrosciences and Engineering, p. 476–481.

Simon, A., and Hupp, C.R., 1986, Channel evolution in modified Tennessee channels, in Proceedings of the Fourth Federal Interagency Sedimentation Conference, Las Vegas, Nevada, Vol. 2, p. 5-71–5-82.

Simon, A., Curini, A., Darby, S.E., and Langendoen, E.J., 2000, Bank and near-bank processes in an incised channel: Geomorphology, v. 35, no. 3–4, p. 193–217, doi:10.1016/S0169-555X(00)00036-2.

Tschantz, B.A., and Wright, K.R., 2011, Hidden dangers and public safety at low-head dams: The Journal of Dam Safety, v. 9, no. 1, p. 8–17.

U.S. Bureau of the Census, 1841, Brooklyn: Compendium of the Enumeration of the Inhabitants and Statistics of the United States, 1840: As Obtained at the Department of State, from the Returns of the Sixth Census, by Counties and Principal Towns... to Which Is Added an Abstract of Each Preceding Census: Washington, D.C., Thomas Allen; http://www.census.gov/prod/www/abs/decennial/1840.html.

U.S. Bureau of the Census, 1865, Manufactures of the United States in 1860; Compiled from the Original Returns of the Eighth Census under the Direction of the Secretary of the Interior: Washington, D.C., Government Printing Office, p. 695–708; http://www.census.gov/prod/www/abs/decennial/1860.html.

U.S. Bureau of the Census, 1872, Ninth Census of the United States, 1870, Statistics of Population: Tables I to VIII Inclusive..., Vol. 1: Washington, D.C., U.S. Government Printing Office; http://www.census.gov/prod/www/abs/decennial/1870.html.

U.S. Bureau of the Census, 1884, Census Reports Tenth Census. June 1, 1880, Vol. 5: Washington, U.S. Government Printing Office, http://www.census.gov/prod/www/abs/decennial/1880.html.

Verstraeten, G., and Poesen, J., 2000, Estimating trap efficiency of small reservoirs and ponds: Methods and implications for the assessment of sediment yield: Progress in Physical Geography, v. 24, no. 2, p. 219–251.

Walter, R.C., and Merritts, D.J., 2008, Natural streams and the legacy of water-powered mills: Science, v. 319, no. 5861, p. 299–304, doi:10.1126/science.1151716.

Wohl, E., and Merritts, D.J., 2007, What is a natural river?: Geography Compass, v. 1, no. 4, p. 871–900, doi:10.1111/j.1749-8198.2007.00049.x

Wolman, M.G., 1954, A method of sampling coarse river-bed material: Transactions of the American Geophysical Union, v. 35, no. 6, p. 951–956.

Wolman, M.G., 1959, Factors influencing erosion of a cohesive river bank: American Journal of Science, v. 257, no. 3, p. 204–216.

Wynn, T.M., and Mostaghimi, S., 2006a, Effects of riparian vegetation on stream bank subaerial processes in southwestern Virginia, USA: Earth Surface Processes and Landforms, v. 31, no. 4, p. 399–413, doi:10.1002/esp.1252.

Wynn, T., and Mostaghimi, S., 2006b, The effects of vegetation and soil type on streambank erosion, southwestern Virginia, USA: Journal of the American Water Resources Association, v. 42, no. 1, p. 69–82, doi:10.1111/j.1752-1688.2006.tb03824.x.

Wynn, T.M., Henderson, M.B., and Vaughan, D.H., 2008, Changes in streambank erodibility and critical shear stress due to subaerial processes along a headwater stream, southwestern Virginia, USA: Geomorphology, v. 97, no. 3–4, p. 260–273, doi:10.1016/j.geomorph.2007.08.010.

Manuscript Accepted by the Society 29 June 2012